ADVANCES IN ENZYMOLOGY

AND RELATED AREAS OF
MOLECULAR BIOLOGY

Volume 74

LIST OF CONTRIBUTORS

Frank M. Raushel
Department of Chemistry
Texas A&M University
College Station, TX 77843-2128

Hazel M. Holden
Department of Biochemistry
University of Wisconsin
Madison, WI 53706

Henry M. Miziorko
Biochemistry Department
Medical College of Wisconsin
Milwaukee, WI 53226

Perdeep K. Mehta
Philipp Christen
Biochemisches Institut der Universitat Zurich
CH-8057
Zurich, Switzerland

Chia-Hui Tai
Department of Chemistry and Biochemistry
University of Oklahoma
620 Parrington Oval
Norman, OK 73019

Dr. Paul F. Cook
Department of Chemistry and Biochemistry
University of Oklahoma
620 Parrington Oval
Norman, OK 73019

Paul F. Fitzpatrick
Department of Biochemistry and Biophysics
Department of Chemistry
Texas A&M University
College Station, TX 77843-2128

Ronald E. Viola
Department of Chemistry
University of Akron
Akron, OH 44325-3601

ADVANCES IN ENZYMOLOGY

AND RELATED AREAS OF MOLECULAR BIOLOGY

Founded by F.F. NORD

Edited by DANIEL L. PURICH

UNIVERSITY OF FLORIDA COLLEGE OF MEDICINE
GAINESVILLE, FLORIDA

WILEY

2000

AN INTERSCIENCE® PUBLICATION

New York • Chichester • Weinheim • Brisbane • Singapore • Toronto

This book is printed on acid-free paper. ∞

Published simultaneously in Canada.

Library of Congress Cataloging-in-Publication Data:

Library of Congress Cataloguing-in-Publication Data is available.
0-471-34921-6

Printed in the United States of America.

10 9 8 7 6 5 4 3 2 1

CONTENTS

PREFACE

Today's fast-paced growth of mechanistic information on enzyme reactions even astonishes those who have personally conducted decades of research on biological catalysis. The confluence of structural and dynamic information now provides a highly efficient means for discerning subtle features of enzyme catalysis, and the lessons learned from studies on a particular enzyme can no longer be viewed in isolation. Instead, these lessons serve as beacons that will illuminate and guide future explorations, often on entire classes of enzyme-catalyzed reactions. Part B in this Advances in Enzymology subseries entitled "Mechanism of Enzyme Action" continues the mission to provide new and valuable information about the nature of enzyme intermediates, the stepwise organochemical transformations of substrate to product, as well as the nature of barriers to interconversion of the various enzyme-bound species. Presented here are authoritative accounts by leading scientists who have taken this opportunity to share insights that are the product of highly imaginative and productive research. As has become the custom for Advances in Enzymology, these authors were encouraged to acknowledge the relevant findings from many research groups, while focusing greater emphasis on the scaffolding of logic that guided the research accomplished in their own laboratories. In this respect, the editor bears full responsibility for discouraging authors from providing exhaustive treatments that recapitulate findings already well described in earlier reviews on a particular topic.

Daniel L. Purich
Gainesville, Florida
February 2000

ABSTRACTS

Phosphotriesterase: An Enzyme in Search of Its Natural Substrate

The phosphotriesterase from *Pseudomonas diminuta* catalyzes the hydrolysis of a broad range of organophosphate nerve agents. The kinetic constants, k_{cat} and k_{cat}/K_m, for the hydrolysis of paraoxon, diethyl p-nitrophenyl phosphate, are 2200 s^{-1} and 4×10^7 $M^{-1}s^{-1}$, respectively. The utilization of substrates that are chiral at the phosphorus center has demonstrated that the overall hydrolytic reaction occurs with inversion of configuration. The native enzyme contains Zn^{2+} bound to the active site and catalytic activity is retained upon substitution with Mn^{2+}, Ni^{2+}, Cd^{2+}, or Co^{2+}. NMR and EPR spectroscopy has revealed the presence of an antiferromagnetically coupled binuclear metal center. X-ray crystallography shows the enzyme to exist as a homodimer and the protein adopts an $\alpha\beta_8$ folded structure. The binuclear metal center is ligated to the protein via four histidine residues, an aspartate, and a carbamoylated lysine residue that serves as a bridge between the two metal ions. The overall structure is very similar to the binuclear nickel center in urease. A naturally occurring substrate has not been identified for this enzyme but this protein has received considerable attention as a possible catalyst for the detoxification of organophosphate insecticides and military chemical weapons.

Phosphorivulokinase: Current Perspectives on the Structure/Function Basis for Regulation and Catalysis

Phosphoribulokinase (PRK), an enzyme unique to the reductive pentose phosphate pathway of CO_2 assimilation, exhibits distinctive contrasting properties when the proteins from eukaryotic and prokaryotic sources are compared. The eukaryotic PRKs are typically dimers of –39 kDa subunits while the prokaryotic PRKs are octamers of –32 kDa subunits. The enzymes from these two classes are regulated by different mechanisms. Thioredoxin f mediated thiol-disulfide exchange interconverts eukaryotic PRKs between reduced (active) and oxidized (inactive) forms. Allosteric effectors, including activator NADH and inhibitors AMP and phosphoenolpyruvate, regulate activity of prokaryotic PRK. The effector binding site has been identified in the high resolution structure recently elucidated for prokaryotic PRK and the apparatus for transmission of the allosteric stimulus has been identified. Additional contrasts between PRKs include marked differences in primary structure between eukaryotic and prokaryotic PRKs. Alignment of all available deduced PRK sequences indicates that less that 10% of the amino acid residues are invariant. In contrast to these differences, the mechanism for ribulose 1,5-bisphosphate synthesis from ATP and ribulose 5-phosphate (Ru5P) appears to be the same for all PRKs. Consensus sequences associated with M^{++}-ATP binding, identified in all PRK proteins, are closely juxtaposed to the residue proposed to function as general base catalyst. Sequence homology and mutagenesis approaches have suggested several residues that may potentially function in Ru5P binding. Not all of these proposed Ru5P binding residues are closely juxtaposed in the structure of unliganded PRK. Mechanistic approaches have been employed to investigate the amino acids which influence $K_{m\ Ru5P}$ and identify those amino acids most directly involved in Ru5P binding. PRK is one member of a family of phospho or sulfo transferase proteins which exhibit a nucleotide monophosphate kinase fold. Structure/function correlations elucidated for PRK suggest analogous assignments for other members of this family of proteins.

The Molecular Evolution of Pyridoxal-5'-Phosphate-Dependent Enzymes

The pyridoxal-5′-phosphate-dependent enzymes (B_6 enzymes) that act on amino acid substrates are of multiple evolutionary origin. The numerous common mechanistic features of B_6 enzymes thus are not historical traits passed on from a common ancestor enzyme but rather reflect evolutionary or chemical necessities. Family profile analysis of amino acid sequences supported by comparison of the available three-dimensional (3-D) crystal structures indicates that the B_6 enzymes known to date belong to four independent evolutionary lineages of homologous (or more precisely paralogous) proteins, of which the α family is by far the largest. The α family (with aspartate aminotransferase as the prototype enzyme) includes enzymes that catalyze, with several exceptions, transformations of amino acids in which the covalency changes are limited to the same carbon atom that carries the amino group forming the imine linkage with the coenzyme (i.e., Cα in most cases). Enzymes of the β family (tryptophan synthase β as the prototype enzyme) mainly catalyze replacement and elimination reactions at Cβ. The D-alanine aminotransferase family and the alanine racemase family are the two other independent lineages, both with relatively few member enzymes. The primordial pyridoxal-5′-phosphate-dependent enzymes apparently were regio-specific catalysts that first diverged into reaction-specific enzymes and then specialized for substrate specificity. Aminotransferases as well as amino acid decarboxylases are found in two different evolutionary lineages. Comparison of sequences from eukaryotic, archebacterial, and eubacterial species indicates that the functional specialization of most B_6 enzymes has occurred already in the universal ancestor cell. The cofactor pyridoxal-5′-phosphate must have emerged very early in biological evolution; conceivably, organic cofactors and metal ions were the first biological catalysts. In attempts to simulate particular steps of molecular evolution, oligonucleotide-directed mutagenesis of active-site residues and directed molecular evolution have been applied to change both the substrate and reaction specificity of existent B_6 enzymes. Pyridoxal-5′-phosphate-dependent catalytic antibodies were elicited with a screening protocol that applied functional selection criteria as they might have been operative in the evolution of protein-assisted pyridoxal catalysis.

O-Acetylserine Sulfhydrylase

0-Acetylserine sulfhydrylase (OASS) is a pyridoxal 5'-dependent enzyme that synthesizes L-cysteine in enteric bacteria, such as *Salmonella typhimurium* and *Escherichia coli,* and plants. OASS is a member of the β-family of PLP-dependent enzymes that specifically catalyze β-replacement reactions. Enzymes in this class include the β-subunit of tryptophan synthase (β-TRPS), cystathionine β-synthase, β-cyanoalanine synthase, and cysteine lyase. Other than OASS, only β-TRPS has been extensively studied, and thus mechanistic comparisons will be limited to it. This review focuses on the structure that has been solved recently, kinetic and acid-base chemical mechanisms, and spectroscopic studies using ^{31}P NMR, UV-visible, rapid-scanning stopped-flow phosphorescence, static and time-resolved fluorescence techniques. In addition, kinetic isotope effects and stereochemistry of the OASS reaction are discussed.

The Aromatic Amino Acid Hydroxylases

The enzymes phenylalanine hydroxylase, tyrosine hydroxylase, and tryptophan hydroxylase constitute the family of pterin-dependent aromatic amino acid hydroxylases. Each enzyme catalyzes the hydroxylation of the aromatic side chain of its respective amino acid substrate using molecular oxygen and a tetrahydropterin as substrates. Recent advances have provided insights into the structures, mechanisms, and regulation of these enzymes. The eukaryotic enzymes are homotetramers comprised of homologous catalytic domains and discrete regulatory domains. The ligands to the active site iron atom as well as residues involved in substrate binding have been identified from a combination of structural studies and site-directed mutagenesis. Mechanistic studies with nonphysiological and isotopically substituted substrates have provided details of the mechanism of hydroxylation. While the complex regulatory properties of phenylalanine and tyrosine hydroxylase are still not fully understood, effects of regulation on key kinetic parameters have been identified. Phenylalanine hydroxylase is regulated by an interaction between phosphorylation and allosteric regulation by substrates. Tyrosine hydroxylase is regulated by phosphorylation and feedback inhibition by catecholamines.

L-Aspartase: New Tricks From an Old Enzyme

The enzyme L-aspartate ammonia-lyase (aspartase) catalyzes the reversible deamination of the amino acid L-aspartic acid, using a carbanion mechanism to produce fumaric acid and ammonium ion. Aspartase is among the most specific enzymes known with extensive studies failing, until recently, to identify any alternative amino acid substrates that can replace L-aspartic acid. Aspartases from different organisms show high sequence homology, and this homology extends to functionally related enzymes such as the class II fumarases, the argininosuccinate and adenylosuccinate lyases. The high-resolution structure of aspartase reveals a monomer that is composed of three domains oriented in an elongated S-shape. The central domain, comprised of five ¯-helices, provides the subunit contacts in the functionally active tetramer. The active sites are located in clefts between the subunits, and structural and mutagenic studies have identified several of the active site functional groups. While the catalytic activity of this enzyme has been known for nearly 100 years, a number of recent studies have revealed some interesting and unexpected new properties of this reasonably well-characterized enzyme. The non-linear kinetics that are seen under certain conditions have been shown to be caused by the presence of a separate regulatory site. The substrate, aspartic acid, can also play the role of an activator, binding at this site along with a required divalent metal ion. Truncation of the carboxyl terminus of aspartase at specific positions leads to an enhancement of the catalytic activity of the enzyme. Truncations in this region also have been found to introduce a new, non-enzymatic biological activity into aspartase, the ability to specifically enhance the activation of plasminogen to plasmin by tissue plasminogen activator. Even after a century of investigation there are clearly a number of aspects of this multifaceted enzyme that remain to be explored.

CLASSICS IN ENZYMOLOGY

Reprint of Chapter 7 from Advances in Enzymology and Related Subjects, Volume 2, Edited by F.F. Nord and C.H. Werkman. Published by Interscience Publishers, Inc., New York, 1942.

HETEROTROPHIC ASSIMILATION OF CARBON DIOXIDE

By C.H. Werkman and H.G. Wood, Department of Bacteriology, State University of Agriculture, Ames, Iowa

HETEROTROPHIC ASSIMILATION OF CARBON DIOXIDE*

By

C. H. WERKMAN AND H. G. WOOD

Ames, Iowa

CONTENTS

I. Introduction

In 1935 heterotrophic assimilation of carbon dioxide was advanced by Wood and Werkman (1) as a definite and experimentally supported concept. They stated, "It has been established with several species of *Propionibacterium* that the total carbon dioxide liberated during fermentation of glycerol plus that remaining in the form of carbonate is less than the original carbon dioxide added as carbonate. This decrease is believed to result from utilization of carbon dioxide by the bacteria during their dissimilation of glycerol. Carbon and oxidation-reduction balances support this view." The unexpected finding of carbon dioxide utilization by such typically heterotrophic organisms as the propionic acid bacteria had been

* Presented in part at the Seminar of Organic Chemistry, Fordham University, New York, on February 4, 1942.

Advances in Enzymology and Related Areas of Molecular Biology, Volume 74: Mechanism of Enzyme Action, Part B, Edited by Daniel L. Purich
ISBN 0-471-34921-6

first obtained some two years previous, but the unexpected nature of the results led to additional experiments in order to obtain convincing and, if possible, conclusive proof. It was for this reason that the authors in their initial proposal of heterotrophic utilization of carbon dioxide took a definite stand and have remained firm in their pronouncement notwithstanding considerable doubt and criticism expressed in private communications and in print.

The concept of heterotrophic utilization of carbon dioxide was first proposed at the Spring (1935) Meeting of the North Central Branch of the Society of American Bacteriologists in connection with studies on the fermentation of glycerol by bacteria belonging to the genus *Propionibacterium.* These bacteria do not form sufficient carbon dioxide from the glycerol to mask the uptake of carbon dioxide. Therefore, in a medium containing carbonate to neutralize the acids formed from glycerol, *i. e.*, propionic and succinic with a trace of acetic, determination of the carbon balance indicated that the carbon dioxide at the end of the experiment was not equivalent to that of the original medium in the form of carbonate, and that the products of fermentation contained more carbon than was present in the glycerol fermented.

Table I taken from the original work of Wood and Werkman (2) clearly shows that carbon dioxide was utilized by four species of the heterotrophic propionic bacteria used in the experiment. It was pointed out at this time (1936) that:

> "The fact that chemical analysis shows a decrease of carbon dioxide (accountable as carbonate carbon dioxide) is, perhaps, proof enough of carbon dioxide utilization. However, the carbon and oxidation-reduction balances furnish additional evidence."

The authors then continued (1936), "This observation (carbon dioxide utilization) requires a reinterpretation of previous results. Investigators have not considered the possibility of carbon dioxide utilization in constructing schemes of dissimilation. If one considers the limited number of bacteria which have been shown to utilize carbon dioxide and also that such forms (autotrophic) differ markedly from the propionic acid bacteria, failure to consider the possibility of carbon dioxide utilization may be understood." The principle of heterotrophic carbon dioxide utilization was again presented before the Second International Congress of Microbiology, meeting in London during the Summer of 1936. It was not, however, readily accepted and opposing comments were made.

It is significant that the same authors (4) made the following comment in 1938 regarding the utilization of carbon dioxide by animal tissue.

"Krebs and Johnson (1937) have recently shown that citric acid is synthesized by avian tissue from oxalacetic acid and some unknown compound. It is possible that this synthesis involves utilization of carbon dioxide."

TABLE I

DISSIMILATION OF GLYCEROL BY PROPIONIC ACID BACTERIA

Culture	Glycerol fermented per liter, mM.	CO_2 utilized per 100 mM. of fermented glycerol, mM.	Products per 100 mM. of fermented glycerol			Carbon recovery		Oxidation-reduction index	
			Propionic acid, mM.	Acetic acid, mM.	Succinic acid,‡ mM.	Basis-glycerol *plus* CO_2, %	Basis-glycerol only, %	Basis-glycerol *plus* CO_2	Basis-glycerol only
49W	212.6	37.7	55.8	2.9	42.1	101.2	114.0	1.081	2.550
34W	209.0	43.2	59.3	2.0	34.5	93.1	106.6	0.925	2.270
52W*	112.0	20.0	78.4	5.9	8.7	94.6	101.0	0.918	1.386
11W†	218.4	1.1	89.3	2.6	3.9	96.5	96.8	1.135	1.162
15W	176.4	12.3	78.4	5.8	7.8	89.1	92.6	1.047	1.376

* 7.0 mM. of lactic acid produced per 100 mM. of fermented glycerol.
† 0.5 mM. of lactic acid produced per 100 mM. of fermented glycerol.
‡ Succinic acid identified by melting point and mixed melting point.

The experimental proof of carbon dioxide assimilation by animal tissue came in 1940 through the work of Evans and Slotin (3).

Since the isotopes of carbon have become available for use as tracers of fixed carbon dioxide, there has been a tendency to disregard the work done previously. It is true that with the advent of the tracer technique, detection of the fixation of carbon dioxide and its behavior in metabolism have been facilitated; nevertheless, fixation by heterotrophic forms already had been clearly demonstrated by quantitative data obtained with the propionic acid bacteria. Moreover, since all the products were aliphatic carbon compounds of two or more carbon atoms, fixation in a carbon to carbon linkage was shown to occur. Location of the fixed carbon among the products and its position within the molecule was a matter of speculation at that time. The isotopic investigations have been of particular service in clearing up these latter points.

Wood and Werkman (4) showed an equimolar relationship between the carbon dioxide fixed and the succinic acid formed, and found that inhibition of fixation by sodium fluoride (5) resulted in a corresponding reduction in succinic acid. As a result the proposal was made that the succinic acid was the result of a C_3 and C_1 synthesis. Pyruvic acid was suggested as the possible C_3 compound since it could be isolated from the fermentation (6).

This was essentially the situation at the beginning of 1940 when isotopes of carbon first became available.

Autotrophism and Heterotrophism

It is desirable at this point, and before detailed consideration of the phenomenon, to define in the light of present knowledge the expression "heterotrophic carbon dioxide assimilation." The photosynthetic utilization of carbon dioxide has been known for nearly a hundred years but less generally known to occur is the process referred to as chemosynthesis in which carbon dioxide is utilized by an organism employing "chemical energy" in contradistinction to "radiant energy," to reduce the carbon dioxide to form a product of assimilation. Previous to the discovery of the utilization of carbon dioxide by heterotrophic forms, chemosynthesis referred to the utilization of carbon dioxide by a group of organisms known as chemo-autotrophs, discovered in 1890 by the eminent bacteriologist, Sergius Winogradsky (7). Winogradsky established the existence of chemo-autotrophic bacteria which grow and reproduce in a wholly inorganic medium in the dark, *i. e.*, they contain no photosynthetic pigment. The energy required to build cell substance and to carry on metabolism is obtained from relatively simple chemical reactions involving the oxidation of such inorganic compounds as ammonia and nitrite in the case of the nitrifying bacteria (*Nitrosomonas* and *Nitrobacter*, respectively) or sulfur compounds in the case of the non-photosynthetic sulfur bacteria. Hydrogen sulfide is oxidized to free sulfur by *Beggiatoa*, and sulfur to sulfate by *Thiobacillus thio-oxidans*. Hydrogen gas is oxidized by *Carboxydomonas oligocarbophilia*, methane by *Methanomonas methanica* and certain organisms such as *Didymohelix* and *Crenothrix* may oxidize Fe^{++} or Mn^{+} to Fe^{+++} or Mn^{++}. In all cases carbon dioxide is, of course, reduced, generally along with oxygen of the air. The autotrophs are, in large measure, aerobic forms or reduce nitrate. Their nitrogen is generally obtained from ammonium salts or other inorganic salts such as nitrates or nitrites, but the important point is that carbon requirements of these bacteria are satisfied wholly by carbon dioxide.

The metabolism of the autotrophs is either (*a*) relatively simple or, more likely, (*b*) the organisms are able to synthesize the essential complex substances of the nature of vitamins (coenzymes) which must be supplied to the heterotrophs.

The existence of chemo-autotrophic forms of life is frequently not appreciated, to wit, the following statement,

"It is generally known to scientists that photosynthesis is the synthesis of organic matter in green plants with the help of sunlight, and that *this process is the only source of organic matter existing on earth*" (8).

In similar relationship stands heterotrophic carbon dioxide fixation. Thus the idea has become fixed in mind that the assimilation of carbon dioxide is a process uniquely limited to photosynthesis by green plants.

Autotrophs may be photosynthetic, rather than chemosynthetic, such as certain of the sulfur bacteria (*Thiobacteriales*) which contain chlorophyll. Bacterial photosynthesis was confirmed by van Niel (9) after considerable controversy initiated by the original work of Engelmann (10) in 1883. These investigations constitute probably the first evidence for the existence of a photosynthetic process among bacteria. Engelmann found that the red pigmented (purple sulfur) bacteria possess a well-defined absorption spectrum and congregate in portions of the spectrum identical with those absorbed. Engelmann concluded that the pigment plays an essential role in the metabolism of these bacteria.

Winogradsky's original conception of autotrophism envisaged only the chemosynthetic aspects. He could not explain satisfactorily the role of hydrogen sulfide and light required for the growth of the photosynthetic purple sulfur bacteria (*Thiorhodaceae*). It was difficult to explain the light energy requirement in view of the oxidation of hydrogen sulfide to sulfur or sulfate, inasmuch as no rational reason was at hand for the two apparently independent sources of energy. Molisch's (11) discovery of the *Athiorhodaceae*, organisms which require organic substances as hydrogen donators to replace hydrogen sulfide or water in the case of the *Thiorhodaceae* or typical green plants, respectively, further confused the problem of bacterial photosynthesis until the investigations of van Niel offered a rational explanation and cleared the way for a better understanding of bacterial photosynthesis. Kluyver and Donker (12) in 1926 had suggested that with the purple sulfur bacteria hydrogen sulfide functions as a hydrogen donator replacing water in the typical green plant photosynthesis.

Van Niel (13) represents bacterial photosynthesis by the following equations:

$$\begin{array}{lll} 4(H_2O + h) & \longrightarrow 4(H + OH) & \text{Light reaction} \\ 4H + CO_2 & \longrightarrow (CH_2O) + H_2O & \left.\right\} \text{Dark reaction} \\ 2OH + H_2A & \longrightarrow 2H_2O + A & \end{array}$$

H_2A is usually hydrogen sulfide although molecular hydrogen, organic H_2-donators, such as fatty acids, or sulfur oxides may serve, depending on the species of purple bacterium used. In all photosynthesis water is the original source of hydrogen which is ultimately responsible for the reduction of carbon dioxide. In the case of the typical green plants the two "OH" groups form a peroxide which is decomposed into oxygen and water. With bacteria the "OH" is reduced, for example, by hydrogen sulfide by the purple sulfur bacteria to form sulfur and water. For further details the reader is referred to the review by van Niel (13).

The general concept of photosynthesis as portrayed by van Niel is important in that the explanation of the dark reaction which results in the reduction of carbon dioxide may involve the same type of reduction as described by Wood and Werkman (4) for the heterotrophic assimilation of carbon dioxide and represented by: $CO_2 + CH_3{\cdot}CO{\cdot}COOH \longrightarrow COOH{\cdot}CH_2{\cdot}CO{\cdot}COOH$. This point will be further discussed. The function of light is to form active hydrogen; from here on photosynthesis may bear a close analogy to heterotrophic assimilation of carbon dioxide as proposed by Wood and Werkman.

In contrast to the autotrophic bacteria are the heterotrophic organisms which require a source of carbon more complex than carbon dioxide, *i. e.*, they are unable to utilize carbon dioxide (and now must be added) as a sole source of carbon. In view of the work of Wood and Werkman (1, 2, 14), Slade, *et al.* (15), Carson and Ruben (16), Ruben and Kamen (17), Barker, *et al.* (18), and others, the distinction between autotrophs and heterotrophs is becoming less evident; however, the terms are useful and carry a practical meaning. It is not possible at present to express in clear terms an explanation of the implied difference between an autotroph and a heterotroph; it remains a question just why heterotrophic forms require complex carbon sources when they are able to utilize carbon dioxide. It may be that inability of heterotrophic bacteria to use carbon dioxide as a sole source of carbon is linked with an inability to synthesize a certain molecular structure essential in their metabolism. On the other hand, this suggestion may be questioned, inasmuch as no such specific structure seems required by heterotrophs.

The differentiation of heterotrophs and autotrophs on the basis of carbon assimilation is difficult to apply in practice. The first difficulty is to determine whether the carbon (*e. g.*, from carbon dioxide) is assimilated. What constitutes assimilation? It is generally defined as the incorporation or conversion of nutrient material into body substance. If, then, the carbon of carbon dioxide is found at one stage in the protoplasm of the organism, assimilation clearly has taken place; if the carbon is found in a molecule of an enzyme active in metabolism, this would be accepted, probably by all, as assimilation. On the other hand, if the carbon is found in an excretion product, a question may be raised. Possibly a thermodynamic approach is to be preferred, but here again certain difficulties may arise in differentiating exergonic from endergonic reactions.

From the standpoint of mechanism, the difference between a typical heterotroph and a typical autotroph, as judged from present studies, is that the heterotroph can bring about a carbon to carbon linkage if one component of the linkage is organic, but it cannot repeat the process to form a linkage in which both components originate from inorganic carbon. In the case of the autotroph this can be done, however, and adjacent carbons can be inorganic. This only tells what the difference is but does not answer the fundamental question of why there is this difference. There are, however, certain borderline cases. For example, it is not to be concluded that certain organisms will not be found which require the presence in the medium of a substance of the nature of a vitamin which is essential for an autotrophic type of synthesis. An organism, *Clostridium aceticum*, isolated and described by Wieringa (19), is able to live an essentially chemoautotrophic existence when an unknown organic constituent of Dutch mud, which certainly cannot furnish appreciable energy, is added to the inorganic medium. Since the organism apparently reduces carbon dioxide by molecular hydrogen to form acetic acid, sufficient energy is available and only a coenzyme-like substance is required, and this is provided by the mud. In this case we have the picture of a typically autotrophic organism losing the property of synthesizing an essential organic constituent which must be supplied in the medium. The loss of the property in this case is a first step toward heterotrophism. The organism can, in fact, use sugars and function as a heterotroph.

It is likely that various organisms manifest heterotrophism for different reasons. One may be heterotrophic because it is unable to oxidize an inorganic substrate to provide the energy required to assimilate carbon dioxide, whereas inability to synthesize essential growth factors may force heterotrophism on another. It is not important and quite impossible to draw a sharp line of demarcation. An example of an essentially heterotrophic form requiring organic compounds as a source of carbon, and yet possessing the synthetic properties of an autotroph, is the organism recently isolated by Barker (18), *Clostridium acidi-urici*, which attacks purine compounds such as uric acid, xanthine and hypoxanthine anaerobically to form cell substance, ammonia, carbon dioxide and acetic acid. By the use of radioactive carbon dioxide (18) it was shown that the acetic acid was synthesized from carbon dioxide and that the fixed carbon occurred in both the methyl and carboxyl groups. While it was not possible to prove that individual molecules contained the fixed carbon in both groups since in any one molecule fixation may have occurred in only one group, nevertheless it seems quite likely that fixation does occur in both groups in individual molecules. This organism may be an example of a heterotroph forming a carbon chain from C_1 compounds, *i. e.*, an autotrophic property.

Both *Clostridium acidi-urici* and *Clostridium aceticum* are examples of intermediate forms and manifest their autotrophism particularly by their ability to synthesize a carbon to carbon linkage from the C_1 compound, carbon dioxide. The heterotrophic utilization of carbon dioxide has not been proved, as yet, to be an essential step in cellular metabolism and may be no more than a vestige of autotrophism incapable of providing the carbon requirements of heterotrophic bacteria. Later discussion, however, will point out the probable importance of the utilization in the metabolism of the cell.

There is no evidence to indicate that chemo-autotrophic bacteria do not utilize carbon dioxide by the same or similar mechanism employed by the heterotrophs. Inasmuch as carbon dioxide is the sole source of carbon for the autotroph, it is to be expected that all the carbon to carbon linkages formed will comprise atoms originating from carbon dioxide, whereas probably only a small fraction of the typically heterotrophic assimilation employs carbon from carbon dioxide. Urgently needed are studies on autotrophic bacteria employing tracer carbon dioxide, preferably $C^{13}O_2$. The use of isotopes has opened new methods of attack and the whole problem of intermediary metabolism needs intensive study.

Probably the earliest suggestion that carbon dioxide plays an active role in heterotrophic metabolism was the result of the work of Novak (20) and Smith (21). The necessity for carbon dioxide became apparent in the case of *Brucella abortus* which was found to grow more readily, particularly if freshly isolated, when grown in the presence of an aerobe (*Bacillus subtilis*); although the effect was first attributed to decrease in oxygen tension, it was soon found to be due to an increased tension of carbon dioxide. Rockwell and Highberger (22) from a study of bacteria, yeasts and fungi, ventured the suggestion that heterotrophic microorganisms utilize carbon dioxide in their metabolism. Winslow, *et al.* (23), and Gladstone, *et al.* (24), examined the subject more closely. Carbon dioxide-free air bubbled continuously through liquid cultures of a large number of aerobes, and anaerobes prevented or greatly retarded growth, whereas when ordinary air was used growth was normal. The early work was largely qualitative and the effect of the carbon dioxide was frequently ascribed to its physical behavior. Thus until 1935 our ideas regarding the role of carbon dioxide in cellular metabolism were largely limited to the photosynthetic and chemosynthetic processes of the autotrophs.

It was mentioned that with the discovery of heterotrophic carbon dioxide assimilation, differentiation between the autotrophs and the heterotrophs became less distinct. It is not, however, implied that formerly a sharp line of demarcation was drawn between the groups. It was early recognized that intermediate forms occurred, and they were called facultative heterotrophs, if they were able to grow in an inorganic medium but tolerated the presence of organic constituents, especially in small concentration, and facultative autotrophs if organic material was preferred but growth could take place in its absence in a mineral medium. Even the early distinction between the autotroph and the heterotroph was appreciated to be a matter of convenience and rested in considerable degree on the assumption that heterotrophs were unable to assimilate carbon dioxide.

Inasmuch as the differentiation of autotrophs and heterotrophs is based on nutritional requirements of the organisms, it is convenient to visualize a spectrum (Fig. 1) in which the autotrophs and heterotrophs represent the

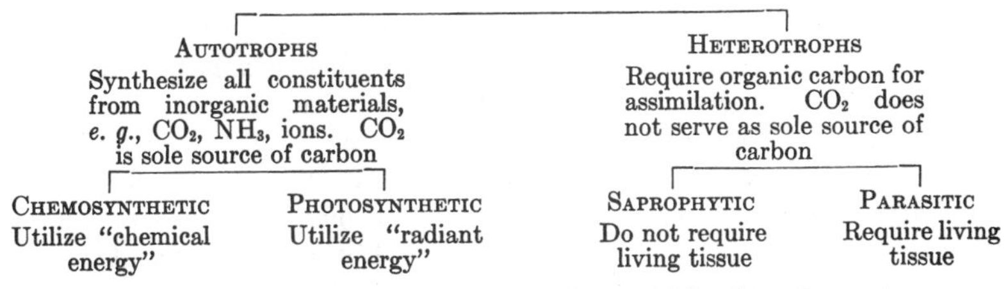

Fig. 1. Classification of organisms based on nutritional requirements.

two extremes and between these occur an indeterminate number of intermediate forms. The *Athiorhodaceae*, photosynthetic bacteria discovered by Molisch (11), apparently are unique in that they require organic compounds but in this case the organic substance serves simply as a hydrogen donator in a manner not clear, for the photosynthetic reduction of carbon dioxide. The carbon chain is not dissimilated and the fractions then assimilated to form cellular material. These bacteria are essentially autotrophic; however, in the present system of classification, their requirement of organic compounds would place them nearer the heterotrophs. A finer distinction might be made on the basis of function performed by the organic compound required. Further examples of intermediate forms are the organisms discovered by Wieringa (19) and by Barker, *et al.* (18).

There are, however, discernible differences in the type of function performed by carbon dioxide in cellular physiology. It may serve (1) simply as a hydrogen acceptor, or (2) it may be incorporated to form more complex carbon compounds in the cell, *i. e.*, assimilated by the creation of a carbon to carbon linkage. The two types have not been differentiated heretofore. The studies of Woods (25), Barker (26) and Hes (27) illustrate the first type, and although a utilization of carbon dioxide by heterotrophic organisms is involved, the same implications do not result as in investigations in which heterotrophic assimilation has been shown to occur, *i. e.*, creation of a carbon to carbon linkage. This point will be discussed in further detail.

II. Mechanism of Heterotrophic Carbon Dioxide Fixation by Bacteria

No attempt will be made to give detailed consideration to all articles that have appeared. Only those that have a direct bearing on the mechanism of fixation will be discussed. The recent reviews (28, 29, 30) should be consulted for additional information. It is clear that the present consideration of the mechanism of heterotrophic fixation of carbon dioxide by bacteria will not offer a final solution to the problem raised by the discovery. Investigation of the mechanism has only recently begun in earnest, and it is true that future observations may change the general picture. Nevertheless, an outline of the present position should be of value in realizing shortcomings of present ideas, and in pointing out the next steps to be taken. It is hoped that the present review will be of value in investigations on animal physiology, a field in which the results of bacterial fixation of carbon dioxide have become only recently so pertinent. It is likely that extensive application of the fundamental facts discovered in bacterial physiology will be made to animal physiology. In fact, some applications have already been made, and these will be discussed. Thus far, most of the work on the actual mechanism of fixation of carbon dioxide has been done with bacteria, and particularly with the propionic acid bacteria. For this reason studies on the linkage of carbon atoms with carbon dioxide have been largely confined to the C_3 and C_1 addition. There are fixations in a number of other heterotrophic fermentations which fundamentally may be examples of C_3 and C_1 addition but the evidence is not clear at present. The mechanisms of bacterial fixation of carbon dioxide, therefore, will be considered under the following heads: (1) fixation of carbon dioxide not involving carbon to carbon linkage; (2) fixation of carbon dioxide involving carbon to carbon linkage. Under the latter will be discussed (*a*) C_3 and C_1 addition and (*b*) miscellaneous fixation reactions.

A. Fixation of Carbon Dioxide Not Involving Carbon to Carbon Linkage

Formation of Formic Acid.—The reduction of carbon dioxide to formic acid by gaseous hydrogen with *Escherichia coli* has been clearly shown (25) and it presumably occurs with other bacteria that contain hydrogenylase, *i. e.*, produce carbon dioxide and hydrogen from formic acid. The exact kinetics of the reaction are not known, perhaps:

$$\mathrm{HO{\cdot}\overset{\overset{H}{O}}{\overset{\cdot}{C}}{:}O + H_2 \rightleftharpoons HO{\cdot}\overset{\overset{H}{O}}{\overset{\cdot}{\underset{\underset{H}{\cdot}}{C}}}{\cdot}OH \rightleftharpoons HO{\cdot}\overset{O}{\overset{\cdot\cdot}{\underset{\underset{H}{\cdot}}{C}}} + H_2O}$$

The enzymes have not been isolated or purified so as to permit a detailed study of possible coenzymes or carriers that may be involved. The reaction may be more complicated than pictured and, perhaps, include phosphorylation.

A problem of some importance is the mechanism of formation of formic acid, particularly whether it is usually formed by reduction of carbon dioxide. It is generally accepted that formic acid originates from pyruvic acid by a hydroclastic split:

$$\mathrm{\underset{\underset{HOH}{+}}{CH_3{\cdot}CO{\cdot}COOH}} \longrightarrow \mathrm{CH_3{\cdot}\underset{\underset{OH}{\cdot}}{C}{:}O + HCOOH} \qquad (1)$$

and that carbon dioxide and hydrogen are formed from the formic acid. It seems just as likely that the reaction may occur stepwise as follows:

$$\mathrm{CH_3{\cdot}CO{\cdot}COOH + HOH + 2}A \longrightarrow \mathrm{CH_3{\cdot}COOH + CO_2 + 2H}A \qquad (2)$$

$$\mathrm{CO_2 + 2H}A \longrightarrow \mathrm{HCOOH + 2}A \qquad (3)$$

$$\mathrm{HCOOH \rightleftharpoons CO_2 + H_2} \qquad (4)$$

A in the above reactions functions as a hydrogen carrier. Accordingly formic acid would not be a direct intermediary product of pyruvic acid breakdown but would result by reduction of carbon dioxide. The hydroclastic reaction thus would be only a special case of pyruvic acid breakdown in which carbon dioxide was the hydrogen acceptor. The donator function of the pyruvic acid would be the same as in the dissimilation of pyruvic acid by a number of organisms to form acetic acid and carbon dioxide, *e. g.*, the aerobic dissimilation studied by Lipmann (31) and Barron (32), the dismutations of Nelson and Werkman (33) and Krebs (34), and the formation of acetic acid from pyruvate in the propionic acid fermentation (35, 36). The hydroclastic reaction was originally proposed to account for formation of formic acid from pyruvate before it was known that carbon dioxide is an active metabolite. Therefore, consideration has not been given to the reduction of carbon dioxide. Reactions (2) and (3) appear to account for the facts as readily as does reaction (1). For example, *Eberthella typhi* would be able to reduce carbon dioxide to formic acid with the proper hydrogen donator but not with gaseous hydrogen. In all fermentations in which formic acid is formed in the presence of labelled carbon dioxide, the acid should contain isotopic carbon, even though reaction (4) did not occur. This point should be investigated further.

Formation of Methane.—Foster and his co-workers (28) have described the methane fermentation as

> "One of the most interesting and up to now best understood of the biological processes utilizing carbon dioxide. . ."

Whether the process is considered well understood depends largely on the point of view. If one is comparing the old concept of the fermentation with the new concept, then our present understanding appears favorable. According to the present concept (26), methane fermentations of all types may be considered as a process of oxidation in which carbon dioxide acts as a hydrogen acceptor:

$$4H_2A + CO_2 \longrightarrow 4A + CH_4 + 2H_2O$$

H_2A is the oxidizable molecule and A the oxidation product. H_2A may be ethyl alcohol and be oxidized to acetic acid, *e. g.*, *Methanobacterium omelianskii*, or it may be acetic acid and be oxidized to carbon dioxide as by *Methanosarcina methanica*. The reaction has been proved in the case of *Methanobacterium omelianskii* but with *Methanosarcina methanica* the proof of the reduction of carbon dioxide to methane as a result of oxidation of acetic acid is yet considered inconclusive (37) although quite likely.

On the other hand, when one views the methane fermentation from the standpoint of understanding the actual mechanism of the reduction of carbon dioxide to methane, it must be admitted that we are almost completely in the dark. The intermediate steps are completely unknown, and a set of reactions for a possible explanation of the process has not been ventured. According to Barker (38) formic acid is not an intermediary in the process since it would not replace carbon dioxide in the fermentation by *Methanobacterium omelianskii*. It is evident that much important work is yet to be done before we have any understanding of the reactions which take place in the conversion of the most highly oxidized form of carbon to its state of greatest reduction.

B. Fixation of Carbon Dioxide Involving Carbon to Carbon Linkage

1. C_3 and C_1 Addition

Evidence That Succinate Is Formed by Fixation of Carbon Dioxide.—The first evidence of fixation of carbon dioxde by C_3 and C_1 addition was obtained by fermentation of glycerol with the propionic acid bacteria (2, 4). It was established that for each mole of succinic acid formed from the glycerol one of carbon dioxide was fixed, *i. e.*, there was a net decrease in carbon dioxide of the system and in amounts equivalent to the succinic acid formed. When the fermentation was conducted in phosphate buffer in the absence of carbon dioxide little or no succinic acid was formed. It was, therefore, clear that carbon dioxide played an important role in succinic acid formation. Since the succinic acid (C_4) was formed from the glycerol (C_3) in amounts equivalent to the consumed carbon dioxide (C_1), the synthesis by C_3 and C_1 addition seemed the most probable mechanism. Pyruvic acid was proposed as the probable intermediary C_3 compound of the fixation reaction because it could be easily isolated from the fermentation liquor by the addition of sulfite to fix carbonyl compounds (6).

$$CO_2 + CH_3{\cdot}CO{\cdot}COOH = COOH{\cdot}CH_2{\cdot}CO{\cdot}COOH$$

$$COOH{\cdot}CH_2{\cdot}CO{\cdot}COOH + 4H = COOH{\cdot}CH_2{\cdot}CH_2{\cdot}COOH + 2H_2O$$

It was recognized that the equivalence of succinic acid formed and the carbon dioxide utilized might be the result of the stoichiometric demands of the oxidation-reduction balance, succinic being oxidized and carbon dioxide reduced. But this explanation did not seem correct because other acceptors such as oxygen (39) did not increase the yield of succinic acid; likewise increased yields of the reduced compound propyl alcohol were not accompanied by an increase in succinic acid.

The fact that Elsden (40) had shown that the rate of succinic acid formation by *Escherichia coli* is a function of the concentration of carbon dioxide in the medium, further supported the role of carbon dioxide in succinic acid synthesis, and indicated that the phenomenon was not limited to the propionic acid bacteria.

Further investigations (41, 5) on the propionic acid fermentation did not bring forth much additional information as to the actual mechanism of the fixation. Carbon dioxide fixation was demonstrated with several substrates, and such inhibitors as malonate, azide, arsenite, cyanide and pyrophosphate were found to have no effect on the fixation, whereas sodium fluoride and iodoacetate did. The inhibition by fluoride caused an equivalent decrease in succinic acid formation in accordance with the concept of C_3 and C_1 addition. Some indication was obtained that phosphate had a function in carbon dioxide fixation.

The availability of carbon isotopes provided a great impetus to further investigation. Heretofore, studies on the mechanism of fixation were of necessity largely speculative, for although one could prove that carbon dioxide was fixed and actually entered into a carbon to carbon linkage, there was no method of determining its location in the carbon compound.

With isotopes the fixed carbon dioxide may be located in the compound. Work on the propionic acid fermentation with radioactive carbon was immediately initiated at the University of California and simultaneously a cooperative investigation between Iowa State College and the University of Minnesota was started with heavy carbon (C^{13}). Fixation by the propionic acid bacteria fermenting glycerol was confirmed with both the radioactive (16) and with heavy carbon dioxide (42, 14). The picture was not as simple, however, as previously thought, for the isotopic carbon was found in both the propionic acid and in the succinic acid in contrast to predictions made on the basis of the stoichiometric relationship between carbon dioxide fixed and succinic acid formed.

The fixation of carbon dioxide by coliform bacteria likewise was investigated by Wood *et al.* (42, 14), and with the C^{13} isotope a clear-cut fixation was demonstrated in the fermentation of galactose, pyruvate and citrate. Previous investigations, not using the isotope, had led to the proposal that these bacteria fix carbon dioxide but the results could not be conclusive, since there was always a net production of carbon dioxide in these fermentations. It is virtually impossible to prove conclusively, without isotopes, that carbon dioxide is both produced and assimilated in experiments involving a net increase in carbon dioxide, but with labelled carbon dioxide this can be accomplished. Fixation of carbon dioxide by the coliform bacteria occurred solely in the succinic and formic acids. The indications were, therefore, that all fixation in these fermentations was by C_3 and C_1 addition, except that formed by reduction of carbon dioxide to formic acid as described by Woods (25).

Nishina, Endo and Nakayama (43) by use of radioactive carbon have demonstrated the synthesis of malic acid and fumaric acid from pyruvic acid and carbon dioxide in fermentations by *Escherichia coli.* These authors prepared recrystallized derivatives of the acids; thus there is every reason to believe they were dealing with pure compounds. The results are significant because they demonstrate the presence of fixed carbon in compounds proposed to occur as intermediates in the conversion of oxalacetate to succinate. Radioactive fumarate also was demonstrated from a fermentation of glucose, and when ammonium chloride was added to the dissimilation of pyruvate, aspartate containing fixed carbon could be isolated. The scheme first proposed by Wood and Werkman (5) involving oxalacetate synthesis from pyruvate and carbon dioxide and a subsequent stepwise conversion to malate, fumarate and succinate, is suggested as the mechanism involved in these fixations. Krebs and Eggleston (44) have shown that this set of reversible reactions from oxalacetate to succinate occurs in the propionic acid fermentation, demonstrating that such a set of reactions is feasible. No new data were supplied, however, on the actual fixation of carbon dioxide.

Other Possible Mechanisms of Succinate Formation.—Whether succinic acid is ever formed from carbohydrate by a mechanism other than fixation, *e. g.*, by acetic acid condensation or by decarboxylation of α-ketoglutaric acid as in the Krebs cycle, must await further investigation. The authors have presented evidence on several occasions which they believed to indicate a formation of succinic acid in the propionic acid fermentation by acetic or pyruvic acid condensation. The fact that fluoride had only a partial inhibitory effect on succinate formation from glucose, at concentrations which in glycerol fermentations almost completely inhibited the fixation of carbon dioxide as measured manometrically by the net decrease in carbon dioxide, led to this suggestion (5). These results together with other evidence made it appear that there was a fluoride insensitive mechanism of succinic acid formation which was independent of carbon fixation. This point has now been examined with heavy carbon dioxide, on the supposition that if there were such a mechanism, succinate formed in the presence of fluoride would not contain heavy carbon. The results (unpublished) show that fluoride does not prevent fixation of carbon dioxide in either the succinate or propionate formed from glucose by the propionic acid bacteria or in succinate formed from galactose, arabinose or mannose by *Escherichia coli*. This is apparent, for the C^{13} concentration was practically the same in the succinate from fermentations with or without addition of fluoride, and there was substantial fixation of carbon dioxide in each case. The yield of succinic acid per mole of substrate fermented was reduced in the presence of fluoride but apparently the succinate which was formed involved a fixation reaction. There was thus no indication by this method that succinate formation occurred by any other than the fixation reaction.

An explanation is now available for this incomplete inhibition of carbon dioxide fixation by fluoride. Studies with heavy carbon have shown that there is a formation of carbon dioxide from glycerol by propionic acid bacteria even in those experiments in which there is a net decrease in carbon dioxide in the system (14). This fact is apparent, for the C^{13} of the sodium bicarbonate and carbon dioxide was diluted during the fermentation. That the dilution was caused by $C^{12}O_2$ produced from the substrate is certain, inasmuch as the possibility of dilution by a miscellaneous exchange reaction was eliminated since all the C^{13} of the original system was accounted for in the products, residual sodium bicarbonate and carbon dioxide, at the conclusion of the experiment. Actually the carbon dioxide fixed in the glycerol fermentation by propionic acid bacteria is not just that observed manometrically as a net decrease in carbon dioxide but it is this quantity plus that produced from the glycerol by fermentation. The inhibition observed by fluoride was only that fraction of the total carbon fixation measured manometrically by a net decrease in gas. Thus the fixation in the glycerol fermentations was never completely inhibited by fluoride.

The carbon dioxide fixed in the presence of fluoride is almost entirely in propionic acid since there is little succinic acid formed under these conditions. There is reason to believe, however, that this carbon dioxide is initially fixed by C_3 and C_1 addition, and propionate is formed from the resulting C_4 dicarboxylic acid. This point will be considered later.

Recently Krebs and Eggleston (44) have rejected the idea that succinate may be formed by acetic acid condensation in the propionic acid fermentation. This reaction (or pyruvic acid condensation prior to formation of acetic acid) was proposed (45, 6) to account for low yields of acetic acid frequently found. Since in some cases the yield of succinate was not sufficient to account for the carbon dioxide formed on the basis that carbon dioxide equals the acetate plus twice the succinate, it was proposed that part of the succinate was in turn broken down to propionic acid and carbon dioxide. Krebs and Eggleston reject this explanation on the basis of the following points:

(*a*) They could not demonstrate an anaerobic breakdown of succinate and conclude that it is not metabolized by *Propionibacterium shermanii* anaerobically.

(*b*) They observed the simultaneous formation of both fumarate and succinate and state concerning this observation:

"Since the reaction, succinate → fumarate, does not occur under anaerobic conditions, fumarate formed anaerobically cannot have arisen from succinate; if the formation of succinate by reduction is rejected, the improbable assumption of two separate mechanisms for the formation of succinate and fumarate must be made."

They suggest that the acetic acid is oxidized to carbon dioxide anaerobically by an unknown mechanism, and it is this reaction that causes the low yields of acetate and high yields of CO_2.

The points raised in rejection of the Wood and Werkman scheme are not conclusive for the following reason. Shaw and Sherman (46), Hitchner (47), Wood *et al.* (6), and Fromageot and Bost (48) have all reported that succinate is fermented by propionic acid bacteria. In view of these findings by so many different investigators, and extended over a period of eighteen years, it can be concluded that Krebs and Eggleston's conclusion that succinate is not fermented by *Propionibacterium shermanii* is incorrect. Erb (49) in this laboratory has studied the anaerobic breakdown of succinate by these bacteria (*Propionibacterium shermanii* included) rather extensively. He has found that washed cell suspensions from a five-day culture grown on a medium of yeast extract (Difco) 0.3 per cent, peptone (Difco) 0.2 per cent and glucose 1.0 per cent decarboxylate succinate anaerobically. The optimum *p*H for the reaction is 5.2; at *p*H 6.4 there was hardly any activity. The inactivity in Krebs and Eggleston's experiments may have been caused by the alkaline *p*H of the $NaHCO_3$ buffer.

Wood and Werkman (unpublished) have determined the products of this anaerobic breakdown of succinate. The rate of succinate fermentation falls off rapidly after the first six hours so that a large conversion was not obtained (15.7 mM per liter were fermented in two days by *Propionibacterium arabinosum*). After applying a correction for endogenous values (1.07 mM propionic acid, 1.36 mM acetic acid and 2.93 mM carbon dioxide), 15.84 mM propionic acid, 16.93 mM carbon dioxide and 3.6 mM acetic acid were obtained for the products. No other products were detected. The results show that the succinate was appreciably decarboxylated to propionic acid and carbon dioxide.

The experiments quoted by Krebs and Eggleston from Wood and Werkman (45) as direct evidence that succinate is not formed by condensation are satisfactorily explained, if one assumes a subsequent breakdown of succinate.

With regard to the second point three objections may be raised.

1. Under the proper experimental condition, *i. e.*, with methylene blue as an acceptor, the reaction succinate → fumarate does occur. Whether it occurs in a natural dissimilation of glucose is a question that awaits further experimentation.

2. It is doubtful whether anyone would agree to rejection of the formation of succinate by reduction, for few would deny the possibility of succinate formation by carbon dioxide fixation yielding oxalacetate which is subsequently reduced to succinate.

3. Apparently a more crucial objection of Krebs and Eggleston to succinate formation by condensation is that this gives two mechanisms for formation of dicarboxylic acids, the second being by the fixation reaction. There is, however, a clear-cut example of the formation of succinate by two mechanisms. In the aerobic dissimilation of pyruvate by pigeon liver with addition of malonate as an inhibitor, succinate is

formed which contains no fixed heavy carbon, and in the same dissimilation fumarate is formed containing fixed heavy carbon (50). Obviously these two dicarboxylic acids arise by different reactions. Anaerobically in the absence of malonate, both succinate and fumarate are formed from pyruvate by pigeon liver. In this latter case, in which no inhibitor is added, there is every reason to believe that these acids arise by both the fixation reaction and a modified Krebs cycle. These reactions have not been shown to occur in the propionic acid fermentation but they certainly serve to illustrate that there is no reason *a priori* to conclude there cannot be two mechanisms. The authors have always held the view that propionic acid fermentation is more complex than generally believed, and that some modification is necessary in the scheme, be it inclusion of acetic acid condensation, pyruvic acid condensation or a modified Krebs cycle. The point is that Krebs and Eggleston have not furnished the necessary information to warrant rejection of any given proposal as yet—not that their proposals may not be correct. It may be of some interest in this connection to point out that the aerobic oxidation of propionic acid with liberation of carbon dioxide by propionic acid bacteria is more rapid than acetic acid oxidation.

Of some significance is the fact that in the large number of dissimilations studied with a variety of bacteria (*Propionibacterium*, *Escherichia*, *Aerobacter*, *Citrobacter*, *Proteus*, *Staphylococcus* and *Streptococcus paracitrovorus*) (14, 15) there has been fixation of carbon dioxide in succinate without exception, whenever it was formed. This indicates the general occurrence of the fixation reaction but does not necessarily eliminate the possibility that there are two mechanisms of formation of succinate. The fact that the amount of C^{13} fixed in succinate varies over a rather wide range in different experiments may be an indication that there is more than one source of succinate. The formation of succinate by a mechanism not involving C^{13} fixation would dilute the C^{13} of the succinate formed by fixation.

The Location of Fixed Carbon in Succinate.—There is good evidence that carbon dioxide is fixed by a C_3 and C_1 addition; however, assuming that fixation does occur by C_3 and C_1 addition, the elucidation of the actual mechanism is difficult even with the use of isotopes. The evidence that there is fixation by C_3 and C_1 addition will be considered first, and then preliminary studies will be discussed of what is believed to be the initial reaction of carbon dioxide fixation.

The formation of succinic acid, particularly, is believed to involve C_3 and C_1 addition. In this case the fixed carbon dioxide should be in the terminal carboxyl group of succinic acid. That this is the location of the fixed carbon has been definitely shown (51, 52) by degrading succinic acid containing fixed $C^{13}O_2$. Succinic acid isolated from fermentations by *Propionibacterium pentosaceum, Escherichia coli, Proteus vulgaris, Aerobacter indologenes* and *Streptococcus paracitrovorus* has been tested, and in each the methylene carbon atoms have been found to contain the normal complement of C^{13}, whereas the carboxyl groups contain all the fixed heavy carbon. The succinic acid was decarboxylated (51) by converting it to a mixture of fumaric and malic acids with a heart muscle preparation containing succinic dehydrogenase and fumarase. The malic acid was oxidized to two molecules of carbon dioxide and one of acetaldehyde by acid permanganate. The aldehyde is from the methylene groups and the carbon dioxide from the carboxyl groups.

Only by quantitative methods is it possible to prove that a symmetrical dicarboxylic acid contains fixed carbon in both its carboxyl groups. If the concentration of C^{13} in the carboxyl groups is greater than the average value calculated on the basis that one carboxyl group has a normal complement of C^{13} and the other a complement equivalent to that of the $C^{13}O_2$ available for fixation, obviously fixation has occurred in both carboxyl groups of at least part of the dicarboxylic acid molecules. If the C^{13} concentration in the carboxyl groups is less than or equal to this average value there is no means available at present to determine whether any one molecule has more than one carboxyl containing fixed carbon. In none of the fermentations so far examined (51, 52) has there been a concentration of C^{13} in the carboxyl carbons of the succinate sufficiently high to prove fixation in both carboxyl groups. In the propionic acid fermentation of glycerol the concentration of C^{13} in the carboxyl groups of succinate approaches rather closely the value estimated for fixation in only one carboxyl group. This fact may be interpreted as evidence that succinate is formed in this fermentation solely by C_3 and C_1 addition, the C_3 compound being formed from the glycerol with little or no exchange of its carbon with $C^{13}O_2$.

The Initial Reaction in C_3 and C_1 Addition.—As a working hypothesis the reaction $CO_2 + CH_3{\cdot}CO{\cdot}COOH \rightleftharpoons COOH{\cdot}CH_2{\cdot}CO{\cdot}COOH$ was proposed (4, 42) as the possible initial conversion in fixation of carbon dioxide by C_3 and C_1 addition. It should be clearly understood that at best the reaction represents merely the over-all conversion. It is quite possible and even probable that phosphorylated intermediates are involved and that the reaction is more complex than represented. The proposal that the fixation of carbon dioxide by C_3 and C_1 addition is a reaction of general biochemical importance has been advanced by the authors for the past several years (4) without general acceptance until recently. The present tendency is, perhaps, to the other extreme in that suggestions involving some speculation may now be too readily accepted. This results, in part, from the great general interest in isotopic investigations and a consequent focussing of attention on studies on fixation of carbon dioxide. The fixation reaction has recently been suggested by Solomon, *et al.* (53), as a part of a possible mechanism of fixation of carbon dioxide in glycogen synthesized from lactate by liver; also Meyerhof (54) has used the reaction to explain the Pasteur effect. These authors are fully aware that they are dealing with complex phenomena for which a number of possible explanations may be offered, and, furthermore, that since the mechanism of the fixation reaction has not been fully elucidated, there can be no certainty as to how it fits into a scheme. It is desirable, however, that such ideas be advanced as working hypotheses. Inasmuch as the fixation reaction is used frequently in a wide variety of schemes, there is a tendency to accept the reaction as fact and to forget that the actual mechanism of the reaction is not completely established.

Krebs and Eggleston (55, 44) in two articles entitled, "The Biological Synthesis of Oxalacetate from Pyruvic Acid and Carbon Dioxide," imply that they show carbon dioxide fixation with pyruvic acid to form oxalacetate. In no case, however, was a synthesis of oxalacetate shown; furthermore, fixation of carbon dioxide was not demonstrated and in one of the investigations pyruvic acid was not used as a substrate. Their principal experimental contribution, in so far as carbon dioxide fixation is concerned, was to show that an increased carbon dioxide concentration stimulates succinate formation from pyruvate by pigeon liver. In addition they provided evidence that cocarboxylase has a function in fixation (*cf.* page 158).

The present status of evidence relative to the actual mechanism of the fixation reaction is outlined below. It should be emphasized that the uncertainty as to the mechanism of C_3 and C_1 addition in no way alters the evidence (51, 52) that there is such a fixation. The question is not so much whether it occurs, but how it occurs. Krebs' (56) statement:

> "This (Evans and Slotin's demonstration of fixed carbon in α-ketoglutarate) completes the proof of the occurrence of reaction (9) ($COOH{\cdot}CO{\cdot}CH_3 + CO_2 = COOH{\cdot}CO{\cdot}CH_2{\cdot}COOH$) in pigeon liver,"

is hardly accurate. The position of the fixed carbon in α-ketoglutarate was not known at this time, and even now that the position of the fixed carbon is known, it can only be said that the results do not conflict with the proposed mechanism of fixation. Fixation in the carboxyl group of fumarate or malate, for example, will give the same result as fixation in oxalacetate since these compounds are convertible to oxalacetate. For that matter, it has not been proved that oxalacetate, as such, is the compound that reacts with pyruvic acid in the Krebs' cycle. The only way to prove that carbon dioxide can be fixed by addition to pyruvic acid is to use isolated enzyme systems and to isolate the oxalacetic acid under acceptable conditions. The enzyme system must be such as to eliminate miscellaneous reactions which confuse the picture. Even with evidence such as this, objection can be raised that the fixation was not conducted under natural conditions.

The fixation reaction as represented in the above equation is a carboxylation of pyruvic acid and implies that the decarboxylation of oxalacetic acid is reversible. Until recently the only enzymes known to decarboxylate oxalacetic acid were carboxylase and the thermostable enzyme from muscle studied by Breusch (57). There is no certainty that carboxylase acts on the carboxyl next to the methylene carbon in oxalacetate. Its action may well be on the carboxyl next to the carbonyl to yield malonic aldehyde, which breaks down spontaneously to acetaldehyde and carbon dioxide.

$$COOH{\cdot}CH_2{\cdot}CO{\cdot}COOH \xrightarrow{\text{Carboxylase}} COOH{\cdot}CH_2{\cdot}CHO + CO_2$$

$$COOH{\cdot}CH_2{\cdot}CHO \longrightarrow CO_2 + CH_3CHO$$

In this case carboxylase would not be a component of the fixation reaction since it activates the carboxyl adjacent to the carbonyl group instead of the one adjacent to the methylene group. The thermostable enzyme studied by Breusch has a very low activity and is probably not concerned in the fixation reaction.

Krampitz and Werkman (58) have recently discovered a new enzyme which offers promise of being one involved in the fixation reaction. The enzyme is heat-labile and catalyzes the decarboxylation of oxalacetate to pyruvate. An acetone and alkaline phosphate washed preparation of *Micrococcus lysodeikticus* is used to demonstrate the presence of the enzyme. Such preparations are free from cocarboxylase and Mg^{++} and will not decarboxylate or oxidize either oxalacetate or pyruvate. When Mg^{++} is added, the enzyme decarboxylates the oxalacetate, and there is no further action on the resulting pyruvate. When cocarboxylase is then added, the resulting pyruvate is oxidized to acetic acid and carbon dioxide. It is therefore evident that the decarboxylation of oxalacetate by this enzyme is dependent on Mg^{++} but independent of cocarboxylase.

On the basis that this enzymic reaction is reversible, and that the equilibrium is not too far to the side of pyruvic acid, the synthesis of oxalacetate from pyruvate and carbon dioxide should be possible with this enzyme.

Krampitz and his co-workers (59) attempted this synthesis but were unable to demonstrate any formation of oxalacetate. With $C^{13}O_2$ they have obtained evidence, however, that the enzyme is involved in the carboxylation or fixation reaction.

A decarboxylation of oxalacetic acid was conducted in the presence of $C^{13}O_2$ with the deficient preparation to which Mg^{++} was added. The decarboxylation was allowed to proceed until it was approximately 50 per cent complete, and then the C^{13} concentration in the carboxyl adjacent to the methylene group was determined in the residual oxalacetate. The necessary carboxyl was obtained by decarboxylation with citric acid and aniline (60). The resulting carbon dioxide contained a concentration of C^{13} (1.4 per cent) significantly above the normal. On the contrary, similar experiments conducted without addition of the enzyme contained substantially no fixed carbon. Clearly the enzyme catalyzed the induction or exchange of carbon dioxide with the carboxyl group of oxalacetate. Essentially this exchange is C_3 and C_1 addition, for the products of the enzymic conversion are solely C_3 and C_1 compounds and there is no reason to believe the intermediate steps involve other carbon chains. The following reactions are proposed as a possible explanation of the observed exchange during decarboxylation of oxalacetate and the failure to obtain a synthesis of oxalacetic acid from pyruvic acid and carbon dioxide.

$$COOH \cdot CH_2 \cdot CO \cdot COOH \rightleftharpoons CO_2 + (C_3 \text{ compound})$$

$$(C_3 \text{ compound}) + C^{13}O_2 \rightleftharpoons C^{13}OOH \cdot CH_2 \cdot CO \cdot COOH$$

$$(C_3 \text{ compound}) \longrightarrow CH_3 \cdot CO \cdot COOH$$

It is suggested that the C_3 compound in parentheses is a derivative of pyruvic acid, possibly a phosphorylated compound, and that it is this compound rather than pyruvic acid that is a component of the fixation reaction. In addition it is suggested that the unknown C_3 compound is converted to pyruvic acid, and that in the system as employed by Krampitz *et al.*, this reaction is not reversible. Pyruvic acid thus would not be suitable for demonstration of the synthesis of oxalacetic acid. Furthermore, the decarboxylation of oxalacetic acid may not be by a direct splitting out of carbon dioxide but may involve a preliminary phosphorylation. Additional studies are necessary before these possibilities can be evaluated fully. Nevertheless, the above results offer the first direct evidence that oxalacetic acid is a component in the fixation reaction. No previous investigation has shown a fixation in oxalacetic acid, assumed to be the initial product of C_3 and C_1 addition.

The question arises whether or not an exchange of carbon dioxide with the carboxyl group occurs during the action of carboxylases in general or is specific for the particular carboxylase which acts on the carboxyl beta to the keto group of oxalacetic acid. If there is a similar exchange with other carboxylases, the possibility exists of fixation of carbon dioxide by a rather large number of reactions, and the theory that C_3 and C_1 addition is the principal path for fixation of carbon dioxide by typical heterotrophs would probably have to be revised. The evidence for fixation of carbon dioxide by other mechanisms will be considered in a later section, but usually such mechanisms have not been found in typical heterotrophs. It is true that some exchange might be expected when any carboxylase is active but for most carboxylases the equilibrium may be so far to the side of decarboxylation that from a practical standpoint the reaction is irreversible. The extent of exchange of carbon dioxide with the carboxyls of pyruvic, lactic and α-ketoglutaric acids during aerobic decarboxylation by *Micrococcus lysodeikticus* has been determined by Krampitz *et al.* (59). The decarboxylation of the acids was allowed to proceed in the presence of $C^{13}O_2$ until somewhat over 50 per cent of the acid was converted, then the C^{13} concentration in the carboxyl group was measured. The results showed that there was practically no exchange in the carboxyl group of these acids under the conditions of the experiments. Pyruvate arising from lactate was, likewise, tested with negative results. The absence of exchange in this pyruvate is, perhaps, of greater significance than in the direct use of pyruvate since it was formed within the cell and in all probability came in contact with the active enzyme centers, whereas there is no assurance that such are the conditions when starting with pyruvate and determining C^{13} in the unfermented portion. More extensive studies are needed before a definite conclusion can be drawn, but at present the possibility of a general fixation of carbon dioxide through the action of carboxylases as a group seems remote. Fixation by carboxylases may be limited largely to the enzyme studied by Krampitz *et al.* Evans (61), likewise, has found no evidence of exchange during the action of carboxylase on pyruvate. The exchange was studied by use of radioactive carbon dioxide at pressures as high as 300 atmospheres. Ruben and Kamen (17) also consider the reaction irreversible.* Their evidence is hardly conclusive. Experiments were conducted in which yeast suspensions, with no added substrate, fixed radioactive carbon dioxide. By use of pyruvate as a carrier, a fraction was isolated as the hydrazone. This hydrazone contained but a small per cent of the total fixed radioactive carbon. There was no proof provided, however, that there actually was

* However, *cf. Proc. Natl. Acad. Sci. U. S.*, **27**, 475 (1941).

any pyruvate formed in the dissimilation (none was added apparently). The negative result, therefore, could have been caused either by the absence of an exchange or because there was not enough pyruvate present to give a detectable amount of fixed carbon, even though an exchange did occur.

Proof that there is no exchange in pyruvate is of considerable importance because, both in bacterial fermentations (15) and in the dissimilation of pyruvate by animal tissue (50), lactate is formed containing fixed carbon in the carboxyl group. Since pyruvate is generally believed to be the precursor of lactate, fixation by exchange in the pyruvate is one of the possible mechanisms of fixation in lactate. Other mechanisms for this fixation will be considered later.

Relationship of Cocarboxylase to the Fixation Reaction.—In contrast to the findings of Krampitz and Werkman (58) and Krampitz *et al.* (59), Krebs and Eggleston (55) and Smyth (62) believe that cocarboxylase is essential for the fixation of carbon dioxide in oxalacetate. Krebs and Eggleston (55), reasoning from indirect evidence with pigeon liver and from analogy with bacterial fixation of carbon dioxide (5), concluded that there is a synthesis of oxalacetic acid from pyruvic acid and carbon dioxide in pigeon liver. The evidence will be considered in greater detail in the section on animal fixation of carbon dioxide, but it is entirely indirect since a net uptake of carbon dioxide was not demonstrated. The proposed function of thiamin in the fixation reaction is based on the observation that thiamin, on addition to muscle and liver suspensions from thiamin-deficient pigeons, causes an increased dissimilation of pyruvate by liver but not by muscle. The vitamin, therefore, was presumed to take part in a reaction present in liver but absent in muscle. The fixation reaction was believed by Krebs and Eggleston to meet this requirement, and they, therefore, concluded that the vitamin acts in this fixation reaction. Evans and Slotin (63) have since presented evidence that there is, in fact, no fixation of carbon dioxide by pigeon breast muscle. The action of carboxylase in pyruvate oxidation, according to Krebs and Eggleston's concept, is to synthesize oxalacetate which is necessary as a hydrogen carrier and as a component of the Krebs cycle. The stimulating action of cocarboxylase in apparently dissimilar reactions, *i. e.*, the non-oxidative decarboxylation of pyruvate by yeast and oxidative decarboxylation by animal tissue and bacteria, was believed to be explained on the basis that decarboxylation of pyruvate and carboxylation of oxalacetate were enough alike so that they would be catalyzed by similar or identical enzymes, both requiring cocarboxylase. The usual assumption of a dual function of a carboxylase as a catalyst was thus avoided.

Although the above explanation of an analogous function of carboxylase in oxidative and non-oxidative decarboxylation of pyruvate may seem attractive, there are several facts that argue against these proposals. The experimental evidence presented by Krebs and Eggleston in support of the proposed function of cocarboxylase is hardly conclusive. They presented no evidence to show that the breast muscle preparation was actually deficient in cocarboxylase. If such were not the case, addition of cocarboxylase would not increase the rate of pyruvate oxidation by the muscle. Ochoa and Peters (64) have shown that there is a marked difference in the vitamin content of the different organs of an animal showing effects of vitamin deficiency, and that liver usually has a lower content of the vitamin than the muscle. The observed difference in the effect of the vitamin on addition to muscle and liver may not have resulted from a stimulation of a reaction present in liver and absent in muscle but, instead, because the liver was deficient in cocarboxylase and the muscle was not. It also has been pointed out by Barron (65) that the difference may have resulted because minced muscle is unable to phosphorylate thiamin rapidly (66). The necessary cocarboxylase, in this case, would not be formed in muscle, but would be formed in liver.

Aside from these weaknesses in experimental proof, and the evidence of Krampitz and Werkman that cocarboxylase is not necessary in the reaction in the case of *Micrococcus lysodeikticus*, the concept is not convincing, for it is not in agreement with known facts of pyruvate oxidation. Banga and her co-workers (67) have shown that besides cocarboxylase, a C_4 dicarboxylic acid is an essential component of the pyruvate oxidizing system. Clearly in these oxidations, the function of the carboxylase is not to synthesize C_4 dicarboxylic acids as proposed by Krebs and Eggleston.

Smyth's evidence (62) that thiamin is a component of the fixation reaction was obtained from a study of the dismutation of pyruvate by thiamin deficient *Staphylococcus*. The dismutation could be stimulated either by the addition of thiamin or oxalacetate. In accordance with the Krebs and Eggleston concept, it was proposed that the function of the thiamin in the pyruvate dismutation was to promote synthesis of oxalacetate which is necessary as a hydrogen carrier. When oxalacetate is present, the need of thiamin, therefore, is removed. At present there seems to be no other ready explanation of Smyth's results. The system is complex, however, and it is by no means certain that Smyth's explanation is correct. Krebs and Eggleston (44) have reported recently that addition of oxalacetate, fumarate or malate catalytically accelerated the fermentation of glycerol by propionic acid bacteria. There is no reason to believe the fixation reaction was impaired in any way in these fermentations, yet oxalacetate stimulated the dissimilation. This serves to illustrate that stimulation with oxalacetate as observed by Smyth may not have resulted because of a weak fixation reaction, but for some other reason, as in the propionic acid fermentation. At any rate it can be concluded that the above-proposed action is not the function of cocarboxylase in the dissimilation of pyruvate by *Micrococcus lysodeikticus* (58) for in this case both Mg^{++} and cocarboxylase are necessary for the oxidation of pyruvate even when oxalacetate is added.

On the basis of present results it is apparent that there is fairly direct evidence (58) that cocarboxylase is not a component of the fixation reaction as such. It is conceivable that in the complex reactions studied by Krebs and Eggleston and by Smyth there may be an indirect connection between carboxylase activity and carbon dioxide fixation, but if there is, the mechanism is obscure at present.

The Location of Fixed Carbon in Propionic Acid.—The only compound other than C_4 dicarboxylic acids that definitely has been proposed to arise by C_3 and C_1 addition is propionic acid. Carson and Ruben (16) and Wood *et al.* (42, 14), independently with isotopic carbon found that propionic acid formed in the fermentation of glycerol by *Propionibacterium* contains fixed carbon dioxide. Both proposed that the carbon is initially fixed by C_3 and C_1 addition and that the propionate arises from a dicarboxylic acid. Accordingly, the fixed carbon would be in the carboxyl group. Carson and his co-workers (68), degraded propionate obtained from the fermentation of glycerol, and erroneously concluded that the fixed carbon is located not only in the carboxyl group but is probably equally distributed among all three carbon atoms of the molecule. Wood *et al.* (69), then isolated propionic acid containing fixed C^{13} from a fermentation of glycerol and degraded the acid by a different set of reactions. They found the fixed carbon to be exclusively in the carboxyl group. The degradation was accomplished by α-bromination to give bromo-propionic acid, conversion of this compound to lactate with silver hydroxide, and oxidation of the lactic acid with acid permanganate to acetaldehyde and carbon dioxide. The carbon dioxide is formed from the carboxyl group of the propionate, the acetaldehyde from the α- and β-carbon atoms. The aldehyde contained a normal per cent of C^{13}, and the carbon dioxide a high per cent, *i. e.*, all the fixed carbon. In order to make certain that the conflicting observations were not the result of a difference in the synthetic reactions of the bacteria under the respective experimental conditions, the reliability of the reactions used by Carson *et* *al.*, was determined with propionic acid synthesized by the following reaction:

$$CH_3{\cdot}CH_2{\cdot}MgBr + C^{13}O_2 \longrightarrow CH_3{\cdot}CH_2{\cdot}C^{13}OOMgBr \xrightarrow{H_2O} CH_3{\cdot}CH_2{\cdot}C^{13}OOH$$

The resulting acid was degraded by the reactions used by Carson *et al.* (68), *i. e.*, alkaline permanganate oxidation and dry distillation of the barium salt. Oxalate and carbonate are formed in the permanganate oxidation. It was assumed by Carson *et al.*, that the carbonate arose from the carboxyl group and oxalate from the α- and β-carbon atoms. The results (70) from the degradation of the synthetic acid showed definitely that the reaction does not occur as assumed, for the carboxyl carbon was found in both the oxalate and carbonate. In the dry distillation of barium propionate, diethyl ketone and carbonate are formed. According to the accepted mechanism of the reaction 50 per cent of the carboxyl carbon should be in the carbonate, the other 50 per cent in the ketone. This distribution of carboxyl carbon was found with the synthetic acid (70). Carson *et al.* (71), and Nahinsky and Ruben (72) independently have reinvestigated the problem. They likewise have found with synthetic propionic acid containing C^{14} radioactive carbon in the carboxyl group, that the alkaline permanganate oxidation is not reliable. The dry distillation of the barium salt apparently was not checked with the synthetic acid to determine the reliability of this degradation. Judging from the results of Wood *et al.* (70), however, it seems probable that the reaction is reliable, and a faulty experimental procedure was used in the original experiments of Carson *et al.* At any rate Carson and his co-workers (71) on reinvestigation with the biological acid have found the fixed carbon only in the carboxyl group.

This series of investigations serves to illustrate the difficulty faced in determining the location of isotopic carbon in a compound, and also in synthesizing compounds which will contain isotopic carbon in certain positions. The exact mechanism of the chemical reactions which are employed are often not known. Under the circumstances it is necessary to study the mechanism of the reaction before use. Much experimental work is needed in this field before the full benefits of isotopes can be realized.

The Mechanism of Fixation of Carbon Dioxide in Propionic Acid.—With the position of the fixed carbon definitely shown to be in the carboxyl group of the propionate, the problem of the mechanism of fixation can be considered more accurately. The following mechanism has been proposed by Carson, *et al.* (71), and Krebs and Eggleston (44) for formation of propionate containing fixed carbon:

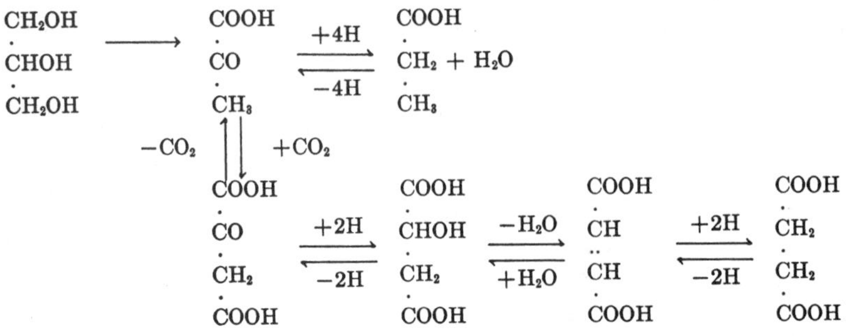

$$\begin{array}{c} CH_2OH \\ \cdot \\ CHOH \\ \cdot \\ CH_2OH \end{array} \longrightarrow \begin{array}{c} COOH \\ \cdot \\ CO \\ \cdot \\ CH_3 \end{array} \underset{-4H}{\overset{+4H}{\rightleftharpoons}} \begin{array}{c} COOH \\ \cdot \\ CH_2 \\ \cdot \\ CH_3 \end{array} + H_2O$$

$$-CO_2 \; \uparrow\downarrow \; +CO_2$$

$$\begin{array}{c} COOH \\ \cdot \\ CO \\ \cdot \\ CH_2 \\ \cdot \\ COOH \end{array} \underset{-2H}{\overset{+2H}{\rightleftharpoons}} \begin{array}{c} COOH \\ \cdot \\ CHOH \\ \cdot \\ CH_2 \\ \cdot \\ COOH \end{array} \underset{+H_2O}{\overset{-H_2O}{\rightleftharpoons}} \begin{array}{c} COOH \\ \cdot \\ CH \\ \cdot\cdot \\ CH \\ \cdot \\ COOH \end{array} \underset{-2H}{\overset{+2H}{\rightleftharpoons}} \begin{array}{c} COOH \\ \cdot \\ CH_2 \\ \cdot \\ CH_2 \\ \cdot \\ COOH \end{array}$$

Fig. 2.

Viewed critically there is not a great deal of evidence to support the scheme, other than the fact that it provides a mechanism for the fixation of carbon dioxide in the carboxyl group of propionic acid. Krebs and Eggleston (44) have shown that the reactions from oxalacetic acid to succinic acid are reversible with the propionic acid bacteria. It has not been shown conclusively, however, that under the conditions of a glycerol fermentation a significant amount of fumarate or succinate is converted to propionate. According to the scheme (Fig. 2) it is essential that the reaction proceed as far as a symmetrical molecule (fumarate or succinate) and then reverse itself so that pyruvate may be formed containing fixed carbon. It is questionable whether much fumarate would be oxidized to oxalacetate in the presence of glycerol which is a good hydrogen donator.

If there was a rapid shifting back and forth from pyruvate to fumarate, part of the dicarboxylic acids should contain fixed carbon in both carboxyl groups, since some pyruvic acid containing fixed carbon would be present and could re-enter the fixation reaction. According to the scheme those dicarboxylic acids not containing two fixed carbons would contain at least one, therefore the average should be well above fixation of one carbon. Actually the quantitative data of Wood *et al.* (14), indicate that the succinate formed in the glycerol fermentation contains only one fixed carbon atom. This fact is a strong argument against any scheme that implies a part of the succinate is to contain two fixed carbons. The evidence supporting the contention that there is approximately only one fixed carbon in each succinate molecule has been arrived at by the following calculation. It has been assumed that the concentration of C^{13} in the carbon dioxide fixed by the cell is equal to that of the medium at the conclusion of the fermentation. It is difficult to estimate the concentration of C^{13} in the carbon dioxide available to the cell, since there is no assurance that the carbon dioxide produced within the cell comes to equilibrium with that dissolved in the medium. The above assumption probably gives a minimal value for the available $C^{13}O_2$, for at the start of the experiment the C^{13} concentration is higher than at the conclusion when dilution with $C^{12}O_2$ from the substrate has occurred. At any rate, on this basis and the assumption that only one $C^{13}O_2$ is fixed in each succinate molecule, the calculated values for the C^{13} in the succinate of two experiments are 1.72 and 1.34 per cent. The observed values were 1.65 and 1.28 per cent, respectively. Calculated on the basis that both carboxyls contain fixed carbon, the values are 2.71 and 1.77 per cent. The calculations are approximations but, nevertheless, they seem to indicate that fixation occurs in only one carboxyl group.

Accordingly, the scheme in Fig. 2 involves the splitting of a carboxyl from a symmetrical dicarboxylic acid containing fixed carbon in only one position to form propionate containing fixed carbon. Wood and his co-workers (14) have presented quantitative evidence to show that practically every molecule of propionate formed from glycerol would have to pass through fumarate if the above scheme holds. If every molecule of propionate was formed *via* decarboxylation of a symmetrical dicarboxylic acid and none directly from the glycerol, the C^{13} concentrations in the carboxyl group of the succinic and propionic acids should be equal. In the two fermentations so far examined the values were 2.21 and 1.47 per cent C^{13} for the carboxyl of propionic acid as compared to 2.29 and 1.54 per cent, respectively, for the carboxyls of succinate. The implication is that none of the propionate formed in the fermentation of glycerol has arisen by direct reduction of pyruvate prior to its conversion to fumarate and back again. This argument holds whether it is accepted that succinate contains only one fixed carbon or not. It seems unlikely on the basis of Fig. 2 that not any of the propionate will be formed by reduction of pyruvate before it passes through fumarate and back again. Furthermore, it is unlikely that once the propionate is formed there will be any extensive passage back to pyruvate. There is every reason to believe that if there is an equilibrium between pyruvate and propionate, it is far to the side of propionate, for in a propionic acid fermentation, pyruvate can be detected only by special methods.

Carson and his co-workers (71) have offered evidence which they suggest may show that there is interconversion of propionate and succinate. They admit that the rates of interconversion may be too slow to account for the radioactivity usually found in propionic acid formed from glycerol. In fact, on addition of radioactive succinate (obtained from the propionic fermentation) to glycerol or pyruvate fermentations containing no other source of C^{11}, the radioactivity of the formed propionate was found to be 2 ± 2 and 1 ± 1 units, respectively, on the addition of 100 units of succinate. In the case of added propionate the recovered succinate was 5 ± 2 and 9 ± 2 units. Even if these figures are considered significant, there is no assurance that the succinate and propionate are formed by the reversible series of reactions of Fig. 2. There is a possibility that the succinate may be decarboxylated directly to propionate; also that the propionate is oxidized with liberation of radioactive carbon dioxide which then is fixed in the succinate. In one fermentation in which radioactive propionate was added 8 ± 1 units of carbon dioxide were formed.

Most of the above discussion relative to the mechanism of formation of propionate has been of a negative type, in that it has pointed out only the weaknesses of existing scheme of Fig. 2. What constructive suggestions can be offered? Frankly, sufficient data are not available with which to formulate a defensible scheme. It is likely that the mechanism of succinate formation from glycerol by the propionic acid bacteria is substantially as shown in the above scheme with the exception that in the presence of glycerol the series of equilibria between pyruvate and succinate are largely shifted to succinate, owing to the reducing intensity of the glycerol. This would account for the succinate containing only one fixed carbon per molecule, a conception which has been our basic assumption since 1938. Elucidation of the mechanism of propionic acid formation has always been a difficult problem. Of several mechanisms proposed, that involving removal of water from lactic acid to form acrylic acid which is then reduced to propionic acid has seemed the most probable.

$$CH_3{\cdot}CHOH{\cdot}COOH \xrightarrow{-H_2O} CH_2{:}CH{\cdot}COOH \xrightarrow{+2H} CH_3CH_2{\cdot}COOH$$

Up to the present, however, the reduction of acrylic acid by propionic acid bacteria has not been demonstrated. As shown in the previous discussion, there is some basis for assuming that all the propionic acid formed in the fermentation of glycerol may arise by decarboxylation of a symmetrical dicarboxylic acid containing one fixed carbon. Accordingly, 50 mM of $C^{12}O_2$ will be formed for each 100 mM of propionate, since there is an equal chance of splitting out the carboxyl of the dicarboxylic acid which originates from the glycerol and the carboxyl that is formed by fixation of $C^{13}O_2$. If the C^{13} of the sodium bicarbonate be calculated on the basis of this dilution by $C^{12}O_2$, there is reasonable agreement between the experimentally observed and calculated values (calculated: 3.76 and 2.23; observed: 3.62 and 2.10 per cent C^{13}). This is further evidence that the propionic acid is formed in the glycerol fermentations exclusively by decarboxylation of a symmetrical dicarboxylic acid. It is the mechanism of the decarboxylation that particularly is uncertain. The evidence that the mechanism does not involve the reversible reactions of Fig. 2 has already been discussed. There is some evidence that the propionic acid bacteria can decarboxylate succinate anaerobically ($COOH \cdot CH_2CH_2 \cdot COOH \rightarrow CO_2 + CH_3CH_2 \cdot COOH$) but it is questionable whether or not the rate of this reaction is high enough to be of any considerable importance.

Whether or not propionate may be formed from glucose and pyruvate exclusively by the fixation reaction, *i. e.*, by C_3 and C_1 addition with subsequent decarboxylation of a dicarboxylic acid is uncertain, and there is some evidence to the contrary. Carson and his co-workers (71) in the fermentation of pyruvic acid by the propionic acid bacteria found 5 per cent of the fixed radioactive carbon in the volatile acids and 95 per cent in the non-volatile acid. Since the yield of volatile acids, and especially of propionic acid, was not given, no definite idea can be reached as to how much of the total propionic acid was formed by the fixation reaction. If the yield of volatile acid was at all normal, however, and the propionic acid was formed through the fixation reaction, there would have been more fixed carbon in the volatile acid fraction than was observed.

In a preliminary unpublished experiment by the authors, it has been found, contrary to the results of Carson *et al.*, that a large part of the fixed carbon is in the volatile acid from a pyruvate fermentation. However, by the same method of calculation that was employed in the glycerol fermentation, only 60 per cent of the propionic acid is indicated to have arisen by the decarboxylation of a symmetrical dicarboxylic acid. In this calculation it is assumed that all the C_4 dicarboxylic acids are formed by the fixation reaction. In the fermentation of glucose and pyruvate there is some indication that more than one mechanism of succinate formation occurs. Therefore, there is a possibility that all the propionate is formed by decarboxylation of a symmetrical dicarboxylic acid.

The experiment by Carson *et al.* (71), was of short duration (50 minutes), whereas that of the authors ran for 18 hours, *i. e.*, until the pyruvate was all fermented. This experimental difference may account for the discrepancy in the amount of fixed carbon in the volatile acids. Carson and his co-workers also reported the formation of a non-volatile keto acid which contained 70 per cent of the fixed carbon of the non-volatile acid fraction. The acid was not pyruvic acid and was not identified. Its identification is of interest because of the light that might be thrown on the mechanism of carbon dioxide fixation. The compound can hardly be oxalacetic acid, for this acid is rapidly decomposed on heating in acid solution and very likely would have been converted to pyruvate during the steam distillation.

It is evident from the above discussion that propionic acid is formed by a fixation reaction and probably by decarboxylation of a C_4 dicarboxylic acid that is formed by C_3 and C_1 addition. The mechanism of decarboxylation is not known, neither is it known whether or not this reaction is the general mechanism by which propionate is formed in all propionic acid fermentations.

2. *Miscellaneous Fixation Reactions*

There have been a number of other fixation reactions demonstrated, but information is as yet too meager to allow any definite idea of their mechanisms. In some cases it is probable that the mechanism is simply the cleavage of a C_4 compound in which carbon dioxide has been fixed by C_3 and C_1 addition. The mechanism, thus, is fundamentally the same as described under C_3 and C_1 addition, in so far as the fixation itself is concerned. In others it is evident that the fixation reactions may involve the formation of a carbon chain entirely from C_1 compounds. Particularly this latter type of synthesis is not well understood. It is here that information is needed to solve the mechanism of the strict autotrophs and of photosynthesis.

Examples of fixations that may occur by C_3 and C_1 addition or, on the other hand, may be proved to involve a mechanism quite different, have recently been demonstrated by Slade *et al.* (15, 52). They have investigated fixation of carbon dioxide by several of the typical heterotrophic bacteria through use of $C^{13}O_2$. The compounds found in the different fermentations have been isolated and their C^{13} content determined. The species used were *Staphylococcus candidus, Aerobacter indologenes, Streptococcus paracitrovorus, Clostridium welchii, Clostridium acetobutylicum, Proteus vulgaris, Lactobacillus plantarum* and *Streptococcus lactis*. Glucose and citrate were used as substrates with cell suspensions in most cases. The last two species are homo-lactic acid bacteria, *i. e.*, bacteria that form substantially nothing but lactic acid. There was no evidence of fixation by either of these bacteria but there was fixation by all the other cultures. It has been pointed out previously that all the genera forming succinate (*Staphylococcus, Aerobacter, Streptococcus* and *Proteus*) fixed carbon in this compound.

Fixation in Lactate and Acetate.—Of particular interest is the observed fixation in lactic acid and acetic acid (15, 52). There was fixation in lactic acid by the following species: *Staphylococcus candidus, Streptococcus para-*

citrovorus, Clostridium welchii, Clostridium acetobutylicum, Proteus vulgaris and *Aerobacter indologenes.* The carbon fixed in the lactic acid has been found in each case to be exclusively in the carboxyl group as located by acid permanganate oxidation of the lactic acid to carbon dioxide and acetaldehyde. The carbon dioxide arises from the carboxyl carbon, the acetaldehyde from the α- and β-carbons. The fact that the homo-lactics do not fix carbon dioxide, as do the hetero-lactics, may be of some significance in indicating that a cleavage of the C_3 chain is necessary before fixation of carbon dioxide in lactate can occur. On the other hand, the failure of the homo-lactic acid bacteria to fix carbon dioxide in lactic acid may result from an inability to fix carbon dioxide by a direct C_3 and C_1 addition. A possible mechanism of fixation of carbon dioxide in lactic acid by C_3 and C_1 addition is as follows:

$$C^{13}O_2 + CH_3{\cdot}CO{\cdot}COOH = C^{13}OOH{\cdot}CH{:}COH{\cdot}COOH = C^{13}OOH{\cdot}COH{:}CH{\cdot}COOH$$
$$= C^{13}OOH{\cdot}CO{\cdot}CH_3 + CO_2$$

Pyruvic acid containing heavy carbon is formed in this reaction by shifting the hydroxyl and hydrogen of enol-oxalacetic acid and subsequently decarboxylation. Meyerhof (54) has proposed this reaction and suggests that the shifting of hydroxyl with hydrogen occurs spontaneously. The formation of pyruvate containing fixed carbon thus can be explained without passage through a symmetrical C_4 dicarboxylic acid. The pyruvate is then reduced to lactate. A similar scheme might apply to the formation of propionate. The same objection may be raised to this scheme as with others, *i. e.*, if heavy carbon pyruvate is present, part of the dicarboxylic acids should contain two fixed carbons. In no case has the content of fixed carbon in succinate been sufficient to indicate such an occurrence. This may be due, however, to the simultaneous formation of succinate by a non-fixation reaction.

It is significant that, in contrast to the fermentation of glucose, the fermentation of citrate by *Streptococcus paracitrovorus* yields succinate which contains fixed carbon and lactate which does not contain it. This fact may indicate that carbon is not fixed in lactate by the above equilibria for if such were the case, it would be expected that whenever C_4 dicarboxylic acids were formed containing heavy carbon, heavy carbon pyruvate and lactate would likewise occur. It is apparent that more information is needed before a decision can be reached on the mechanism of fixation of carbon in lactate. The possibility must be left open that heavy carbon lactate is formed by C_2 and C_1 addition but there is no evidence to support this idea at present.

Acetic acid containing fixed carbon was formed by *Aerobacter indologenes* and by *Clostridium welchii*. The fixed carbon has been shown to be exclusively in the carboxyl group. The procedure of degradation (*cf.* Barker *et al.* (18)) involved dry distillation of the barium salt to yield acetone and barium carbonate.

$$(CH_3C^{13}OO)_2Ba \longrightarrow CH_3{\cdot}C^{13}O{\cdot}CH_3 + BaC^{13}O_3$$

The acetone was then degraded by the iodoform reaction.

$$CH_3{\cdot}C^{13}O{\cdot}CH_3 + 3I_2 + 4NaOH \longrightarrow CHI_3 + CH_3C^{13}OONa + 3NaI + 3H_2O$$

The barium carbonate contained heavy carbon, whereas there was none in the iodoform. The mechanism of this fixation is unknown. It is possible that the acetic acid is formed by cleavage of a C_4 dicarboxylic acid. No detectable amount of C_4 dicarboxylic acid was formed in the fermentation by *Clostridium welchii* but this does not necessarily mean that it did not occur as an intermediate. Further studies are being made on the mechanism of this fixation, for it offers possibilities of representing a type differing from any heretofore investigated.

It is noteworthy that in the fermentation by *Aerobacter*, ethyl alcohol was formed which did not contain fixed carbon. This fact indicates that the alcohol was not from the same source as the acetate containing fixed carbon. These bacteria under proper conditions reduce acids to alcohols (73) but apparently did not do so in the present experiment.

Carbon to Carbon Linkage, Both Components from Carbon Dioxide.—None of the typical heterotrophic bacteria studied by Slade *et al.* (15, 52), formed a carbon to carbon linkage in which both members of the link were from carbon dioxide. It may be a characteristic of most heterotrophs that one member of the link must be organic. In contrast, the typical autotroph uses inorganic carbon for both members of the link. That there are intermediate types is not unexpected. Barker and his co-workers (18) have made an interesting contribution concerning this intermediate group. With radioactive carbon it was shown that the formation of acetic acid by *Clostridium acidi-urici* involves fixation of carbon dioxide. By use of the degradation reactions described above 67 per cent of the radioactive carbon was demonstrated in the methyl and 33 per cent in the carboxyl groups. The unequal distribution of the fixed carbon in the two groups of the molecule may indicate that the acid is not synthesized entirely from carbon dioxide. There is no information available on the mechanism of this fixation. Its solution, obviously, is of fundamental importance.

The organism studied by Wieringa (19) apparently falls in the same category as that of Barker *et al.*, only it is even more autotrophic. It reduces carbon dioxide with molecular hydrogen to acetic acid.

Barker and his co-workers (18, 37) claim to have shown fixation of carbon dioxide in cell protoplasm. The mechanism of this fixation has not been investigated, and the type of linkage is unknown.

III. Mechanism of Carbon Dioxide Fixation by Animal Tissue

Understanding of the mechanism of fixation of carbon dioxide by bacteria is admittedly inadequate but the situation is even less satisfactory for animal tissue. It has been known since 1935 that even the more fastidious heterotrophs can utilize carbon dioxide. Although the true significance and the fundamental importance of the phenomenon were not generally recognized, there was considerable attention devoted to it by a number of bacteriologists. It is, however, only within the last year that any extensive consideration has been given to the potentialities of carbon dioxide assimilation by animals. True, Krebs and Henseleit (74) in 1932 offered proof of the participation of carbon dioxide in the formation of urea, but the significance of the conversion was largely overlooked. The possibilities of carbon dioxide fixation by animals, therefore, have been investigated to only a very limited extent. Much has been accomplished on the fixation of carbon dioxide by pigeon liver in the oxidation of pyruvate, because the groundwork had been laid by bacterial studies, and the course of action to be followed was apparent. The scope of fixation by animal tissues has by no means been completely probed, and it is very probable that carbon dioxide will be found to have a function in a number of physiological processes.

The same outline will be followed in considering the mechanism of fixation by animal tissue as was used for bacteria, *i. e.*, (1) fixation of carbon dioxide not involving a carbon to carbon linkage; (2) fixation of carbon dioxide involving a carbon to carbon linkage, under which will be considered (*a*) C_3 and C_1 addition, and (*b*) miscellaneous fixation reactions.

A. Fixation of Carbon Dioxide Not Involving a Carbon to Carbon Linkage

Two examples of such a synthesis are known, formation of carbaminohemoglobin (75)

$$HbNH_2 + CO_2 \rightleftharpoons HbNHCOOH$$

and of urea (74). The demonstration by Krebs and Henseleit of urea synthesis (Fig. 3) from carbon dioxide by liver tissue was the first clearly defined example of heterotrophic utilization of carbon dioxide. Conclusive

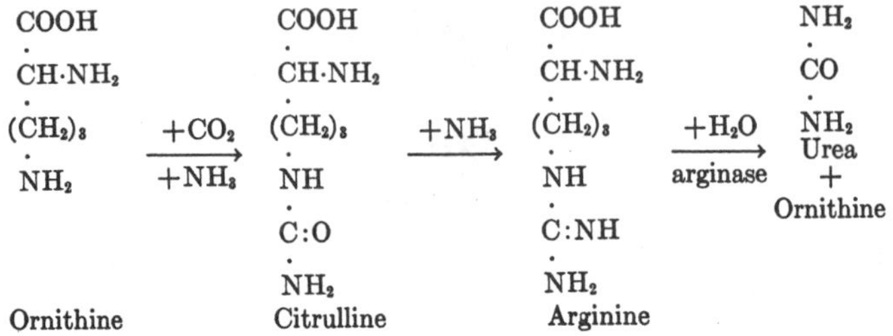

COOH		COOH		COOH		NH_2
·		·		·		·
$CH \cdot NH_2$		$CH \cdot NH_2$		$CH \cdot NH_2$		CO
·		·		·		·
$(CH_2)_3$	$+CO_2$ → $+NH_3$	$(CH_2)_3$	$+NH_3$ →	$(CH_2)_3$	$+H_2O$ → arginase	NH_2 Urea + Ornithine
·		·		·		
NH_2		NH		NH		
		·		·		
		C:O		C:NH		
		·		·		
		NH_2		NH_2		
Ornithine		Citrulline		Arginine		

Fig. 3.

proof that carbon dioxide is fixed in urea by liver has been provided by Rittenberg and Waelsch (76) by use of C^{13} and by Evans and Slotin (77) with C^{11}. Hemingway (private communication) has demonstrated urea synthesis *in vivo* with mice by use of C^{13}. In this connection it is interesting to note that the demonstration of Ruben and Kamen (17) of fixation of carbon dioxide by liver has been cited (53, 55) as evidence for mechanisms of carbon dioxide fixation by liver which involve a carbon to carbon linkage. It is evident that the observed fixation could have been due to urea formation. Therefore, the demonstration of fixation, unaccompanied by identification of the compound concerned, does not give reliable evidence for the suggested reactions.

B. Fixation of Carbon Dioxide Involving a Carbon to Carbon Linkage

1. *C_3 and C_1 Addition*

The evidence for the occurrence of C_3 and C_1 addition in bacterial metabolism has already been considered and particularly the evidence for the reaction:

$$CO_2 + CH_3 \cdot CO \cdot COOH = COOH \cdot CH_2 \cdot CO \cdot COOH$$

The reader is referred to this discussion (p. 153) for a more detailed account concerning the above fixation reaction, for despite Krebs' statement, no direct evidence has been obtained as yet with animal tissue which permits the definite conclusion that pyruvic or oxalacetic acid is a component of the fixation reaction.

Apparently carbon dioxide fixation with formation of a carbon to carbon linkage has been demonstrated in only one tissue—liver. This may be because other tissues have not been examined; it is known, however, that there is no fixation of carbon dioxide during the dissimilation of pyruvate by pigeon breast muscle (63). In liver fixed carbon has been demonstrated in glycogen (53) and in products of pyruvate oxidation (3, 78, 63, 50).

Fixation of Carbon Dioxide in the Dissimilation of Pyruvate by Liver Tissue.—Attention was focussed on this fixation particularly by the experiments of Evans (79), who found that pyruvate is oxidized by pigeon liver even in the presence of malonate, and is converted to C_4 dicarboxylic acids, α-ketoglutaric acid and carbon dioxide. No theory was advanced by Evans (79) to explain the mechanism but viewed in the light of the experiments of Krebs and Eggleston (80) on pyruvate oxidation by pigeon breast muscle, it was evident that pigeon liver very likely possessed a malonate-insensitive mechanism for formation of C_4 dicarboxylic acids and dissimilated pyruvate by the Krebs cycle. With pigeon breast muscle it was necessary to add a C_4 dicarboxylic acid to the malonate-inhibited reaction in order to get oxidation of pyruvate. Apparently breast muscle cannot synthesize C_4 dicarboxylic acids under the conditions as can liver. The only malonate-insensitive reaction that had been described in the literature for synthesis of C_4 dicarboxylic acids is the fixation reaction studied by Wood and Werkman (41, 42). The fact that oxalacetic acid is an intermediate in the Krebs cycle further added to the attractiveness of the hypothesis that the oxalacetate is formed by the fixation reaction occurring in the dissimilation of pyruvate by pigeon liver. Independently Evans and Slotin (3) and Krebs and Eggleston (55) and shortly afterwards Wood *et al.* (78, 50), presented evidence that carbon dioxide may be fixed by C_3 and C_1 addition in pigeon liver.

Krebs and Eggleston, handicapped by not having an available source of carbon isotopes, were forced to rely on indirect methods of demonstrating the role of carbon dioxide in the pyruvate oxidation. They showed that the rate of oxidation of pyruvate and formation of α-ketoglutarate, citrate, malate and fumarate was stimulated by the presence of carbon dioxide. This specific effect of carbon dioxide in connection with other considerations led them to propose that oxalacetate is synthesized by the Wood and Werkman reaction and then is metabolized by the Krebs cycle. The proposed role of thiamin in this reaction has been considered in connection with results obtained with bacteria (page 158).

Evans and Slotin (3) independently provided conclusive evidence that carbon dioxide is fixed during the oxidation of pyruvate by isolation of radioactive α-ketoglutarate from experiments in which $C^{11}O_2$ was used as a tracer. Following this Wood *et al.* (78, 50), with heavy carbon, and independently Evans and Slotin (63) with radioactive carbon, determined the position of the carbon fixed in isolated α-ketoglutaric acid. The isolated acid was degraded by acid permanganate oxidation to succinic acid and carbon dioxide.

$$COOH \cdot CH_2 \cdot CH_2 \cdot CO \cdot C^{13}OOH = COOH \cdot CH_2 \cdot CH_2 \cdot COOH + C^{13}O_2$$

All the fixed carbon was in the carbon dioxide. The identical results of both investigations prove conclusively that the fixed carbon is exclusively in the carboxyl group alpha to the keto group. Krebs (56), convinced that pyruvate was dissimilated in pigeon liver by a combination of oxalacetate synthesis through the fixation reaction and the Krebs cycle, had predicted that the fixed carbon would be found in both carboxyls of the α-ketoglutarate. This, in fact, would be the location of the fixed carbon if pyruvate were fermented by pigeon liver according to the Krebs cycle, since it includes the symmetrical citrate molecule as an intermediate. Wood *et al.* (78), and Evans and Slotin (63), therefore concluded that citrate is not an intermediate in the dissimilation of pyruvate by pigeon liver.

Evans and Slotin (63) have provided further proof that citrate is not an intermediate. Non-radioactive citrate was added to a dissimilation of pyruvate in a radioactive bicarbonate medium and the α-ketoglutarate was isolated. Neither the yield of ketoglutarate nor the ratio of its activity to that of the medium was affected. If citrate were an intermediate in the formation of α-ketoglutarate, it would be expected that the activity of the ketoglutarate would have been lowered due to dilution by ketoglutarate from the citrate.

The status of the Krebs cycle with respect to liver is placed in doubt since citrate is not an intermediate in the dissimilation. Even as applied to pigeon breast muscle the cycle must be accepted with reservations. Unquestionably Krebs' fine investigations have established the general skeleton of the cycle but further investigation is needed to determine the identity of the intermediates. Furthermore, the proof that α-ketoglutarate contains fixed carbon by no means establishes the fixation of carbon dioxide by pigeon liver through C_3 and C_1 addition.

Further investigations by Wood *et al.* (50), on the dissimilation of pyruvate by pigeon liver have more nearly completed this proof, and additional support has been obtained for a modified Krebs cycle.

Fig. 4. Dissimilation of Pyruvate by Pigeon Liver

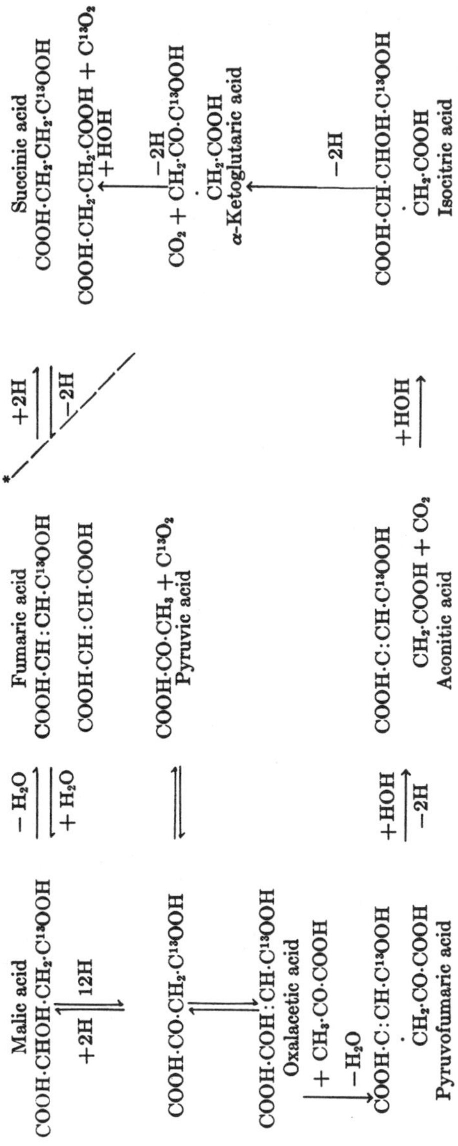

Malic acid
$COOH \cdot CHOH \cdot CH_2 \cdot C^{13}OOH$ $\underset{+ H_2O}{\overset{- H_2O}{\rightleftharpoons}}$ Fumaric acid
$COOH \cdot CH{:}CH \cdot C^{13}OOH$ $\underset{-2H}{\overset{+2H}{\rightleftharpoons}}$* Succinic acid
$COOH \cdot CH_2 \cdot CH_2 \cdot C^{13}OOH$

$COOH \cdot CH{:}CH \cdot COOH$

$COOH \cdot CH_2 \cdot CH_2 \cdot COOH + C^{13}O_2$

Malic acid $\underset{12H}{\overset{+2H}{\rightleftharpoons}}$ $COOH \cdot CO \cdot CH_2 \cdot C^{13}OOH$

$COOH \cdot CO \cdot CH_2 \cdot C^{13}OOH$ $\rightleftharpoons$ $COOH \cdot CO \cdot CH_3 + C^{13}O_2$
Pyruvic acid

$COOH \cdot CO \cdot CH_2 \cdot C^{13}OOH$ $\rightleftharpoons$ $COOH \cdot COH{:}CH \cdot C^{13}OOH$
Oxalacetic acid

$\downarrow$ $+ CH_3 \cdot CO \cdot COOH$, $-H_2O$

$COOH \cdot C{:}CH \cdot C^{13}OOH$
$\cdot CH_2 \cdot CO \cdot COOH$
Pyruvofumaric acid $\xrightarrow[-2H]{+HOH}$ $COOH \cdot C{:}CH \cdot C^{13}OOH$
$CH_2 \cdot COOH + CO_2$
Aconitic acid $\xrightarrow{+HOH}$ $COOH \cdot CH \cdot CHOH \cdot C^{13}OOH$
$\cdot CH_2 \cdot COOH$
Isocitric acid

Isocitric acid $\xrightarrow{-2H}$ $CO_2 + CH_2 \cdot CO \cdot C^{13}OOH$
$\cdot CH_2 \cdot COOH$
α-Ketoglutaric acid $\xrightarrow[+HOH]{-2H}$ $COOH \cdot CH_2 \cdot CH_2 \cdot COOH + C^{13}O_2$

* Reaction is inhibited by malonate.

Figure 4 presents a tentative mechanism which is adaptable to the observed facts. It is a Krebs cycle from which citrate has been deleted and isocitrate retained. It is recognized that most tissues contain aconitase which would induce formation of citrate, but as a working hypothesis the skeleton of the original cycle may as well be retained until information is available which dictates the proper change. It is probable that phosphorylated intermediate compounds are involved.

Three new facts were established in this study and are related to the scheme as follows:

(*a*) *Carbon dioxide was found to be fixed exclusively in the carboxyl groups of C_4 dicarboxylic acids formed from pyruvate.* Malate, fumarate and succinate were isolated and degraded. The α- and β-carbons were found free from fixed carbon, which was contained in the carboxyl carbons. This observation more nearly completes the proof that the C_4 dicarboxylic acids are formed by C_3 and C_1 addition, since the fixed carbon was located directly in the carboxyls of the C_4 dicarboxylic acids. The fixation reaction is the only malonate-insensitive reaction that has been proposed which accounts for these facts. Acetic acid condensation, for example, could not account for the observations, even if an exchange of carbon dioxide with the carboxyl carbons were assumed, because this conversion passes through succinate prior to malate and fumarate formation. This apparently was not the case, for it was shown that the conversion of succinate to fumarate can be blocked without inhibition of the formation of malate or fumarate. If there were interconversion of succinate and fumarate, both acids would contain fixed carbon. This was not always the case.

(*b*) *In the presence of malonate, succinate was shown to contain little or no fixed carbon, whereas the other C_4 dicarboxylic acids did contain fixed carbon.* This observation is of considerable importance. In fact Krebs (81) has referred to the oxidative formation of succinate from oxalacetate in the presence of malonate as the crucial experiment. It is crucial, for, provided one can assume that the anaerobic formation of succinate by reduction of oxalacetate is inhibited by malonate, it proves that there is an oxidative reaction leading from oxalacetate to succinate. Critics of the Krebs cycle (65) have contended that this assumption is without adequate proof. The second fact shows that in the case of pigeon liver, inhibition by malonate was effective and succinate did not arise by anaerobic reduction over malate and fumarate. If such were the case, the C^{13} concentration would have been approximately the same in each compound, as it was when malonate was not added. It is clear that there are two mechanisms for the formation of C_4 dicarboxylic acids. The one is quite probably by C_3 and C_1 addition and is not inhibited by malonate. The other is by an oxidative process, and the resulting C_4 dicarboxylic acids do not contain fixed carbon. This then removes one of the major criticisms of the Krebs cycle, at least as applied to the dissimilation by liver. Much of the criticism of the Krebs cycle has centered around this point, and whether or not citrate is an intermediate. In so far as liver is concerned, the first criticism is not valid, while the second is. However, in viewing objections to the cycle, the fact must not be overlooked that this scheme does give a fairly logical mechanism for oxidation of pyruvate to carbon dioxide. The scheme should not be considered a substitute for the Szent-Györgyi system of hydrogen transfer, for the latter scheme only attempts to tell where the hydrogen goes and not how the carbon chain is cleaved. No other scheme adequately explains the mechanism of oxidation of pyruvate to carbon dioxide, and although certain details of the Krebs cycle may be in error, it seems probable that the general framework is correct. At any rate the scheme in Fig. 4 accounts for all observed facts in so far as location of the fixed carbon is concerned. Additional information on the fixation reaction as such has been given in the section dealing with bacteria (page 153).

(*c*) *Lactate formed either aerobically or anaerobically contained fixed carbon in the carboxyl group and none in the α- and β-carbons.* Further study is required before the significance of these results can be evaluated fully. Probable mechanisms of the formation of lactate containing fixed carbon have already been considered in the section on bacteria (page 166).

A logical objection can be raised to the scheme in Fig. 4 which assigns an essential role to a fixation reaction solely on the basis of the presence and location of fixed carbon in a compound. Fairly reliable information on the mechanism of fixation can be obtained, but the answer as to whether or not the process as a whole is dependent on the fixation reaction cannot be obtained by this method. The carbon may have been fixed, for example, by a non-essential exchange reaction. Some attempt has been made to answer this criticism. For example, the general occurrence of exchange during decarboxylation has been disproved by Krampitz *et al.* (59). Evans and Slotin (63) have considered the problem in some detail. They have presented evidence that the utilization of carbon dioxide is stoichiometric, one molecule of carbon dioxide being utilized for each molecule of α-ketoglutarate. Wood and his co-workers (14) also have shown that there is a stoichiometric relationship of carbon dioxide to propionic and succinic acids formed in the fermentation of glycerol by the propionic acid bacteria. If there is fixation by exchange, it seems questionable whether the reaction would be stoichiometric.

Evans and Slotin (3) have found that α-ketoglutarate formed by breast muscle in the presence of radioactive carbon dioxide is inactive. In muscle, C_4 dicarboxylic acids cannot be formed from pyruvic acid by a fixation reaction; however, if carbon enters the molecule through miscellaneous exchange reactions, there should have been fixation in α-ketoglutarate even in the absence of the true fixation reactions. These results are indicative but admittedly are not conclusive proof that fixation is an essential reaction. There may be a specific fixation by C_3 and C_1 addition and the reaction still be non-essential. For example, the initial oxalacetate synthesis might occur by some other reaction than fixation, and then the carboxyl groups come to equilibrium with the isotopic carbon dioxide by the enzyme exchange reaction of Krampitz *et al.* (59). The C^{13} concentration in the resulting oxalacetate would be the same as if it arose initially by the fixation reaction, and, furthermore, if one blocked fixation the synthesis of oxalacetate and α-ketoglutarate would still go on. There is ample evidence, however, that carbon dioxide does not function solely in a non-essential exchange reaction for it is known to be necessary for growth of microorganisms (22) and in the reduction of methylene blue by dehydrogenases (27).

In summary, the results obtained on fixation of CO_2 by pigeon liver in pyruvate dissimilation, indicate that the CO_2 is fixed by C_3 and C_1 addition, just as in bacterial metabolism. The fact that the fixed carbon of the C_4 dicarboxylic acids has been shown to be exclusively in the carboxyl group supports this view. Proof of fixation of CO_2 in oxalacetate is needed, however, to confirm the occurrence of the reaction:

$$CH_3{\cdot}CO{\cdot}COOH + CO_2 = COOH{\cdot}CH_2{\cdot}CO{\cdot}COOH$$

in liver. Although it is likely that C_4 dicarboxylic acids can be formed by fixation, it is not certain whether this is an essential part of the dissimilation or not.

2. Miscellaneous Fixation Reactions

Fixation of Carbon Dioxide in Glycogen by Liver Tissue.—Solomon and his co-workers (53) in an interesting investigation have found that carbon dioxide is fixed *in vivo* by fasted rats when lactate is fed and radioactive sodium bicarbonate is injected intraperitoneally. Part of the fixed carbon occurs in the liver glycogen. The following are the more pertinent results.

The liver glycogen derived from fixed carbon dioxide was calculated to vary from 7 to 16 per cent. If one carbon atom in six of the glycogen had such an origin, the value would be 16.6 per cent. The calculation was made (*a*) on the assumption that the $C^{11} : C^{12}$ ratio is the same in tissue fluid in all parts of the body at any particular moment and (*b*) that the ratio of carbon dioxide excretion to carbon dioxide synthesis into glycogen is constant. If these assumptions hold, then the mM of carbon dioxide incorporated into the glycogen are equal to the mM of excreted carbon dioxide multiplied by the ratio of radioactivity in the glycogen to the radioactivity in the excreted carbon dioxide. The truth of these assumptions is of considerable importance because much of the evidence for the proposed mechanism of fixation is based on the incidence of not more than one carbon atom derived from carbon dioxide in six of the glycogen.

The fixation reaction is proposed by Solomon *et al.*, to constitute an essential step in the glycogen synthesis. The steps of glycolysis are all reversible (82) except that involving phosphopyruvate (83). It is suggested that the phosphopyruvic acid is formed through the fixation reaction by decarboxylation of the phosphorylated C_4 dicarboxylic acid (84, 85).

$$C^{11}O_2 + CH_3{\cdot}CO{\cdot}COOH = C^{11}OOH{\cdot}CH_2{\cdot}CO{\cdot}COOH$$

$$C^{11}OOH\cdot\overset{H}{C}:\overset{\overset{H}{O}}{C}{\cdot}COOH \xrightarrow{H_3PO_4} C^{11}OOH{\cdot}\overset{H}{C}:\overset{\overset{PO_3H_2}{O}}{C}{\cdot}COOH \xrightarrow{-CO_2} CH_2:\overset{\overset{PO_3H_2}{O}}{C}{\cdot}COOH$$

$$\updownarrow$$

$$C^{11}OOH\cdot\overset{OH}{C}:\overset{H}{C}{\cdot}COOH \xrightarrow{H_3PO_4} C^{11}OOH{\cdot}\overset{\overset{PO_3H_2}{O}}{C}:CH{\cdot}COOH \xrightarrow{-CO_2} C^{11}OOH{\cdot}\overset{\overset{PO_3H_2}{O}}{C}:CH$$

A random distribution of the C^{11} in the two carboxyl group of the dicarboxylic acid may arise from the shift of hydroxyls in enol oxalacetate (54) or by passage through fumarate. Each pair of phosphopyruvate molecules transformed into glycogen would, therefore, contain one labelled carbon atom. Significance is given to the fact that the experimental value never exceeded this but did approach it rather closely in three of the seven experiments. The above proposal is an attractive explanation of the results but, as Solomon and his co-workers suggest, it should be considered only as a working hypothesis. The fixation reaction, in the first place, has not been elucidated as to details of mechanism; second, the evidence of Kalckar (84) that phosphopyruvate is formed from C_4 dicarboxylic acids is only presumptive since the compound was not isolated; and finally Solomon and his co-workers have indicated a possible inconsistency. They find that about the same amount of carbon dioxide is fixed in liver glycogen after feeding glucose as after feeding lactate. Accordingly it must be assumed that the glucose is broken down to pyruvate before glycogen is synthesized. This seems unlikely for the path of breakdown of glucose would be through the same phosphorylated hexose as would be needed for the glycogen synthesis. It might be expected that part of the glucose after phosphorylation will go directly to glycogen and that, therefore, the amount of carbon dioxide fixed would be less than when lactate was fed.

Several other observations by Solomon *et al.*, are of interest. Substantially no fixation in glycogen was observed when the rats were not fed lactate, and radioactive sodium bicarbonate was injected into non-fasted rats. Apparently the animal must be actually making and depositing glycogen before there is fixation in the glycogen. The muscle glycogen, in contrast to liver glycogen, contained no fixed carbon. It is apparent, therefore, that there was no significant interchange between liver and muscle glycogen and that the muscle glycogen did not have an origin similar to that of the liver glycogen.

Fixation of Carbon Dioxide by Trypanosoma.—Searle and Reiner (87) found that in the dissimilation of glucose by *Trypanosoma lewisi* carbon dioxide is fixed with formation of succinic, lactic, pyruvic and acetic acids. The carbon dioxide assimilated was equivalent to the succinic acid formed. One molecule of carbon dioxide was assimilated for each two molecules of glucose fermented. Glucose was not metabolized in the absence of bicarbonate unless pyruvate was added. The activating effect of pyruvate differed from that of bicarbonate since succinate was not formed in the absence of bicarbonate and the products were principally lactate with some acetate.

It was suggested that two results were not in agreement with the Wood and Werkman reaction: (*a*) pyruvate remains unchanged in the presence of bicarbonate, and (*b*) carbon dioxide is required for the dissimilation of glucose and especially glycerol under aerobic conditions. With glycerol there was no evidence that either pyruvate or succinate was formed as an intermediate or final product; the oxidation goes completely to CO_2.

Although CO_2 may participate in a manner other than through C_3 and C_1 addition in the oxidation of glycerol, the above evidence is not entirely conclusive. It is only by special methods that intermediate products can be demonstrated. This is the case in the oxidation of pyruvate by liver tissue, yet it is very probable that CO_2 functions by C_3 and C_1 addition. Whether a product will accumulate will depend on whether its rate of formation is more rapid than its rate of oxidation to CO_2. In the case of the glycerol oxidation the rates may be such as to cause only CO_2 to accumulate.

Failure to observe dissimilation of pyruvate in the presence of CO_2 may be explained. For example, C_3 and C_1 addition may have occurred but pyruvate may not function as an adequate hydrogen donator to reduce the oxalacetate. The reaction might therefore stop after slight accumulation of oxalacetate. Furthermore, the reaction of pyruvate and CO_2 is merely a picture of the over-all conversion and pyruvate as such may not be an actual component of the fixation reaction.

Probably both the fixation of CO_2 by *Tr. lewisi* and by liver in glycogen, which here are classified as miscellaneous fixation reactions, may with more complete data be shown to fall in the category of C_3 and C_1 addition.

Unidentified Fixation of Carbon Dioxide.—Aside from the specific studies on fixation of carbon dioxide in urea, dicarboxylic acids and glycogen, there have been indications of additional fixations. How many of these are by types already studied is not known. Solomon and his coworkers (53), for example, could not account for 39 per cent of the administered C^{11} and believed that much of it was fixed in organic compounds. Evans and Slotin (63) found that a part of the fixed carbon dioxide that could not be accounted for as α-ketoglutarate was released as carbon dioxide by treatment with ninhydrin and with chloramine T. This fact suggests that there is fixed carbon in amino acids or similar compounds. Such a fixation can readily be explained on the basis of the existence of transaminase in liver and its action on oxalacetic acid and α-ketoglutaric acid (86). There seems little doubt that as our investigations of carbon dioxide fixation are broadened we shall find an ever-widening field of its application.

Bibliography

1. Wood, H. G., and Werkman, C. H., *J. Bact.*, **30**, 332 (1935).
2. Wood, H. G., and Werkman, C. H., *Biochem. J.*, **30**, 48 (1936).
3. Evans, E. A., Jr., and Slotin, L., *J. Biol. Chem.*, **136**, 301 (1940).
4. Wood, H. G., and Werkman, C. H., *Biochem. J.*, **32**, 1262 (1938).
5. Wood, H. G., and Werkman, C. H., *Ibid.*, **34**, 129 (1940).
6. Wood, H. G., Stone, R. W., and Werkman, C. H., *Ibid.*, **31**, 349 (1937).
7. Winogradsky, S., *Ann. Inst. Pasteur*, **4**, 213 (1890); **5**, 92, 577 (1891).
8. Franck, J., *Sigma Xi Quart.*, **29**, 81 (1941).
9. Niel, C. B. van, *Arch. Mikrobiol.*, **3**, 1 (1931).
10. Engelmann, T. W., *Arch. ges. Physiol. Pflügers*, **30**, 95 (1883); *Botan. Z.*, **46**, 661 (1888).

11. Molisch, H., "Die Purpurbakterien nach neuen Untersuchungen," Jena, 1907.
12. Kluyver, A. J., and Donker, H. J. L., *Z. Chem. Zelle Gewebe*, **13**, 134 (1926).
13. Niel, C. B. van, "Advances in Enzymology," 1, 263, New York, 1941.
14. Wood, H. G., Werkman, C. H., Hemingway, A., and Nier, A. O. C., *J. Biol. Chem.*, **139**, 365 (1941).
15. Slade, H. D., Wood, H. G., Nier, A. O., Hemingway, A., and Werkman, C. H., *Iowa State Coll. J. Sci.*, **15**, 339 (1941).
16. Carson, S. F., and Ruben, S., *Proc. Natl. Acad. Sci.*, **26**, 422 (1940).
17. Ruben, S., and Kamen, M. D., *Ibid.*, **26**, 418 (1940).
18. Barker, H. A., Ruben, S., and Beck, J. V., *Ibid.*, **26**, 477 (1940).
19. Wieringa, K. T., *Ant. Leeuwenhoek*, **3**, 263 (1936); **6**, 251 (1940).
20. Novak, J., *Ann. Inst. Pasteur*, **22**, 54 (1908).
21. Smith, T., *J. Exptl. Med.*, **40**, 219 (1924).
22. Rockwell, G. E., and Highberger, J. H., *J. Infectious Diseases*, **40**, 438 (1927).
23. Winslow, C. E. A., Walker, H. H., and Stutermeister, M., *J. Bact.*, **24**, 185 (1932).
24. Gladstone, G. P., Fildes, P., and Richardson, G. M., *Brit. J. Exptl. Path.*, **16**, 335 (1935).
25. Woods, D. D., *Biochem. J.*, **30**, 515 (1936).
26. Barker, H. A., *Arch. Mikrobiol.*, **7**, 404 (1936).
27. Hes, J. W., *Ann. fermentations*, **4**, 547 (1938).
28. Foster, J. W., Carson, S. F., and Ruben, S., *Chronica Botanica*, **6**, 337 (1941).
29. Barker, H. A., *Ann. Rev. Biochem.*, **10**, 553 (1941).
30. Werkman, C. H., and Wood, H. G., *Bot. Rev.*, **8**, 1 (1942).
31. Lipmann, F., *Cold Spring Harbor Symposia Quant. Biol.*, **7**, 248 (1939).
32. Barron, E. S. G., *J. Biol. Chem.*, **113**, 717 (1936).
33. Nelson, M. E., and Werkman, C. H., *Iowa State Coll. J. Sci.*, **10**, 141 (1936).
34. Krebs, H. A., *Biochem. J.*, **31**, 661 (1937).
35. Niel, C. B. van, "The Propionic Acid Bacteria," Thesis. Haarlem, 1928.
36. Wood, H. G., Erb, C., and Werkman, C. H., *Iowa State Coll. J. Sci.*, **11**, 287 (1937).
37. Barker, H. A., Ruben, S., and Kamen, M. D., *Proc. Natl. Acad. Sci.*, **26**, 426 (1940).
38. Barker, H. A., *J. Biol. Chem.*, **137**, 153 (1941).
39. Wood, H. G., and Werkman, C. H., *J. Bact.*, **33**, 119 (1936).
40. Elsden, S. R., *Biochem. J.*, **32**, 187 (1938).
41. Wood, H. G., and Werkman, C. H., *Ibid.*, **34**, 7 (1940).
42. Wood, H. G., Werkman, C. H., Hemingway, A., and Nier, A. O., *J. Biol. Chem.*, **135**, 789 (1940).
43. Nishina, Y., Endo, S., and Nakayama, H., *Sci. Papers Inst. Phys. Chem. Research Tokyo*, **38**, 341 (1941).
44. Krebs, H. A., and Eggleston, L. V., *Biochem. J.*, **35**, 676 (1941).
45. Wood, H. G., and Werkman, C. H., *Ibid.*, **30**, 618 (1936).
46. Shaw, R. H., and Sherman, J. H., *J. Dairy Sci.*, **6**, 303 (1923).
47. Hitchner, E. R., *J. Bact.*, **28**, 473 (1934).
48. Fromageot, C., and Bost, G., *Enzymologia*, **4**, 225 (1938).
49. Erb, C., "Respiratory Behavior of the Propionic Acid Bacteria," Thesis. Iowa State College, 1934.

50. Wood, H. G., Werkman, C. H., Hemingway, A., and Nier, A. O., *J. Biol. Chem.*, **142,** 31 (1942).
51. Wood, H. G., Werkman, C. H., Hemingway, A., and Nier, A. O., *Ibid.*, **139,** 377 (1941).
52. Slade, H. D., Wood, H. G., Werkman, C. H., Hemingway, A., and Nier, A. O., *Ibid.* (in press).
53. Solomon, A. K., Vennesland, B., Klemperer, F. W., Buchanan, J. M., and Hastings, A. B., *J. Biol. Chem.*, **140,** 171 (1941).
54. Meyerhof, O., "Symposium on Respiratory Enzymes," University Wisconsin Press (in press).
55. Krebs, H. A., and Eggleston, L. V., *Biochem. J.*, **34,** 1383 (1940).
56. Krebs, H. A., *Nature*, **147,** 560 (1941).
57. Breusch, F. L., *Biochem. J.*, **33,** 1757 (1939).
58. Krampitz, L. O., and Werkman, C. H., *Ibid.*, **35,** 595 (1941).
59. Krampitz, L. O., Wood, H. G., and Werkman, C. H. (unpublished data).
60. Ostern, P., *Z. physiol. Chem.*, **218,** 160 (1933).
61. Evans, E. A., Jr., "Symposium on Respiratory Enzymes," University Wisconsin Press (in press).
62. Smyth, D. H., *Biochem. J.*, **34,** 1598 (1940).
63. Evans, E. A., Jr., and Slotin, L., *J. Biol. Chem.* **41,** 439 (1941).
64. Ochoa, S., and Peters, R. A., *Biochem. J.*, **32,** 1501 (1938).
65. Barron, E. S. G., *Ann. Rev. Biochem.*, **10,** 1 (1941).
66. Ochoa, S., *Biochem. J.*, **33,** 1262 (1939).
67. Banga, I., Ochoa, S., and Peters, R. A., *Ibid.*, **33,** 1980 (1939).
68. Carson, S. F., Foster, J. W., Ruben, S., and Kamen, M. D., *Science*, **92,** 433 (1940).
69. Wood, H. G., Werkman, C. H., Hemingway, A., and Nier, A. O., *Proc. Soc. Exptl. Biol. Med.*, **46,** 313 (1941).
70. Wood, H. G., Werkman, C. H., Hemingway, A., Nier, A. O., and Stuckwisch, C. G., *J. Am. Chem. Soc.*, **63,** 2140 (1941).
71. Carson, S. F., Foster, J. W., Ruben, S., and Barker, H. A., *Proc. Natl. Acad. Sci.*, **27,** 229 (1941).
72. Nahinsky, P., and Ruben, S., *J. Am. Chem. Soc.*, **63,** 2275 (1941).
73. Mickelson, M. N., and Werkman, C. H., *J. Bact.*, **37,** 619 (1939).
74. Krebs, H. A., and Henseleit, K., *Z. physiol. Chem.*, **210,** 33 (1932).
75. Henriques, O. M., *Biochem. Z.*, **200,** 1, 5, 18, 22 (1928).
76. Rittenberg, D., and Waelsch, H., *J. Biol. Chem.*, **136,** 799 (1940).
77. Evans, E. A., Jr., and Slotin, L., *Ibid.*, **136,** 805 (1940).
78. Wood, H. G., Werkman, C. H., Hemingway, A., and Nier, A. O., *Ibid.*, **139,** 483 (1941).
79. Evans, E. A., Jr., *Biochem. J.*, **34,** 829 (1940).
80. Krebs, H. A., and Eggleston, L. V., *Ibid.*, **34,** 442 (1940).
81. Krebs, H. A., *Ibid.*, **34,** 460 (1940).
82. Cori, C. F., "Symposium on Respiratory Enzymes," University Wisconsin Press (in press).
83. Meyerhof, O., Ohlmeyer, P., Gentner, W., and Maier-Leibnitz, W., *Biochem. Z.*, **298,** 396 (1938).
84. Kalckar, H. M., *Biochem. J.*, **33,** 631 (1939).

85. Lipmann, F., "Advances in Enzymology," **1**, 99, New York, 1941.
86. Braunstein, A. E., and Kritzmann, M. G., *Enzymologia*, **2**, 129 New York, 1941.
87. Searle, D. S., and Reiner, L., *Proc. Soc. Exptl. Biol. Med.*, **43**, 80 (1940); *J. Biol. Chem.*, **141**, 563 (1941).

PHOSPHOTRIESTERASE: AN ENZYME IN SEARCH OF ITS NATURAL SUBSTRATE

By FRANK M. RAUSHEL, *Department of Chemistry, Texas A&M University, College Station, Texas 77843* and HAZEL M. HOLDEN, *Department of Biochemistry, University of Wisconsin, Madison, Wisconsin 53706*

CONTENTS

Advances in Enzymology and Related Areas of Molecular Biology, Volume 74: Mechanism of Enzyme Action, Part B, Edited by Daniel L. Purich
ISBN 0-471-34921-6

I. Introduction

Approximately 25 years ago two strains of unrelated soil bacteria (*Pseudomonas diminuta* and *Flavobacterium* sp.) were isolated that had the ability to hydrolyze, and thus detoxify, a broad range of organophosphate insecticides and military-type nerve agents (Munnecke, 1976). The specific chemical reaction catalyzed by these bacterial strains, as exemplified with the insecticide paraoxon, is shown in Scheme 1. This enzymatic transformation is rather interesting since there are very few, if any, naturally occurring organophosphate triesters that have been isolated and characterized. Moreover, a protein that is capable of hydrolyzing organophosphates may eventually find commercial applications for the detoxification of agricultural insecticides and chemical warfare nerve agents.

Organophosphate triesters and related phosphonate diesters are extremely toxic because of their ability to specifically inactivate the enzyme acetylcholinesterase (AchE). Upon incubation of AChE with activated organophosphates, the enzyme rapidly forms a phosphoenzyme intermediate with an active site serine residue. This intermediate cannot be further processed by AChE at a significant rate and thus the catalytic function of the enzyme is destroyed. Since AChE is critical for nerve function, via the hydrolysis of acetylcholine, this inactivation event is lethal to a variety of cell types. The reactions catalyzed by AChE with acetylcholine and paraoxon are summarized in Scheme 2.

The bacterial phosphotriesterase (PTE) treats this class of organophosphate nerve agents not as potent inactivators of enzyme function, but rather as very good substrates with extraordinary turnover numbers. Since organophosphate triesters were not widely released into the environment as agricultural pesticides prior to World War II, it has been difficult to understand how the catalytic activities discovered for PTE could have rapidly evolved during this very short time interval. If PTE has not evolved to specifically hydrolyze organophosphate nerve agents, then the protein must catalyze another metabolic function that has not been discovered. The complete gene sequence for this enzyme has provided no substantive clues as to the

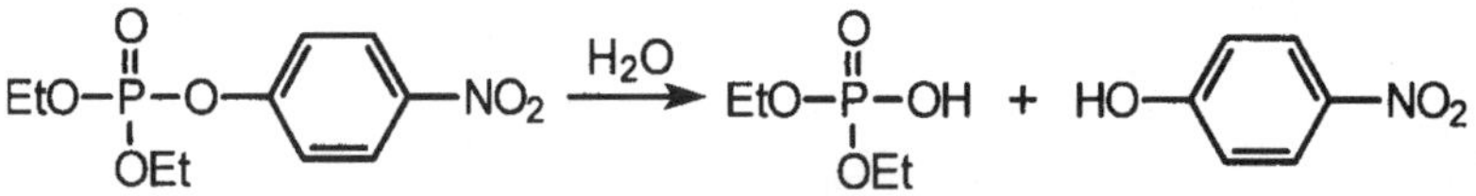

$$(EtO)_2P(=O)\text{-}O\text{-}C_6H_4\text{-}NO_2 \xrightarrow{H_2O} (EtO)_2P(=O)\text{-}OH + HO\text{-}C_6H_4\text{-}NO_2$$

Scheme 1

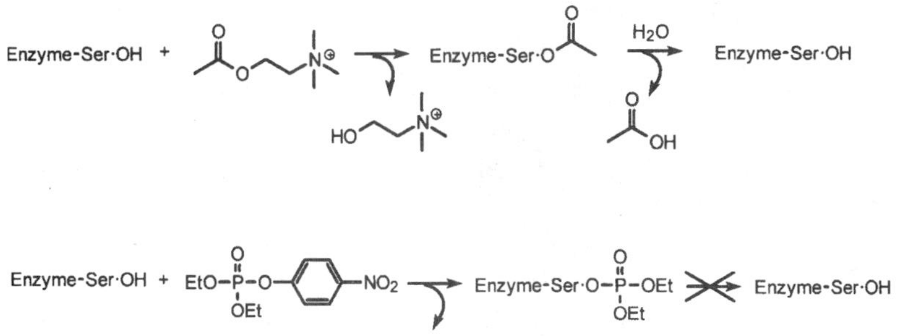

Scheme 2

metabolic role of this enzyme. At the time of discovery, there were no known protein sequences that were homologous to PTE. However, this situation has been clarified somewhat in the intervening period (Holm and Sander, 1997).

Our goal for this review is to provide a detailed account of the relationship between structure, function, and mechanism of this fascinating enzyme and then attempt to correlate the mechanism and structure of this protein with related examples of a growing superfamily of hydrolase enzymes.

II. Structure

A. PROTEIN SEQUENCE

The gene for the bacterial PTE has been sequenced from *P. diminuta* and *Flavobacterium* sp. (Mulbry and Karns, 1989; Serdar et al., 1989). The genes for PTE from both organisms are found within extra chromosomal plasmids. Surprisingly, the DNA sequences are identical, and in both cases the protein is translated as a larger precursor protein prior to the cleavage of a 29-amino acid leader peptide from the N-terminal end of the proenzyme to form the mature enzyme (Mulbry and Karns, 1989). The complete sequence of the unprocessed protein is presented in Figure 1. The constitutively expressed proteins in the native organisms are found as membrane-associated complexes (Brown, 1980; McDaniel, 1985).

The gene for the bacterial PTE has been subcloned into a variety of expression vectors (McDaniel, 1985) and the protein purified to homogeneity using insect cells (Dumas et al., 1989a; Dumas and Raushel, 1990a), *Streptomyces lividans* (Rowland et al., 1991), and *Escherichia coli* (Serdar et al.,

```
1    MQTRRVVLKS  AAAAGTLLGG  LAGCASVAGS  IGTGDRINTV  RGPITISEAG  50
51   FTLTHEHICG  SSAGFLRAWP  EFFGSRKALA  EKAVRGLRRA  RAAGVRTIVD  100
101  VSTFDIGRDV  SLLAEVSRAA  DVHIVAATGL  WFDPPLSMRL  RSVEELTQFF  150
151  LREIQYGIED  TGIRAGIIKV  ATTGKATPFQ  ELVLKAAARA  SLATGVPVTT  200
201  HTAASQRDGE  QQAAIFESEG  LSPSRVCIGH  SDDTDDLSYL  TALAARGYLI  250
251  GLDHIPHSAI  GLEDNASASA  LLGIRSWQTR  ALLIKALIDQ  GYMKQILVSN  300
301  DWLFGFSSYV  TNIMDVMDRV  NPDGMAFIPL  RVIPFLREKG  VPQETLAGIT  350
351  VTNPARFLSP  TLRAS
```

Figure 1. Amino acid sequence for the bacterial phosphotriesterase from *Pseudomonas diminuta*.

1989). Significant improvements in the levels of protein expression have been obtained by deletion of the DNA encoding the 29-amino acid leader sequence (Mulbry and Karns, 1989; Serdar et al., 1989).

B. ACTIVE SITE RESIDUES

In order to catalyze the hydrolysis of an organophosphate triester, an enzyme must function to activate the hydrolytic water molecule for nucleophilic attack and enhance the electrophilic properties of the phosphorus center. Initial speculation on the chemical mechanism of PTE suggested that the hydrolytic water molecule could be activated by general base catalysis and that the leaving group of the organophosphate triester could be activated by general acid catalysis (Donarski et al., 1989; Dumas and Raushel, 1990b). This particular aspect of the reaction mechanism and active site structure was first examined for the purified enzyme via the determination of the pH-rate profile. When paraoxon is used as the substrate, the pH profile versus k_{cat} with PTE is as shown in Figure 2. There is a plateau of catalytic activity above pH ~7 and a loss of activity below pH ~6.5. A fit of the experimental data to the appropriate equation gave a pK_a for a single ionizable group of 6.1 (Dumas and Raushel, 1990b). A similar profile was found for k_{cat}/K_m. Since the substrate cannot lose protons within the pH range examined, it thus appeared that a single group at the active site of PTE must be ionized for maximum catalytic activity. The pH-rate profile was thus consistent with a residue that served to activate the hydrolytic water molecule via general base catalysis. Since there was no loss of activity at high pH there was no evidence for general acid catalysis during the departure of the leaving group phenol. This outcome was not entirely unexpected since the

pK_a of *p*-nitrophenol is approximately 7.0 and there is little to be gained by stabilization of the leaving group phenol. In an effort to determine the identity of the active site residue responsible for the single ionization that was found in the pH-rate profile, the effect of temperature on the pK_a value was measured. The ΔH_{ion} was determined to be 7.9 kcal/mol (Dumas and Raushel, 1990b). The effect of organic solvents indicated that the group was a cationic acid. These data were most consistent with the ionization of a single histidine residue at the active site of PTE.

Confirmation of this assignment was sought through the utilization of group-specific reagents. No significant inactivation could be obtained upon incubation of the enzyme with dithionitrobenzoic acid (DTNB), carbodiimide, pyridoxal, butanedione, or iodoacetate (Dumas and Raushel, 1990b). These results appeared to eliminate cysteine, lysine, arginine, aspartate, and glutamate as accessible residues within the active site. However, the enzyme could be quantitatively inactivated with the histidine-specific reagent diethylpyrocarbonate (DEPC) and the metal chelator, *o*-phenanthroline. Analy-

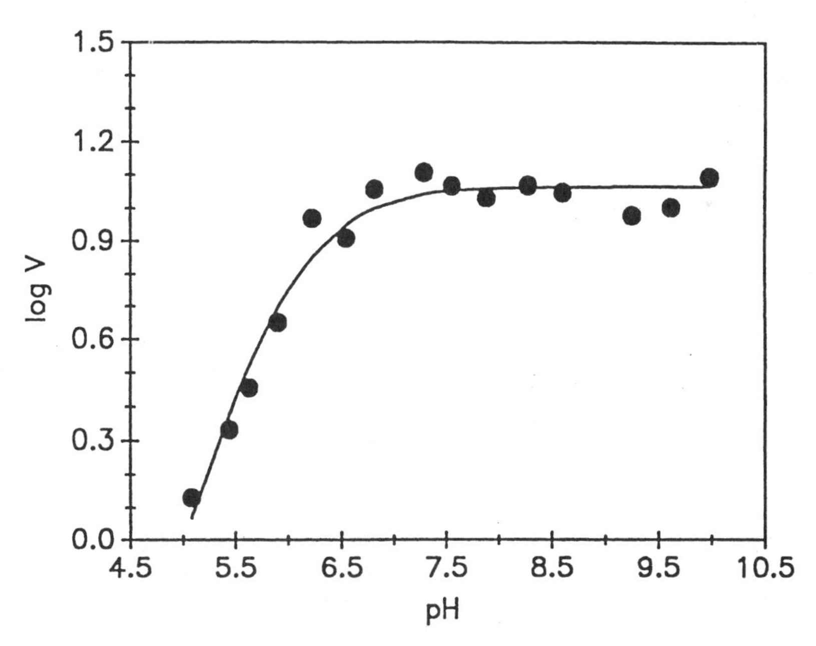

Figure 2. pH-rate profile for the hydrolysis of paraoxon as catalyzed by the bacterial phosphotriesterase.

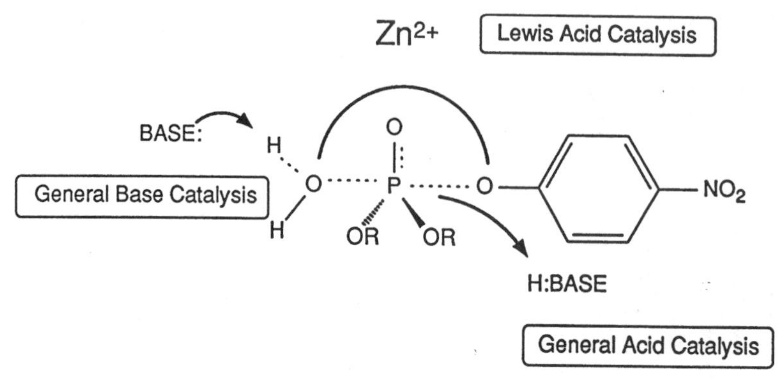

Figure 3. Initial model for the mechanism of organophosphate hydrolysis by phosphotriesterase.

sis of the DEPC inactivation kinetics as a function of pH was consistent with a pK_a value of 6.1 for the histidine that was labeled under the experimental conditions (Dumas and Raushel, 1990b). No loss of activity was observed in the presence of substrate analogs and thus the labeled histidine appeared to be within the active site. Atomic absorption spectroscopy confirmed the presence of zinc bound to the protein. Based upon these results a working catalytic mechanism was proposed as illustrated in Figure 3. In this mechanism the function of the ionized histidine is to remove a proton from the metal-bound water molecule. In addition to serving as a template for the hydrolytic water molecule, the lone zinc ion could serve to polarize the phosphoryl oxygen bond and thus make the phosphorus center more electrophilic.

C. RECONSTITUTION OF APO-ENZYME

The initial discovery that the bacterial phosphotriesterase contained bound zinc prompted a series of experiments designed to substitute other divalent cations into the enzyme active site (Omburo et al., 1992). The most successful of these experiments involved the utilization of a metal chelator to make apo-enzyme and then reconstitution of the metal-free enzyme with a variety of divalent cations. Metal chelators such as *o*-phenanthroline were found to be far more efficient at the removal of bound-metal from the protein than ethylenediaminetetraacetic acid (EDTA). Thus, the second-order rate constant for the inactivation of PTE by *o*-phenanthroline and EDTA are 1.6 $M^{-1} s^{-1}$ and 8×10^{-4} $M^{-1} s^{-1}$, respectively (Omburo et al., 1992). The apo-enzyme has less than 1% of the catalytic activity exhibited by the wild-type enzyme. The apo-enzyme was stable when frozen at –78°C for a period of at least six months and full enzymatic activity could be recovered upon the addition of zinc.

Reconstitution of the apo-enzyme with a variety of divalent cations revealed two very interesting structural and mechanistic features about PTE. Shown in Figure 4 is a plot of the recovered enzymatic activity as a function of metal ion added per equivalent of apo-enzyme. Remarkably, the protein could be reconstituted with Co^{2+}, Cd^{2+}, and Zn^{2+}. The Co^{2+}- and Cd^{2+}-substituted proteins were even more active than the Zn^{2+}-substituted protein. For each of the divalent cations added to the system there was a linear increase in the catalytic activity until the ratio of two metal ions per protein equivalent was achieved. Thus, PTE requires two metal ions for full catalytic activity rather than the single metal ion as originally believed (Omburo et al., 1992). Additional experiments (not shown) demonstrated that Mn^{2+} and Ni^{2+} were also able to activate the enzyme. The optimum pH for reconstitution of apo-enzyme was found to be pH 7–8.5. A summary of the catalytic constants by each of the

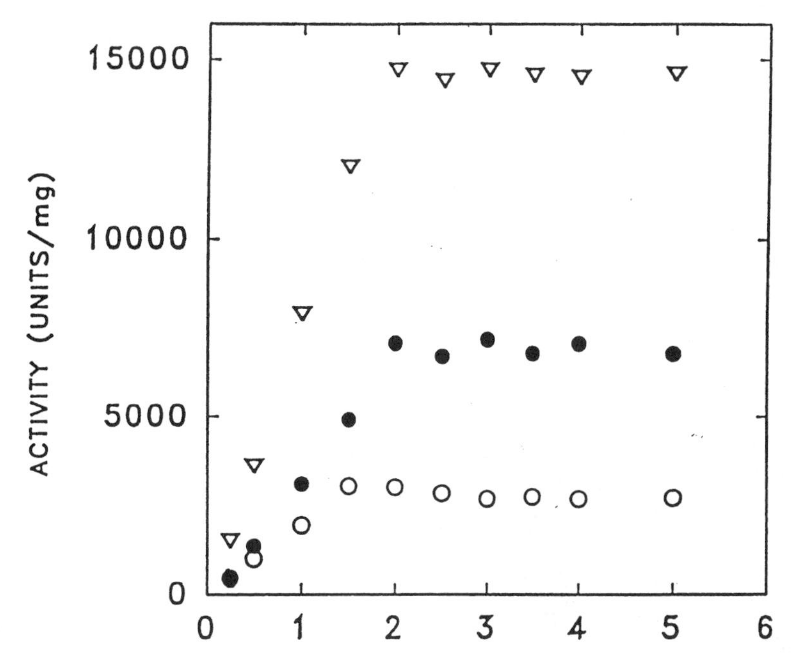

Figure 4. Reconstitution of the apo-phosphotriesterase with various amounts of Co^{2+}, Zn^{2+}, and Cd^{2+}.

TABLE 1
Kinetic Constants for Metal Substituted Phosphotriesterase[a]

Metal	K_m (μM)	k_{cat} (s^{-1})	k_{cat}/K_m ($10^7\ M^{-1}\ s^{-1}$)
Zn^{2+}	90	2400	2.7
Co^{2+}	200	7800	4.0
Mn^{2+}	80	1800	2.2
Cd^{2+}	400	4600	1.2
Ni^{2+}	150	6000	3.9

[a]Determined for paraoxon at pH 9.0.

metal-substituted forms of PTE for the hydrolysis of paraoxon is presented in Table 1. Unfortunately, the titration of apo-enzyme with the various divalent cations provided no clues with regard to whether one or both metal ions were required for catalytic activity. Moreover, it could not be determined whether these metal ions were required for structural or catalytic functions.

D. ^{113}Cd-NMR SPECTROSCOPY

The environment of the metal ion binding site(s) was initially probed with nuclear magnetic resonance (NMR) spectroscopy (Omburo et al., 1993). Cadmium has two isotopes that are NMR active, $^{113}Cd^{2+}$ and $^{111}Cd^{2+}$. Prior investigations with model systems and metal binding sites in proteins of known structure had clearly demonstrated that a great deal of information could be obtained about the direct ligands to each of these metals from a determination of the chemical shift (Summers, 1988). In model systems the chemical shift range for $^{113}Cd^{2+}$ extends in excess of 900 ppm. The most downfield signals are observed for complexes that are composed of all thiolate ligands from cysteine while the most upfield signals are those for complexes with all-oxygen bearing ligands from water or carboxylates (aspartate or glutamate). Nitrogen ligand sets are intermediate in chemical shift between the sulfur and oxygen ligands.

The $^{113}Cd^{2+}$ spectrum of PTE containing two Cd^{2+} per monomeric protein is presented in Figure 5A. The spectrum clearly shows that each of the metal ions is in a slightly different chemical environment. The observed chemical shifts are 116 and 212 ppm. The chemical shift position excluded ligation by even a single cysteine group and strongly suggested that the ligand set consisted of a mixture of oxygen (water, carboxylate) and nitrogen (histidine) ligands. The more downfield signal was expected to arise from Cd^{2+}, which

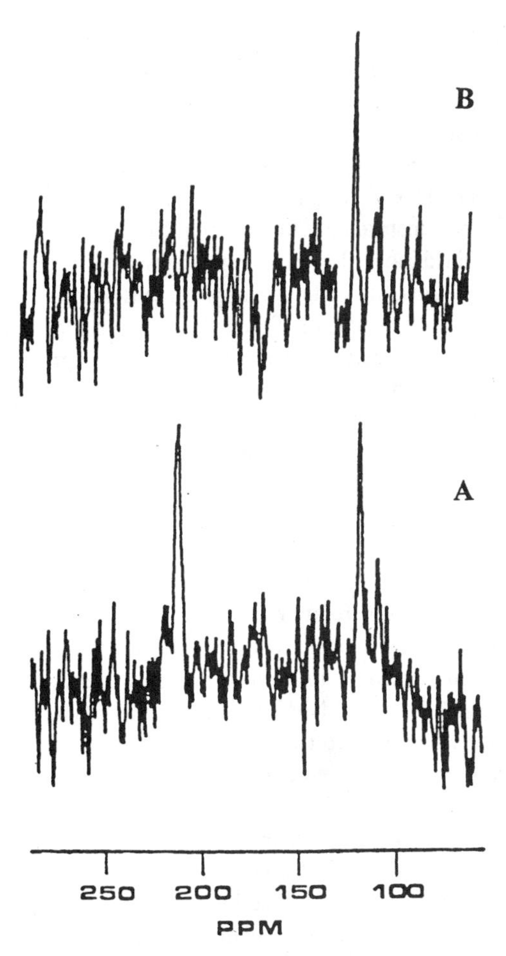

Figure 5. ^{113}Cd-NMR spectra of bacterial phosphotriesterase. **(A)** Apo-enzyme reconstituted with 2 equivalents of Cd^{2+}. **(B)** Apo-enzyme reconstituted with one equivalent each of Zn^{2+} and Cd^{2+}.

had a higher percentage of nitrogen ligands than the more upfield signal. The best estimates, made at that time, proposed that the cadmium with the more downfield signal was bound to three nitrogen ligands and one oxygen ligand while the cadmium exhibiting the upfield signal was bound to two nitrogen ligands and two oxygen ligands. This analysis assumed that the environment for each metal ion was tetrahedral. These conclusions were not entirely correct, as later discovered by the solution of the X-ray crystal structure for PTE. Nevertheless, these studies clearly demonstrated that a majority of the ligands to each of these metal ions originated from the imidazole side chain of histidine residues (Omburo et al., 1993).

^{113}Cd-NMR spectroscopy provided further insight into the structure of an unusual hybrid of PTE. It was found that a hybrid containing one Cd^{2+} and one Zn^{2+} could be constructed (Omburo et al., 1993). The catalytic properties of the Cd/Zn-hybrid more closely resembled those of Zn/Zn–PTE rather than Cd/Cd–PTE, and thus one metal ion appears to dominate the catalytic properties. The ^{113}Cd-NMR spectrum of the hybrid is presented in Figure 5B. Only the single upfield resonance was observed and thus it appeared that cadmium was uniquely occupying only one of the two sites while zinc was located at the other site. This metal hybrid complex may be of great utility in defining the specific functions of the individual metals in binding and catalysis.

E. EPR SPECTROSCOPY

Since the divalent cations bound to PTE could be substituted with Mn^{2+} with high occupancy and relatively high catalytic activity, electron paramagnetic resonance (EPR) spectroscopy was utilized to provide additional information with regard to protein–metal and metal–metal interactions (Chae et al., 1993). The ^{113}Cd-NMR evidence provided no information about whether the protein contained two mononuclear metal centers or if a single binuclear metal center was operable. Shown in Figure 6 is the EPR spectrum of the Mn^{2+}/Mn^{2+}–PTE at 10 K. The X-band EPR spectrum is quite complex. The predominant features were near g-2 and exhibited what appeared to be more than 26 Mn hyperfine splittings at ~45 gauss intervals. The large number of hyperfine splittings, separated by approximately half the magnitude expected for a mononuclear Mn^{2+}-center, suggested the presence of two spin-coupled Mn^{2+} ions. This conclusion was also supported by the Q-band spectrum obtained under similar experimental conditions (Chae et al., 1993). The temperature dependence of the X-band signals provided further evidence that the two manganese ions were antiferromagnetically coupled to one another (Chae et al., 1993). Similar spectra had previously been noted with arginase (Reczykowski and Ash, 1992), enolase (Poyner and Reed, 1992), concanavalin A (Antanaitis et al., 1987), and catalase (Khangulov et al., 1990). These data ruled out the existence of two independent mononuclear sites. Thus, the most likely structure was a coupled binuclear center with a bridging ligand of unknown identity (Chae et al., 1993, 1995). Taken

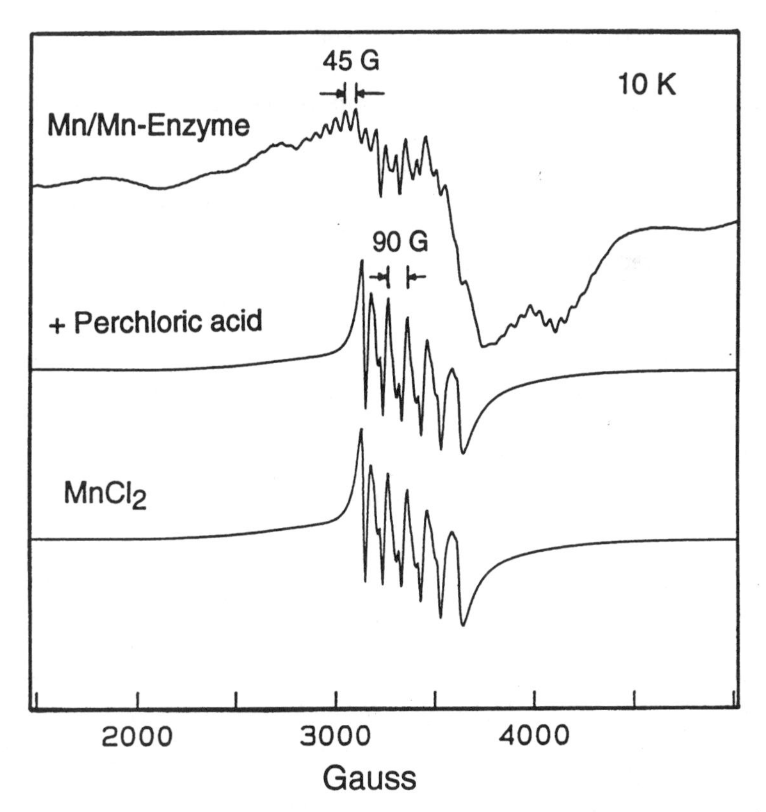

Figure 6. Electron paramagnetic resonance spectrum (X-band) of the apo-enzyme reconstituted with two equivalents of Mn^{2+}.

together the $^{113}Cd^{2+}$-NMR studies and the Mn^{2+}-EPR investigations pointed to the existence of a binuclear metal center where a majority of the direct ligands to the metal ions originated with the imidazole side chains of histidine.

MUTAGENESIS OF ACTIVE SITE RESIDUES.

The spectroscopic and chemical labeling experiments implicated the essential nature of one or more histidine residues at the active site of the bacterial phosphotriesterase. These results prompted two independent investigations toward the identification of those histidine residues required for catalytic activity (Kuo and Raushel, 1994; Lai et al., 1994). In PTE there are only seven histidine residues in the entire molecule and each of these histidine residues was mutated to an asparagine residue. Of the seven residues only His-123 was without measurable effects on the catalytic activities after mutation to asparagine. The remaining six residues, His-55, His-57, His-201, His-230, His-254, and His-257, were all diminished in activity to some extent. Working models, constructed at that time, that attempted to assign individual histidine residues to specific metal ions within the binuclear metal cluster proved later to be slightly incorrect. However, the X-ray structure of PTE clearly demonstrated that His-123 was away from the active site and that the remaining six histidines were in the active site but that only His-55, His-57, His-201, and His-230 had coordinate bonds to to either of the two metals (Benning et al., 1995). All of these histidine residues have also been mutated in a combinatorial library to cysteine (Watkins et al., 1997a)

G. STRUCTURE OF THE Apo-FORM OF PHOSPHOTRIESTERASE

The various structural analyses of phosphotriesterase that were initiated in the spring of 1994 can be aptly described as one surprise after another. Indeed, crystals of the Cd^{2+}-substituted enzyme were first observed with polyethylene glycol 8000, 100 mM bicine (pH 9.0), and 1 M LiCl as the precipitant (Benning et al., 1994). As the X-ray model of phosphotriesterase was being built, however, it became obvious that there were no large peaks in the electron density map corresponding to the bound Cd^{2+} ions of the binuclear metal center and, in fact, the crystallization conditions employed had effectively removed the metals from the protein. While the structural determination of the apo-form of phosphotriesterase was unintentional, this analysis did provide the first three-dimensional glimpse of the enzyme. Indeed, from the packing arrangement of the molecules in the crystalline lattice, it was immediately obvious that, contrary to previous reports (Dumas et al., 1989a), the quaternary structure of the enzyme was dimeric rather than monomeric. Shown in Figure 7 is a ribbon representation of the phosphotriesterase dimer. The total surface area buried upon dimer formation is approximately 3200 Å^2 using a search probe radius of 1.4 Å. The subunit:subunit interface is formed primarily by two regions delineated by Ser-61 to Phe-73 and Met-138 to Phe-149. In particular, there are four aromatic residues, Phe-65, Trp-69, Phe-72, and Phe-73, that participate in various stacking interactions as indicated by the ball-and-stick representations in Figure 7. Apart from these aromatic contacts, there are numerous electro-

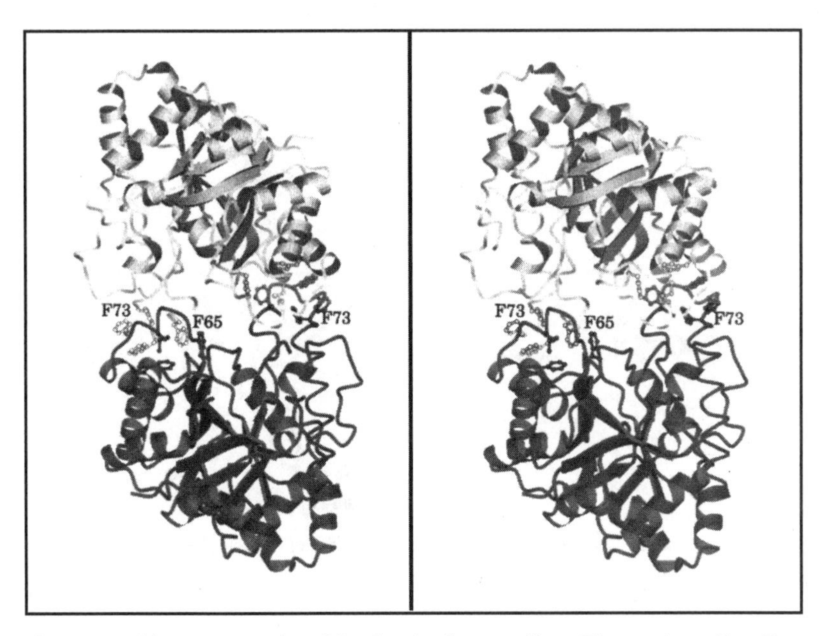

Figure 7. Ribbon representation of the phosphotriesterase dimer. Those amino acid residues forming a hydrophobic patch at the dimeric interface are displayed in ball-and-stick representations.

static interactions that also serve to stabilize the subunit:subunit interface of the dimer.

Each subunit of the dimer is roughly globular with overall dimensions of 51 Å × 55 Å × 51 Å. As can be seen in Figure 7, the fold of each subunit is dominated by eight β-strands that wrap around to form a parallel β-barrel. This β-barrel is flanked on the outer surface by a total of 14 α-helices. In addition to the barrel motif, there are two antiparallel β-strands at the N-terminus. At the time of this investigation, amino acid sequence analyses failed to detect any similarity between phosphotriesterase and other proteins of known structure. Yet, this α/β-barrel motif is a very commonly occurring tertiary structural element in enzymes and is typically referred to as a "TIM-barrel" (Farber and Petsko, 1990). In these TIM-barrel enzymes, the active sites are invariably located at the C-terminal portion of the β-barrel. A close-up view of this region in PTE is shown in Figure 8. As originally predicted from site-directed mutagenesis experiments, His-55, His-57, His-201, His-230, His-254, and His-257 are, indeed, clustered within this active site re-

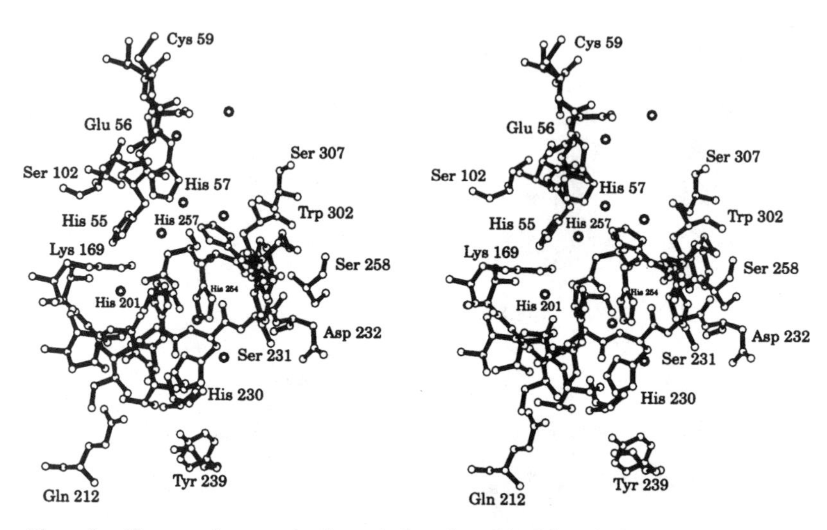

Figure 8. Close-up view near the C-terminal portion of the β-barrel. The *gray spheres* represent ordered solvent molecules.

gion (Kuo and Raushel, 1994; Lai et al., 1994). There is a striking stacking interaction between the imidazole rings of His-201 and His-254. In addition, there is a close contact between the ε-amino group of Lys-169 and $N^{\delta1}$ of His-201 (2.1 Å). Even though this first structural analysis of phosphotriesterase was that of the apo-protein, nevertheless it resulted in the determination of the overall molecular motif of the enzyme, its active site location, and its proper quaternary structure.

H. STRUCTURE OF THE CD^{2+}/CD^{2+}-SUBSTITUTED ENZYME

In an attempt to prepare samples of the native, metal-containing phosphotriesterase, the apo-enzyme crystals were subsequently transferred to various metal-containing solutions such as $CdCl_2$. Unfortunately the X-ray diffraction quality of the crystals deteriorated significantly such that it was not possible to collect X-ray data beyond 3.5 Å resolution. The next set of experiments involved crystallization of the cadmium-containing enzyme with polyethylene glycol but this time in the absence of lithium chloride and with the replacement of the buffer bicine with CHES. Again these experiments met with limited success until the substrate analog, diethyl 4-methylbenzylphosphonate, was added to the precipitant solution. Under these conditions, crystals appeared within one week and diffracted to a nominal resolution of 2.0 Å. The structure of the holoenzyme was subsequently solved by the techniques of molecular replacement, solvent flattening, and molecular averaging (Benning et al., 1995).

A ribbon representation of the Cd^{2+}/Cd^{2+}-containing PTE is displayed in Figure 9A. As predicted from the apo-enzyme structure, the binuclear metal center is, indeed, located at the C-terminal portion of the β-barrel. The substrate analog did not bind within this region, however, but rather at the interface between two symmetry-related molecules in the crystalline lattice. Quite unexpectedly, the bridging ligand for the binuclear metal center was not an aspartate, glutamate, or histidine residue but rather Lys-169, which, as indicated by the electron density, was carbamylated. Of the six histidine residues clustered at the end of the β-barrel, four are directly involved in metal ligation (His-55, His-57, His-201, and His-230). In addition, the carboxylate group of Asp-301 functions as a metal ligand.

A cartoon of the coordination geometry for the binuclear metal center is depicted in Figure 9B. The two cadmium ions are separated by 3.8 Å and are bridged by both the carbamylated lysine and a solvent molecule, most likely a hydroxide ion. The deeper buried cadmium ion is surrounded in a trigonal bipyramidal arrangement by His-55, His-57, Lys-169, Asp-301, and the bridging solvent. Lys-169, His-201, His-230, the bridging hydroxide ion, and two additional waters form a distorted octahedral arrangement around the more solvent-exposed cadmium. The metal:ligand bond lengths range from 2.1 Å to 2.7 Å.

Superpositions of the polypeptide chain backbones for the apo- and holoenzymes are given in Figure 10A. Even though these polypeptide chains are identical in amino acid sequence, they superimpose with a remarkable root-mean-square difference of 3.4 Å. The three-dimensional differences in these polypeptide chains are limited, however, to a few specific regions, the most striking of which occurs near Asp-301. In the apo-enzyme model, Asp-301 adopts dihedral angles of $\phi = -54.2°$ and $\psi = -45.2°$ while in the holoprotein, it assumes torsional angles of $\phi = 58.3°$ and $\psi = 44.0°$. This reversal from a right-handed to a left-handed helical conformation forces the polypeptide chains into completely different directions as can be seen in the close-up view presented in Figure 10B. A superposition of the active site regions for the apo- and holoenzyme models is depicted in Figure 11. In retrospect, it is not surprising that metal-soaking experiments with preformed apo-enzyme crystals met with limited success. As can be seen, apart from the position of His-57, there is a complete restructuring of the

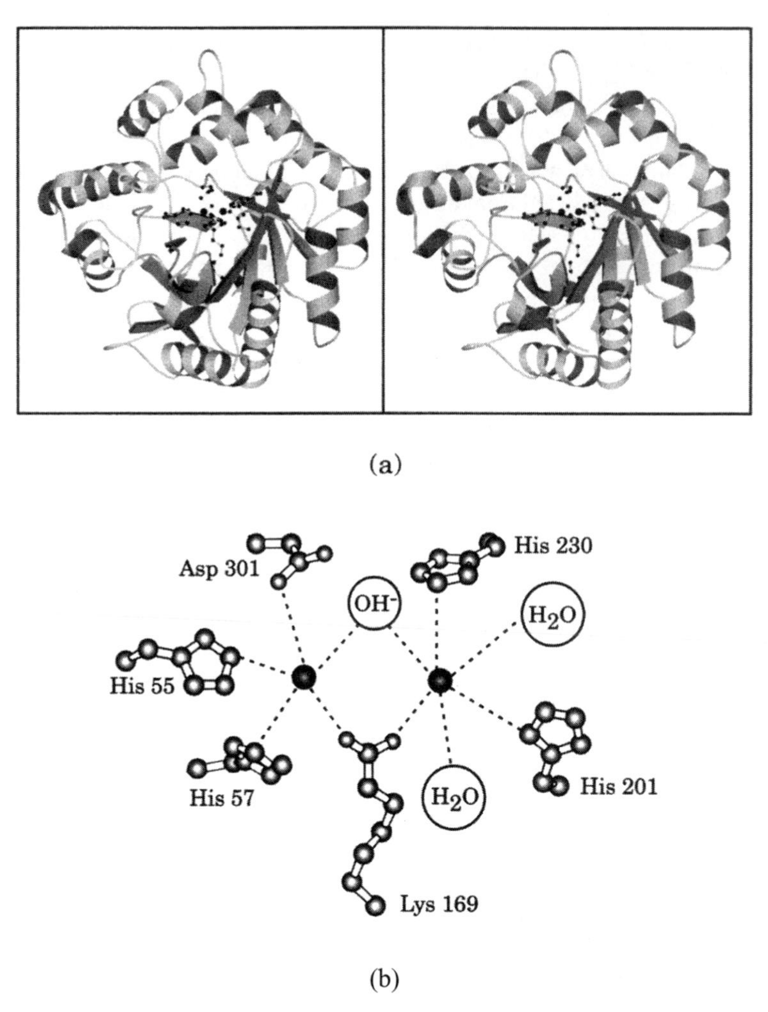

Figure 9. Structure of the Cd^{2+}/Cd^{2+}-containing phosphotriesterase. A ribbon representation of one subunit is given in **(a).** Those residues involved in metal ligation are displayed as ball-and-sticks. A schematic of the coordination geometry about the binuclear metal center is displayed in **(b).**

amino acid side chains upon metal binding and, indeed, these dramatic differences clearly highlight the difficulties in predicting protein structures based simply upon amino acid sequences. In the case of phosphotriesterase, the three-dimensional architecture of the protein is dependent not only upon

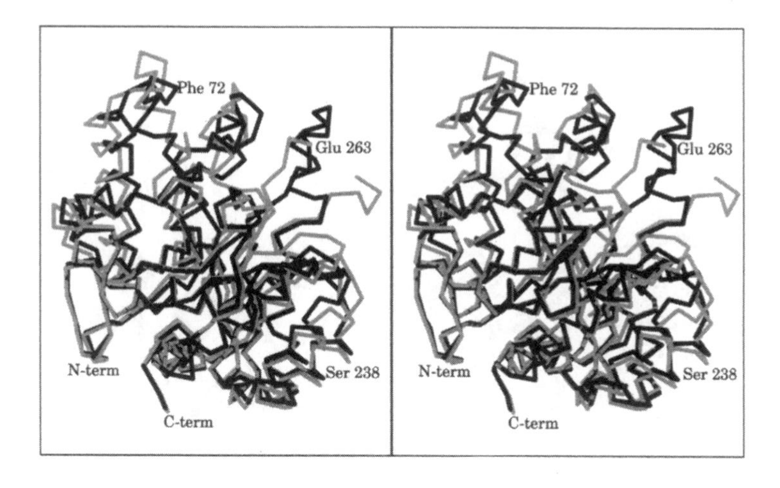

(a)

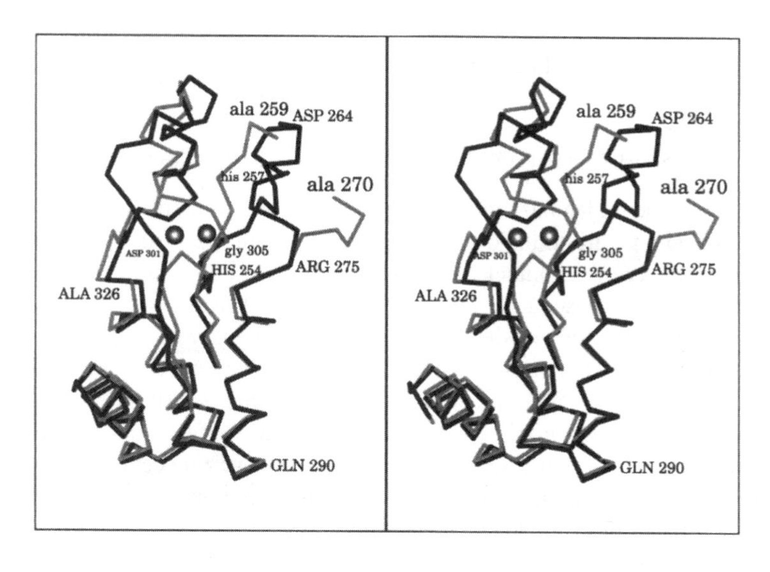

(b)

Figure 10. Comparison of the apo- and holo-enzyme structures. α-Carbon traces for the apo- and holo-enzyme models are displayed in *gray* and *black*, respectively. The two structures differ most in the region shown in (b). Those amino acid residues indicated in lower- and uppercase refer to the apo- and holo-enzymes, respectively. The cadmium ions are represented by the *large spheres* near Asp 301.

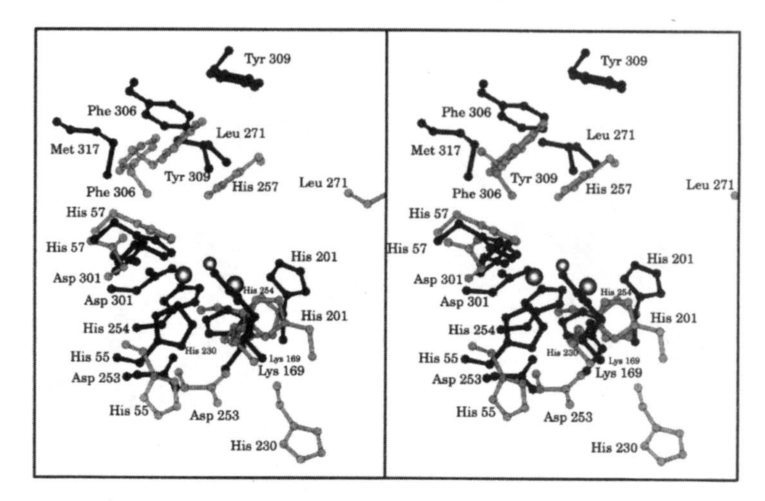

Figure 11. Superposition of the active site regions in the apo- and holo-enzyme models. The apo- and holo-forms of the enzyme are depicted in *gray* and *black*, respectively

its primary structure but also upon the presence or absence of the binuclear metal center.

I. STRUCTURE OF THE Zn^{2+}/Zn^{2+}-CONTAINING ENZYME

While the molecular model of the Cd^{2+}/Cd^{2+}-substituted PTE was informative with regard to the overall fold of the holoenzyme and the coordination geometry of the binuclear metal center, the naturally occurring enzyme contains zinc. Consequently, the next goal in the structural analysis of PTE was to grow crystals of the Zn^{2+}/Zn^{2+}-containing enzyme. Again, crystals were obtained in the presence of the substrate analog, diethyl 4–methylbenzylphosphonate (Vanhooke et al., 1996). This time the inhibitor not only bound within the hydrophobic pocket located between two monomers in the crystalline lattice, but also adjacent to the binuclear zinc center. Note that the α-carbon traces for the zinc- and cadmium-containing proteins superimpose with a root-mean-square deviation of 0.20 Å.

A ribbon representation of the Zn^{2+}/Zn^{2+}-containing enzyme with the bound inhibitor is depicted in Figure 12A and a cartoon of the coordination geometry for the binuclear metal center is displayed in Figure 12B. In the Zn^{2+}/Zn^{2+}-containing protein, the metals are separated by 3.3 Å and the zinc:ligand bond lengths range from 1.8 Å to 2.3 Å. Again, the deeper

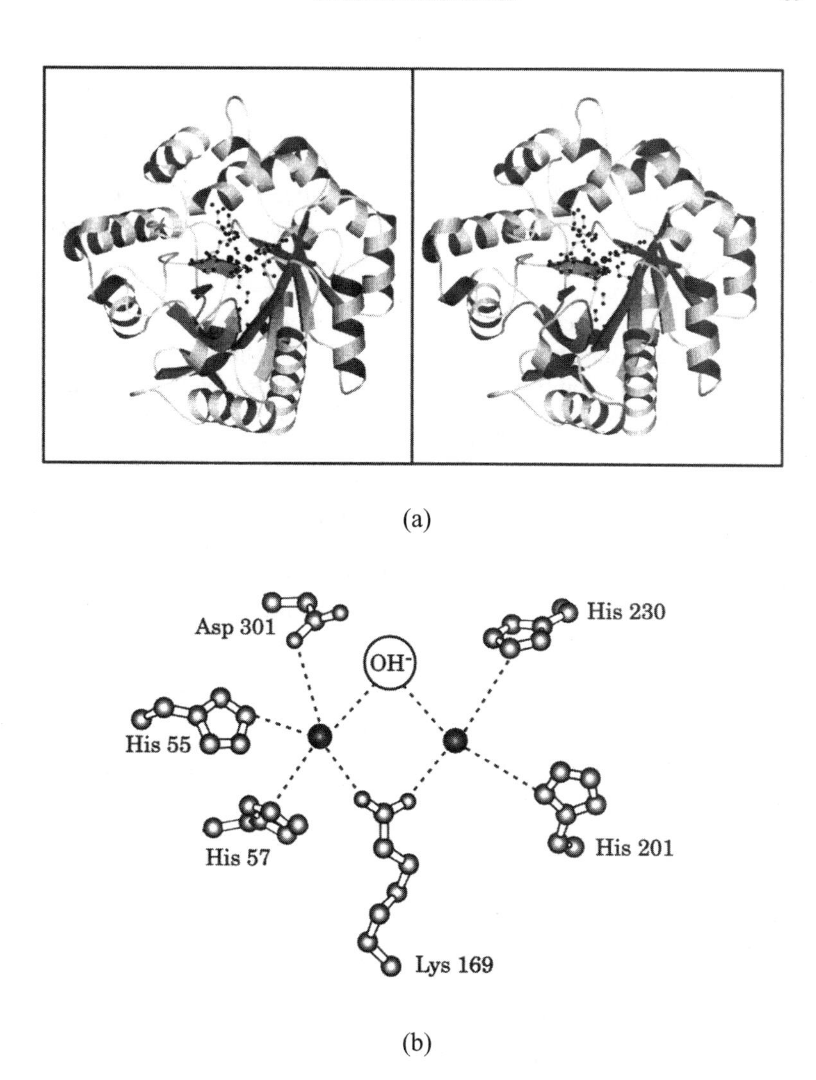

Figure 12. Structure of the Zn^{2+}/Zn^{2+}-containing phosphotriesterase. A ribbon representation of one subunit is given in **(A)** and a cartoon of the coordination geometry is shown in **(B).**

buried metal ion is coordinated to the protein via the side chain functional groups of His-55, His-57, Lys-169, Asp-301, and the bridging solvent in a virtually identical trigonal bipyramidal arrangement as observed for the cadmium-substituted enzyme. The coordination geometry around the more solvent-exposed zinc, however, has changed from octahedral to tetrahedral with the simple removal of water molecules from the coordination sphere. While the atomic radii for Zn^{2+} and Cd^{2+} are 0.74 Å and 0.97 Å, respectively, it appears that the polypeptide chain backbone of phosphotriesterase is ideally suited for accommodating such differences. Indeed, the protein atoms for these two forms of PTE, including side chains, superimpose with a root-mean-square deviation of 0.40 Å.

Perhaps the most notable aspect of the PTE active site is the lack of hydrogen bonding interactions between the protein and the inhibitor. A close-up view of the binding pocket is given in Figure 13. There is only one direct electrostatic interaction between the inhibitor and the protein. Specifically, the phosphoryl oxygen of the substrate analog is located 3.3 Å from $N^{\varepsilon 1}$ of Trp-131 and 3.2 Å from $N^{\delta 1}$ of His-201. In addition, this oxygen is situated within 3.5 Å of the more solvent accessible zinc ion. The methylbenzyl group, as displayed in Figure 13, is wedged into a fairly hydrophobic pocket formed by His-257, Leu-271, Phe-306, and Met-317. Each of the two enantiotopic ethoxy groups of the inhibitor occupy quite distinct chemical environments. The *pro-R* ethoxy group is pointed toward Ile-106, Trp-131,

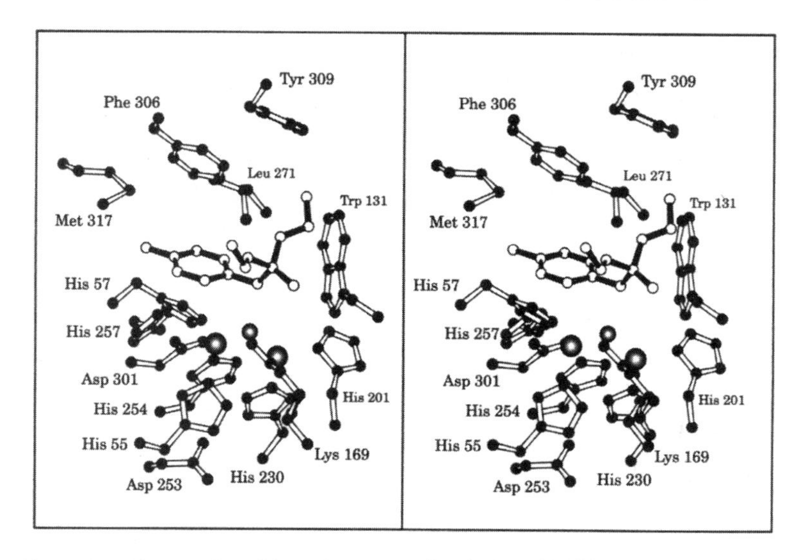

Figure 13. Close-up view of the region surrounding the bound inhibitor diethyl 4-methylbenzyl phosphonate. The protein model is depicted in *open bonds* while the inhibitor is displayed in *filled bonds*.

Leu-303, Phe-306, and Ser-308 while the *pro-S* moiety is directed away from the binuclear metal center toward the solvent and surrounded by Trp-131, Phe-132, Leu-271, Phe-306, and Tyr-309.

J. COMPARISON WITH OTHER ENZYMES

The employment of a binuclear metal center for catalytic activity is, by no means, limited to PTE. Binuclear metal centers are quite common and have been observed, for example, in bovine lens leucine aminopeptidase (Sträter and Lipscomb, 1995), kidney bean purple acid phosphatase (Sträter et al., 1995), *E. coli* alkaline phosphatase (Kim and Wyckoff, 1991), human inositol monophosphatase (Pollack et al., 1994), mammalian protein phosphatase-1 (Goldberg et al., 1995), *Bacillus cereus* phospholipase C (Hough et al., 1989), *E. coli* DNA polymerase I (Beese and Steitz, 1991), *Penicillium citrinum* P1 nuclease (Volbeda et al., 1991), among others. An elegant review of these enzymes can be found in Wilcox (1996). In these abovementioned enzymes, the two metals are typically bridged by either solvent molecules, aspartate and glutamate residues, or both.

Thus far, the only other example of an enzyme that employs a carbamylated lysine as a bridging ligand is urease. This enzyme catalyzes the hydrolysis of urea and employs a binuclear nickel center for activity. Recent X-ray structural analyses of urease from *Klebsiella aerogenes* (Jabri et al., 1995) confirmed the participation of a carbamylated lysine residue in the formation of the binuclear metal center (Park and Hausinger, 1995). As can be seen in the superposition presented in Figure 14, the binuclear metal centers in PTE and urease are remarkably similar. As observed for PTE, the polypeptide chain of one domain of urease folds into a TIM-barrel. Likewise, in urease, the nickels are separated by a comparable distance of 3.5 Å. The more buried nickel ion in urease is surrounded by His-134, His-136, Asp-360, the bridging carbamylated Lys-217, and a hydroxide ion in a distorted trigonal bipyramidal arrangement. The second nickel is ligated to the protein via His-246, His-272, and the bridging lysine residue. Other than the similarities in their overall tertiary structures and the immediate environments surrounding the metal ions, however, the chemical nature of the active sites for PTE and urease are quite different.

Following the three-dimensional analyses of urease and PTE, Scanlan and Reid (1995) demonstrated that a portion of the nucleic acid sequence of the *E. coli* chromosome encodes a protein with 28% amino acid sequence identity to PTE. The three-dimensional structure of this protein, referred to as

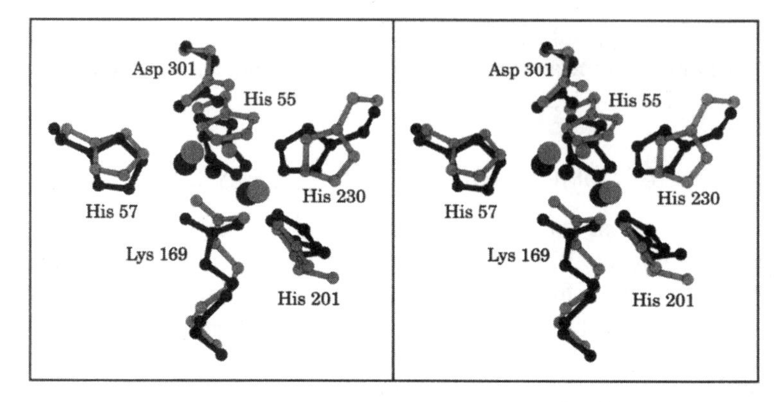

Figure 14. Superposition of the metal binding regions for the Zn^{2+}/Zn^{2+}-containing phosphotriesterase (PTE) and for urease. The PTE model is shown in *black* and the urease model is depicted in *gray*.

PHP (phosphotriesterase homology protein), was subsequently determined in the Fletterick laboratory to 1.7 Å resolution (Buchbinder et al., 1998) and, as expected, the fold is very similar to PTE. The two zinc ions observed binding to PHP are separated by 3.4 Å but in this case the bridging ligand is a glutamate (Glu-125) rather than a carbamylated lysine. As observed in urease and PTE, the deeper buried metal is again surrounded in a trigonal bipyramidal arrangement, specifically by His-12, His-14, Glu-125, Asp-243, and a molecule of unknown identity. The other zinc ion is ligated to PHP in a tetrahedral coordination sphere formed by His-158, His-186, Glu-125, and the unknown solvent molecule. The physiological function of PHP is presently unknown.

On the basis of both amino acid sequence analyses and the structural homologies between urease, PTE, and adenosine deaminase (that only binds one metal) Holm and Sander (1997) have recently identified an even larger set of enzymes that are predicted to have similar active site architectures. Indeed, a characteristic sequence signature of this superfamily includes the aspartate and the four histidine residues involved in metal binding within the active site. Many of the enzymes belonging to this superfamily are involved in nucleotide metabolism, including dihydroorotase, allantoinase, hydantoinases, and the AMP-, adenine, and cytosine deaminases. Also included in this superfamily are proteins involved in animal neuronal development and other enzymes such as imidazolonepropionase, aryldialkylphosphatase, chlorohydrolase, and formylmethanofuran dehydrogenase. Note that some of these enzymes bind only one metal as in the case of adenosine deaminase. Thus far, no invariant lysine residues have been detected for other members of this superfamily, suggesting that urease and PTE are unique in their employment of a carbamylated lysine residue as a bridging ligand.

K. MODIFICATIONS TO CARBAMYLATED LYSINE

The carbamylated lysine residue that serves to bridge the binuclear metal center is unusual. It is not clear why this function could not have been performed by a carboxylate side chain of either glutamate or aspartate. Perhaps this post-translational modification actively participates in the catalytic function of the active site or serves as a regulatory control mechanism for the assembly of the binuclear metal center. Hong et al. (1995) demonstrated that elevated levels of bicarbonate enhanced the rate of formation of the binuclear metal center when the apo-enzyme was mixed with divalent cations. For example, full reconstitution of the metal center in the absence of added bicarbonate is typically observed in 3 hours at room temperature. Preincubation with 100 mM bicarbonate prior to reconstitution with Cd^{2+} enhanced the rate of reactivation by a factor of about six. When the bridging lysine residue is mutated to either methionine (K169M) or alanine (K169A) the catalytic activity is drastically reduced. However, a significant fraction of the wild-type activity can be restored upon the addition of low molecular weight carboxylic acids. The activity of K169A was enhanced 25-fold in the presence of 100 mM propionic acid. Similarly, the activity of K169M was enhanced five-fold in the presence of 100 mM acetic acid (Kuo et al., 1997). The rescue of catalytic activity by the low molecular weight carboxylic acids demonstrate that the carbamate functional group is not absolutely required for the proper functioning of the binuclear metal center. It appears that these carboxylates can substitute for the bridging carbamate as shown in Scheme 3.

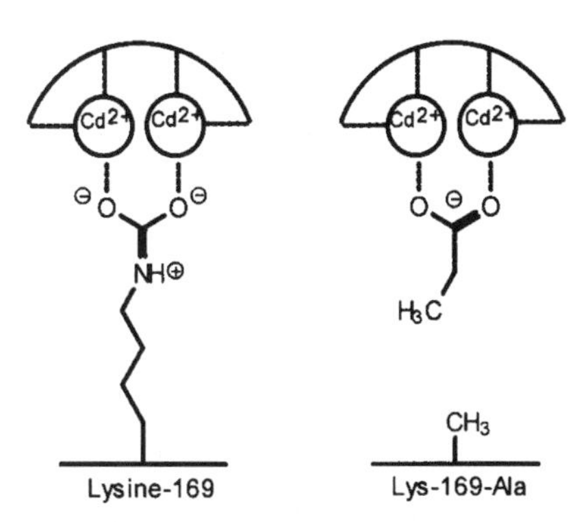

Scheme 3

Although the binuclear metal center in PTE is very similar to the one found in urease there are some notable differences. For example, the binuclear Ni^{2+} center in urease cannot be fully reassembled in vitro (Park and Hausinger, 1995). In fact, there appears to be a collection of accessory proteins that participate in the in vivo assembly of the binuclear metal center in urease.

III. Mechanism of Action

A. STEREOCHEMISTRY AT PHOSPHORUS CENTER

The general mechanism for the enzymatic hydrolysis of paraoxon can be viewed as proceeding by one of three possible routes (see Scheme 4). In mechanism **A** there is a nucleophilic attack at C-1 of the aromatic ring. This mechanism may appear to be highly unlikely but when these experiments were initiated the broad substrate specificity was not known for PTE. Moreover, removal of the electron withdrawing $-NO_2$ substituent diminished the

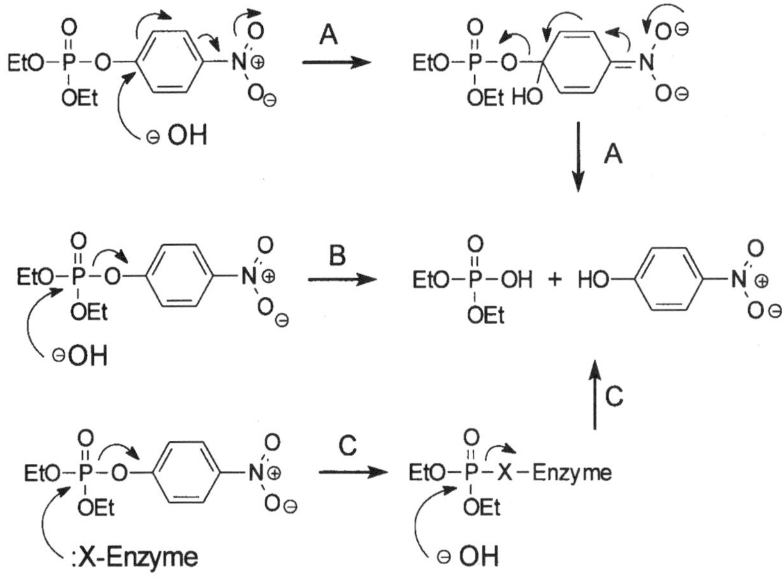

Scheme 4

overall reactivity by over five orders of magnitude. The intermediate or transition state in this particular mechanism could be stabilized by the resonance capability of the nitro group at C-4 of paraoxon. Subsequent C–O bond cleavage would yield the two products.

The second and third mechanisms can be described as variants of an S_N2-like process that differ by the presence of a covalent enzyme-product intermediate. The second mechanism **(B)** involves a single displacement reaction at the phosphorus center. In this mechanism hydroxide, or an activated water molecule, attacks the phosphorus center yielding diethylphosphate and 4-nitrophenol in a single step. In the third mechanism **(C)** there are two in-line displacement reactions at the phosphorus center. The first would occur by a nucleophilic attack at the phosphorus atom by a side chain of a protein residue. This step would be followed by expulsion of 4–nitrophenol to produce the covalent enzyme product complex. Direct nucleophilic attack by water at the phosphorus would then yield diethylphosphate and regenerate free enzyme.

Two rather simple experiments were conducted that were able to differentiate among these three reaction mechanisms. When the reaction was conducted in oxygen-18 water, the heavy isotope was found exclusively in the diethylphosphate, proving that the bond cleavage occurred between the oxygen and phosphorus (Lewis et al., 1988). This result eliminated mechanism **A** from further consideration. In order to distinguish between mechanisms **B** and **C** the stereochemical course of the reaction at phosphorus was determined. If mechanism **C** was operable, then the reaction would proceed with net *retention* of configuration since each of the two steps in this mechanism would proceed with inversion of configuration. Since mechanism **B** has but a single step, the stereochemical course of the reaction would proceed with *inversion* of configuration.

Since neither the substrate nor product of paraoxon hydrolysis is chiral, this substrate cannot be utilized for this endeavor. Fortunately the substrate specificity was found to be promiscuous enough such that the S_P-isomer of the insecticide ethyl-p-nitrophenyl phenylphosphonothiate (EPN) was a good substrate for PTE. When the S_P-isomer of EPN is hydrolyzed by the enzyme, the product, ethyl phenylphosphonothioic acid, was found to have the S_P-configuration. Comparison of the relative stereochemistry of the substrate and product clearly indicated that the overall reaction proceeded with net *inversion* of configuration. Therefore PTE operates via mechanism **B** with a direct attack of an activated water molecule at the phosphorus center (Lewis et al., 1988). See Scheme 5.

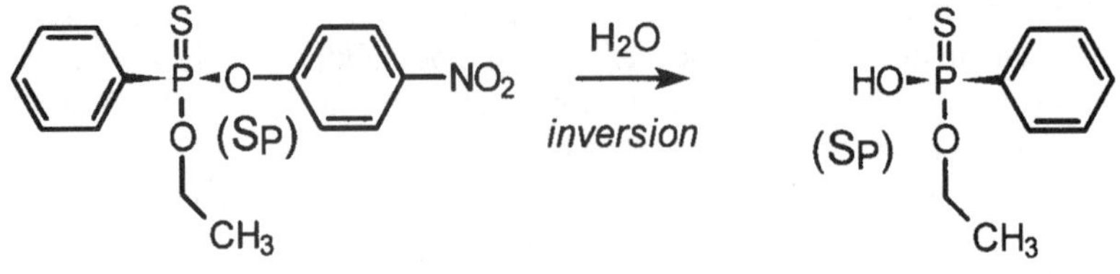

$$\text{PhP(=S)(OCH}_2\text{CH}_3\text{)-O-C}_6\text{H}_4\text{-NO}_2 \text{ (Sp)} \xrightarrow[\textit{inversion}]{H_2O} \text{HO-P(=S)(OCH}_2\text{CH}_3\text{)Ph (Sp)}$$

Scheme 5

B. DETERMINATION OF RATE-LIMITING STEPS

The minimal reaction mechanism that can be written for PTE is presented in Scheme 6. In this scheme there is the simple binding of the triester to form the Michaelis complex (EA), the phosphorus-oxygen bond is cleaved to generate the enzyme-product complex (EPQ), and finally the products dissociate to regenerate the free enzyme for another round of catalysis. The reaction mechanism thus consists of a reversible *association* step, followed by an irreversible *chemical* event, and then the final *dissociation* of the reaction products. The relative magnitude for these three consecutive events was ascertained by a Brønsted analysis of the reaction. The objective of these experiments was to systematically alter the inherent chemical strength of the P–O bond to be cleaved. In the reaction catalyzed by PTE this objective was achieved through the utilization of leaving-group phenols having a range of pK_a values.

Caldwell et al. (1991a) reported the preparation of organophosphate triesters having the general structure shown in Scheme 7. The range of pK_a values spanned from 4.0–10.0. A plot of pK_a of the phenol versus k_{cat} is presented in Figure 15 (a plot of k_{cat}/K_m versus pK_a is similar). The change in the kinetic parameters does not appear to be the result of steric complications and thus the results can be interpreted based only on alterations in the step that involves the cleavage of the phosphorus-oxygen bond. Given this assumption, it is clear that there is a major change in the rate-limiting step as the pK_a of the leaving group phenol increases from 4 to 10.

$$E \underset{k_2}{\overset{k_1A}{\rightleftharpoons}} EA \xrightarrow{k_3} EPQ \xrightarrow{k_5} E + \text{Products}$$

Scheme 6

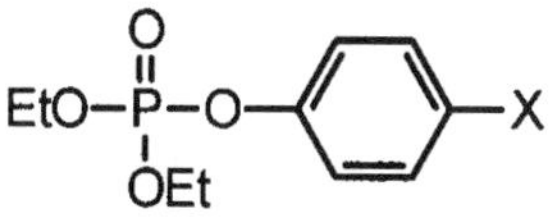

Scheme 7

For the kinetic model that appears in Scheme 6, the relationships for k_{cat} and k_{cat}/K_m are shown in equations 1 and 2.

$$k_{cat} = k_3 k_5 \,/\, (k_3 + k_5) \tag{1}$$

$$k_{cat} \,/\, K_m = k_1 k_3 \,/\, (k_2 + k_3) \tag{2}$$

The value of k_3 will vary according to the magnitude of the pK_a of the leaving group phenol as shown in Equation 3 (Jencks and Carriuolo, 1961)

$$\log k_3 = (\beta \mathrm{p}K_a) + \mathrm{C} \tag{3}$$

These equations predict that as the chemical step, k_3, becomes very fast, the maximum value for k_{cat} will be limited by the step for product release, k_5. Conversely, the relative rates of slow substrates will be inversely proportional to the pK_a of the leaving group phenol. Similarly, the limiting values for k_{cat}/K_m will approach k_1 as the pK_a of the phenol becomes small. A fit of the data that appears in Figure 15 to a combination of equations 1 through 3 provides values of k_1, k_2, and k_5 of 4.1×10^7 M^{-1} s^{-1}, 2600 s^{-1}, and 2200 s^{-1}, respectively (Caldwell et al., 1991a). The value of β is –1.8. There is clearly a change in rate-limiting step from product dissociation to bond cleavage as the pK_a of the phenol increases.

The kinetic analysis of this mechanism yields a rather large β value of –1.8 compared to the values of –1.0 and –0.44 that have been observed for the chemical hydrolysis using water and hydroxide, respectively, as nucleophiles (Khan and Kirby, 1970). Therefore, for the enzymatic reaction, there is a higher sensitivity of the leaving group to the overall rate of hydrolysis. This result is consistent with a significant amount of charge transferred to the phenol in the transition state structure and thus implies that the P–O bond is very nearly completely broken. The observed difference between the β value for chemical and enzymatic hydrolysis may be attributed the hydrophobic environment at the active site.

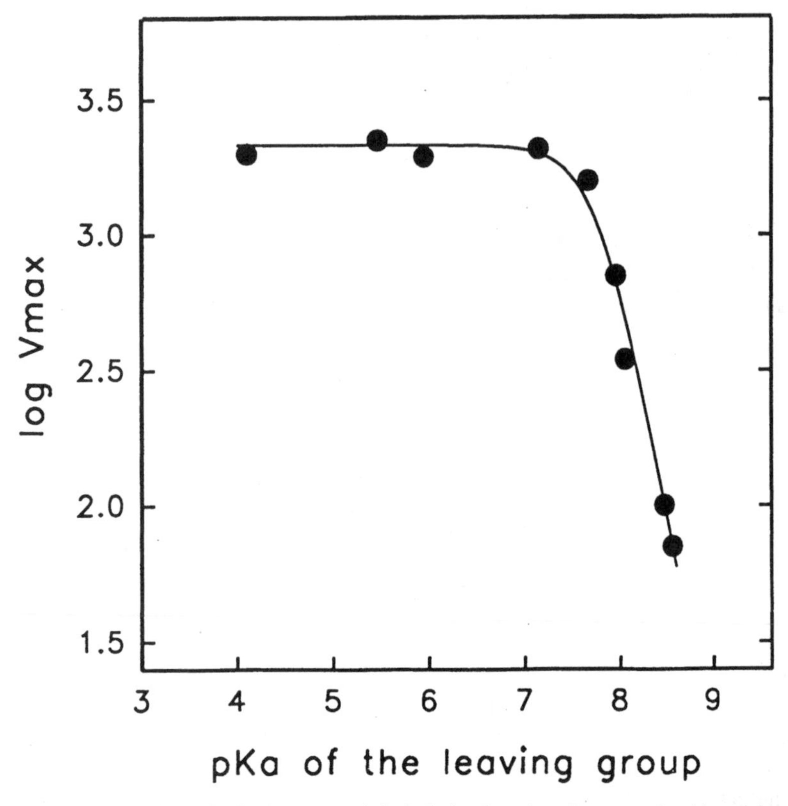

Figure 15. Brønsted plot for the enzymatic hydrolysis of a series of compounds with variable leaving groups.

C. EFFECTS OF SOLVENT VISCOSITY

Alterations in solvent viscosity have also been utilized to determine the relative magnitude of the rate constants presented in Scheme 6. Kirsch demonstrated that alterations in solvent viscosity alter those rate constants for processes that involve either the association or dissociation of enzyme-ligand complexes (Brouwer and Kirsch, 1982). The steps that involve the actual bond making or breaking within the active site are not affected. Thus, in Scheme 6, the rate constants k_1, k_2, and k_5 would be susceptible to changes in solvent viscosity while k_3 would not. The solvent viscosity experiments thus offered a nice complement to the previously described Brønsted analysis where these constants were held constant while k_3 was manipulated by changes in the pK_a

of the leaving group phenol. The effect on the kinetic parameter k_{cat}/K_m with changes in solvent viscosity conforms to Equation 4 (Caldwell et al., 1991a).

$$(k_{cat} / K_m)_o / (k_{cat} / K_m) = (k_3\eta_{rel} + k_2) / (k_2 + k_3) \tag{4}$$

Equation 4 predicts a linear relationship between the relative second-order rate constant, $(k_{cat}/K_m)_o$ / (k_{cat}/K_m) and the relative viscosity. The slope of this line provides a measure of the sensitivity of k_{cat}/K_m to changes in viscosity. From Equation 4, the slope is defined by the relationship shown in Equation 5.

$$(k_{cat} / K_m)_\eta = k_3 / (k_2 + k_3) \tag{5}$$

Therefore, the value of $(k_{cat} / K_m)\eta$ can vary from 0 to 1 depending on the relative magnitude of the rate constants k_2 and k_3. As k_3 becomes infinitely large, $(k_{cat} / K_m)_\eta$ approaches 1, and as k_3 becomes very small, $(k_{cat} / K_m)_\eta$ approaches 0. Shown in Figure 16 is the relationship between the pK_a of the

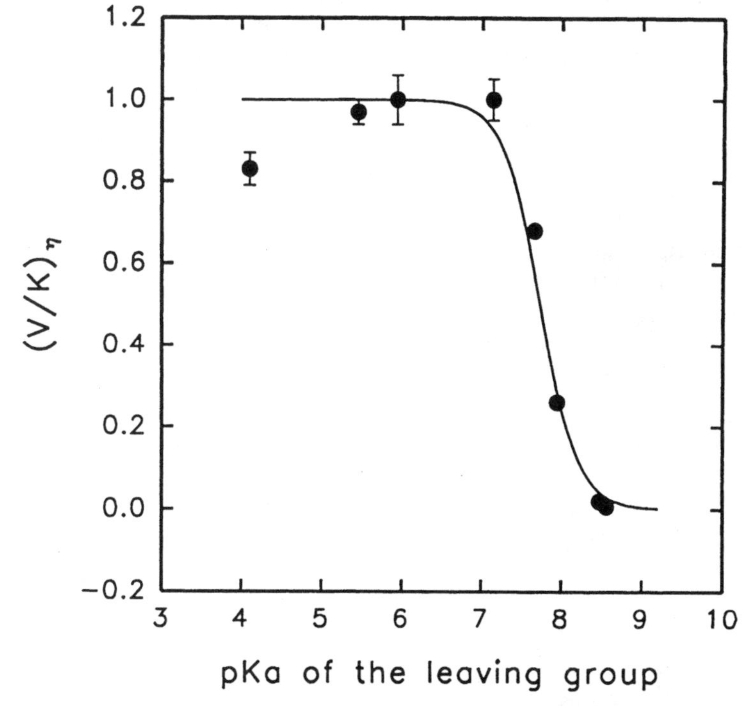

Figure 16. The change in the effect of solvent viscosity on the *V*/*K* values as a function of the pK_a of the leaving group. The Zn/Zn-PTE was utilized for this experiment.

leaving group and $(k_{cat} / K_m)_\eta$. Indeed, those substrates with leaving groups whose pK_a values are <7 are limited by physical diffusion and $(k_{cat} / K_m)_\eta$ approaches a value of unity. In contrast, the substrates with leaving group pK_a >7 are limited by the chemical step k_3 and $(k_{cat} / K_m)_\eta$ approaches a value of 0. The solid line shown in Figure 16 represents a fit to equations 3 and 4 where $k_2 = 2150\ s^{-1}$, $\beta = -1.84$, and C = 17.7. The excellent fit of the viscosity data to the predicted values, based on the rate constants obtained from the Brønsted plot analysis, provides strong experimental support for this kinetic analysis.

D. pH-RATE PROFILES

The utilization of pH-rate profiles can, in certain circumstances, provide information critical to a fuller understanding of the reaction mechanism. Since the substrates hydrolyzed by PTE do not ionize between pH 4 and 10, all of the ionizations observed in such analyses must originate with the protein. Shown in Figure 2 is the pH-rate profile for the hydrolysis of paraoxon by the Zn/Zn-substituted enzyme. Since the catalytic activity does not diminish at high pH, there does not appear to be an enzyme functional group within the active site that offers general acid catalytic assistance to the leaving group. This assertion is supported by the lack of proton-donating residues in the active site from the X-ray crystallographic structure. Also, the extreme value of the Brønsted constant, β, argues against any suppression of the incipient negative charge via proton transfer.

The origin of the ionization that appears at pH ~6 is still unclear. Originally, this ionization was proposed to originate with the protonation of a histidine residue (Dumas and Raushel, 1990b). The histidine residue was postulated to serve as a general base in the abstraction of a proton from the hydrolytic water molecule. The effects of temperature and organic solvents were consistent with this notion. However, in view of the X-ray crystal structure the loss of activity at low pH may arise from the protonation of the bridging hydroxide between the two divalent cations. The variation of the kinetic pK_a values upon alteration of the divalent cation itself is supportive of this proposal. When the active site is substituted with Zn^{2+}, Co^{2+}, Ni^{2+}, Mn^{2+}, and Cd^{2+}, the kinetic pK_a values are 5.8, 6.5, 7.4, 7.0, and 8.1, respectively (Omburo et al., 1992). However, the loss of activity may occur indirectly through protonation of one of the four histidine residues that coordinate the two divalent cations within the active site. The nature the divalent cation may indeed perturb the ionization properties of the direct metal-ligand residues.

E. THE ROLE OF BINUCLEAR METAL CENTER

There are potentially three roles the binuclear metal center may play in the enzymatic hydrolysis of organophosphates. The metals may decrease the pK_a of the bound water molecule and increase the nucleophilic character of the attacking hydroxide. The metal ions can increase the polarization of the P=O bond, and thereby accelerate the approach of an attacking hydroxyl ion by increasing the electrophilic character of the phosphorus center. The metal ions may also neutralize the development of negative charge on the leaving group.

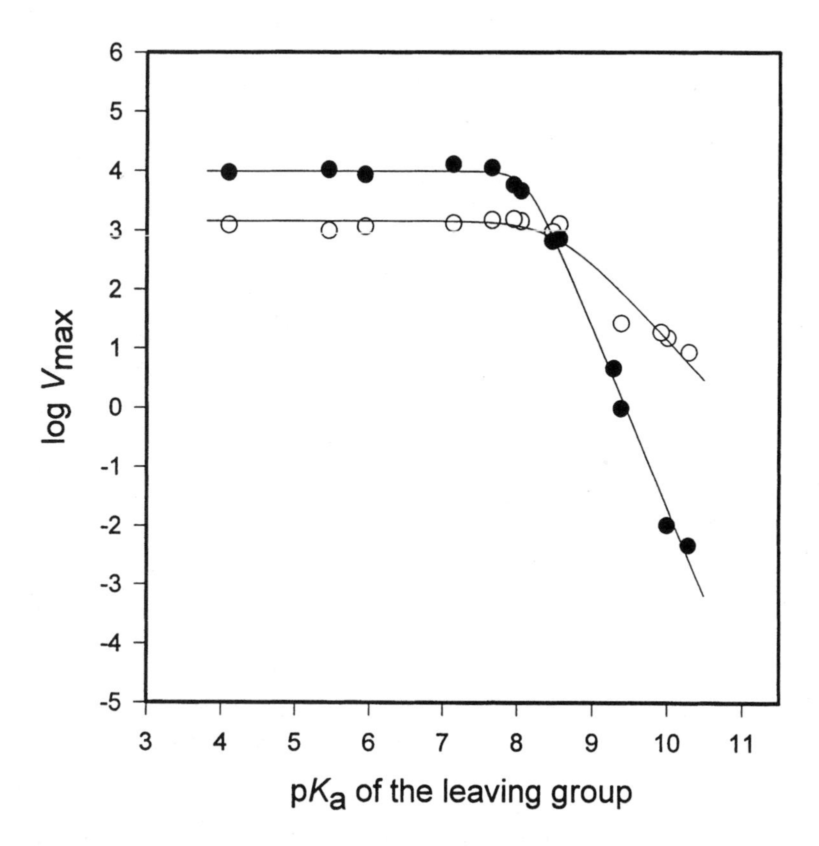

Figure 17. Brønsted plots for the dependence of log V_{max} on the pK_a of the leaving group. *Filled circles* are for a series of organophosphate triesters with leaving group phenols of variable pK_a values. *Open circles* are for a series of organothiophosphate triesters with leaving group phenols of variable pK_a values. The Cd/Cd-PTE was utilized for this experiment.

The direct interaction with the P=O of the substrate was probed by conducting Brønsted-type analyses with different metal-substituted variants of PTE and using a series of phosphate and thiophosphate organophosphate triesters. Shown in Figure 17 is the Brønsted plot for the hydrolysis of phosphate and thiophosphate triesters by Zn/Zn–PTE. Both plots are clearly nonlinear, representative of the change in rate-limiting step as discussed earlier (Caldwell et al., 1991a). It should be stated here that the hydrolysis of organophosphates by OH^- is about 20 times faster than the hydrolysis of the corresponding organothiophosphates (Hong and Raushel, 1996). When the rate-limiting step is dominated by something other than P–O bond cleavage (pK_a values of the leaving group phenols less than ~7) the organophosphate compounds are enzymatically hydrolyzed about one order of magnitude faster than the thiophosphates. For k_{cat} this likely means that the release of diethylthiophosphate is slower than the release of diethylphosphate from the enzyme-product complex. However, when substrates are utilized that are clearly limited by the rate of P–O bond cleavage, the thiophosphate analogs are more reactive than the phosphate series. Thus, there is a crossover in the relative magnitude of the kinetic constants. This observation supports the proposal that one or both of the metal ions in the binuclear metal center directly coordinates the phosphoryl group by polarization of the P=O (or P=S) bond. It has previously been demonstrated that the Cu^{2+}-catalyzed hydrolysis of thiophosphotriesters is much more effective than with the corresponding oxygen analogs (Ketelaar et al., 1956). The faster rate of hydrolysis of the thiophosphate analogs can be explained by the direct coordination of one or both metal ions with the sulfur or oxygen atom of the substrates.

F. HEAVY ATOM ISOTOPE EFFECTS

Oxygen-18 isotope effects have been utilized to probe the transition state structure for the enzymatic and base-catalyzed hydrolysis of organophosphate triesters (Caldwell et al., 1991b, 1991c). The secondary and primary oxygen-18 isotope effects have been measured for PTE using the labeled substrate presented in Scheme 8. The nonbridge, or phosphoryl oxygen, serves as a reporter for the change in bond order in the transition state between this oxygen and the phosphorus core. This provides an indication of the degree of charge delocalization at this site upon attack by the incoming nucleophile. In a fully associative mechanism the attack of the nucleophile leads to a pentavalent phosphorane intermediate followed by the collapse of

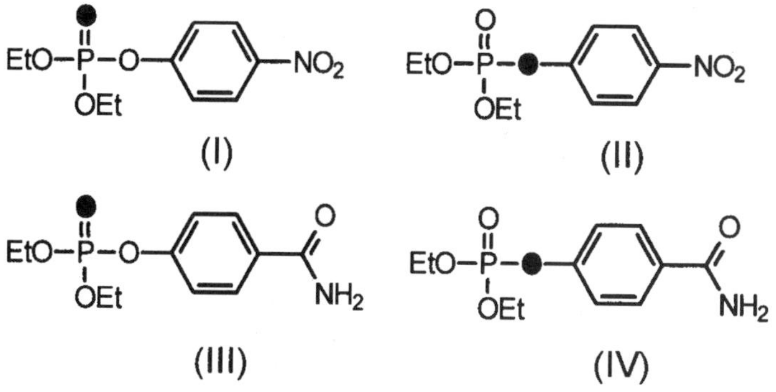

(I) (II)

(III) (IV)

Scheme 8

this structure to products. At the other extreme, the formation of an intermediate is not absolutely required with a direct-displacement mechanism and a single S_N2-like transition state.

In a hybrid mechanism there is simultaneous, but not necessarily synchronous, decrease in bonding to both the phosphoryl oxygen and the leaving group upon the addition of the nucleophile. The relative magnitude of the bond order changes in the transition state at these two atoms is variable and dependent of the nucleophilic character of the attacking ligand and the ability of the leaving group to depart. The magnitude of the primary oxygen-18 effect will depend on the extent of bond cleavage to the leaving group while the secondary oxygen-18 isotope effect will be determined by the change in the bond order to the phosphoryl group.

Two sets of oxygen-18 labeled substrates have been tested with PTE. Paraoxon (**I** and **II**) is a relatively fast substrate for PTE and the kinetic constants k_{cat} and k_{cat}/K_m are limited primarily by diffusional steps rather than bond breaking steps while the other compound (**III** and **IV**) is a relatively poor substrate where the kinetic constants are dominated by bond breaking events. The primary and secondary isotope effects are listed for each of the compounds in Table 2.

The observed primary and secondary isotope effects exhibited by paraoxon are both quite small. This reflects to a significant extent that the chemical step is not rate-limiting. This is not, however, true for the slower substrate where the bond-breaking event is expected to be nearly fully rate-limiting. The extent of phosphorus-phenolic oxygen bond cleavage in the transition state can be estimated from the relative size of the intrinsic

TABLE 2
Primary and Secondary Oxygen-18 Isotope Effects

Compound	$^{16}k/^{18}k$ Primary	$^{16}k/^{18}k$ Secondary
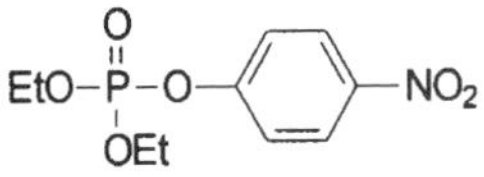 EtO–P(=O)(OEt)–O–C$_6$H$_4$–NO$_2$	1.0020	1.0021
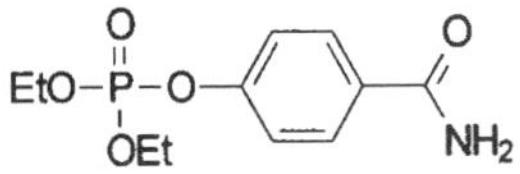 EtO–P(=O)(OEt)–O–C$_6$H$_4$–C(=O)NH$_2$	1.036	1.0181

^{18}O-isotope effect. However, the maximum size of the ^{18}O-isotope effect for phosphotriester hydrolysis is unknown. However, the ^{18}O equilibrium isotope effect for the deprotonation of water and phenol has been estimated to be 1.039 (Rosenberg, 1977). The observed enzymatic effect is very close to this value and thus it appears that the P–O bond is nearly fully broken in the transition state. The secondary ^{18}O-isotope effect of 1.018 suggests that the bond order has changed from 2 to approximately 1.5. Taken together the isotope effects suggest that the reaction proceeds through an associative-type mechanism without any direct evidence for the formation of a phosphorane intermediate. The transition state structure is very product-like.

G. MECHANISM-BASED INHIBITORS

The bacterial PTE was among the first enzymes that catalyze a hydrolytic reaction at a phosphorus center to be inactivated by suicide substrates. The Stang laboratory synthesized a series of alkynyl phosphate esters (Stang et al., 1986). These compounds were constructed with the anticipation that cleavage of the phosphorus–oxygen bond would generate a highly reactive ketene intermediate as illustrated in Scheme 9. The electrophilic carbon center of the ketene intermediate could be subjected to attack by an adjacent nucleophile. Indeed, incubation of PTE with diethyl 1-hexynyl phosphate completely inactivated the enzyme in less than 2 minutes. The catalytic activity could not be recovered upon dialysis (Blankenship et al., 1991).

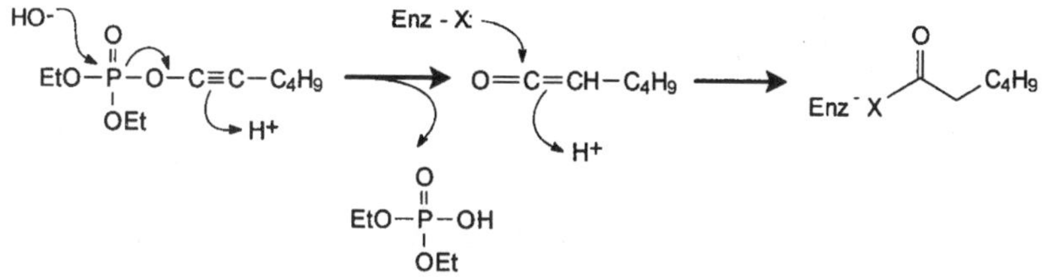

Scheme 9

The efficiency of enzyme inactivation was determined by incubation of a fixed enzyme concentration with variable concentrations of inhibitor. When the fraction of the original enzyme activity was plotted versus the initial ratio of inhibitor to enzyme the intercept on the horizontal axis is 1200. Therefore, approximately 1200 inhibitor molecules are enzymatically hydrolyzed for every enzyme molecule inactivated. The data have been analyzed according to the scheme presented in Scheme 10. In this scheme, **I** is the alkynyl phosphate ester, **Y** is the ketene intermediate, **E-X** in the inactivated enzyme, and **P** is a carboxylic acid after the ketene has reacted with water. The partitioning experiment described indicates that the ratio of k_5/k_7 is approximately 1200.

The inactivation reaction rate was determined using a stopped-flow device. At saturating inhibitor the rate constant for enzyme inactivation was 0.33 s^{-1}. The rate expression for the inactivation of PTE is shown in Equation 6. Since $k_5 >> k_7$, the minimum value of the rate constant (k_7) for the reaction of the putative ketene intermediates is 0.33 s^{-1}. Thus, the minimum value of the rate constants for P–O bond cleavage and product release (k_3 and k_5) is

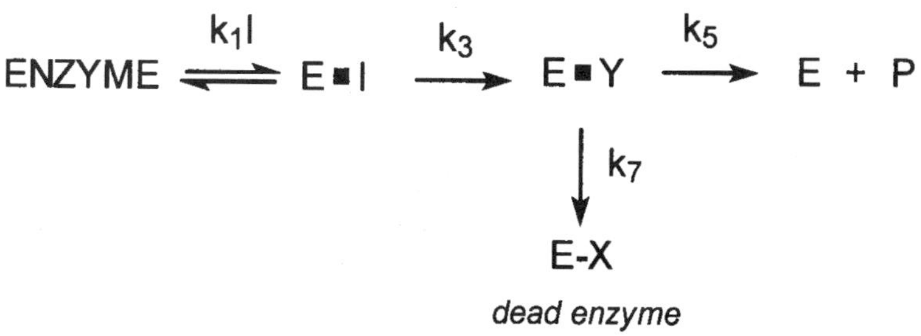

Scheme 10

410 s^{-1}. These values compare quite favorably with the hydrolysis of paraoxon under similar conditions (2100 s^{-1}).

$$k_{in} = (k_3 k_7) / (k_3 + k_5 + k_7) \quad (6)$$

The identity of the residue within the active site that reacts with the putative ketene intermediate was initially probed with a radiolabeled alkynyl phosphate ester (Banzon et al., 1995a). After inactivation of enzyme activity, the stoichiometry of the labeled adduct to protein was 1:1 and the protein could be reactivated within 1.5 h at pH 10.0 (the labeling is conducted at pH 7.0). After the protein was inactivated there was an increase in absorbance at 245 nm ($\varepsilon = 4400\ M^{-1}\,cm^{-1}$). The change in UV absorbance and the reactivation kinetic studies ruled out all other amino acids except for the side chain of histidine.

The identity of the histidine residue that is critical for the maintenance of catalytic activity was probed using the seven histidine to asparagine mutants described earlier (Kuo and Raushel, 1994). Of the seven histidine to asparagine mutants only the H254N mutant was no longer susceptible to inactivation by either alkynyl phosphate esters or the histidine specific reaction, DEPC (Banzon et al., 1995b). The crystal structure of PTE shows that this residue is indeed within the active site but it is not a direct ligand to either of the two metal ions.

Additional ligands have also been designed as potential mechanism-based inactivators and inhibitors of PTE (Hong and Raushel, 1997; Hong et al., 1997). 1–Bromo vinyl- and 4-(bromomethyl)-2-nitro phenyl diethylphosphate (Stowell and Widlanski, 1994; Stowell et al., 1995) were found to be effective inactivators of the bacterial phosphotriesterase. The postulated reaction mechanisms are presented in Scheme 11.

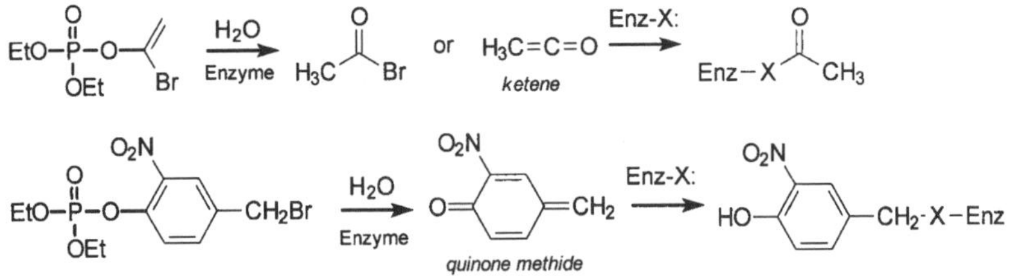

EtO–P(=O)(OEt)–O–C(=CH2)Br —H_2O, Enzyme→ H_3C–C(=O)–Br or H_3C=C=O (ketene) —Enz-X:→ Enz–X–C(=O)–CH_3

EtO–P(=O)(OEt)–O–C6H3(O_2N)–CH_2Br —H_2O, Enzyme→ O=C6H3(O_2N)=CH_2 (quinone methide) —Enz-X:→ HO–C6H3(O_2N)–CH_2-X–Enz

Scheme 11

H. MECHANISM OF ACTION

The chemical probes of the reaction catalyzed by PTE enable a working model of the reaction mechanism to be formulated. The current model is limited by experimental constraints and certain details of the mechanism have not been settled. The working model for this reaction mechanism is illustrated in Scheme 12. The focal point of the model is the binuclear metal center. In order to facilitate this reaction the enzyme must accomplish three objectives. First, the hydrolytic water molecule must be activated for nucleophilic attack toward the phosphorus center of the substrate. The X-ray structure of PTE clearly shows that a solvent molecule bridges the two metal ions within the binuclear metal center. The coordination of the water molecule to the metal center would be expected to reduce the pK_a of the water and thus increase the local hydroxide concentration. There does not appear to be any basic residues within the active site that could serve to assist in the proton removal from any other solvent molecule. Second, the phosphorus center must be made more electrophilic. This can be achieved by coordination of the phosphoryl oxygen to one or both of the metal ions within the cluster. The ligation of the phosphoryl oxygen would polarize this bond and thus reduce the electron density at the phosphorus core. The variation in the magnitude of the β-values from the Brønsted experiments and the crossover in reactivity for the phosphate and thiophosphate substrates demonstrate that metal-substrate interactions are critical for efficient catalytic activity of

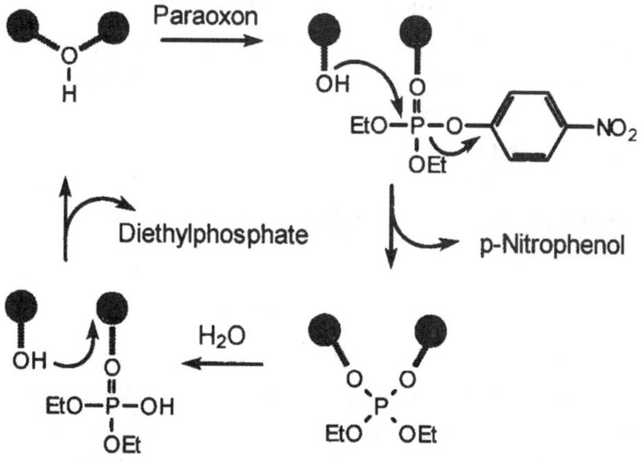

Scheme 12

PTE. Third, the enzyme could activate the leaving group. The activation of the leaving group can potentially be accomplished by general acid catalysis using an appropriately placed residue or, alternatively, by Lewis acid catalysis. However, neither of these effects has thus far been demonstrated for PTE. The best substrates (such as those with *p*-nitrophenolate leaving groups) do not require any significant assistance and there does not appear to be any side chains positioned in the three-dimensional structure of PTE that could serve this function.

The proposed reaction mechanism is initiated by the binding of the substrate to the binuclear metal center. One of the interactions from the metal to the bridging solvent molecule is broken and replaced by ligation to the phosphoryl oxygen of the substrate. Nucleophilic attack is initiated by the bound hydroxide toward the phosphorus center. For substrates such as paraoxon the transition-state is developed quite late. Part of the incoming negative charge is dispersed to the phosphoryl oxygen (via interaction with the metal ion) while the rest is dispersed through the aromatic ring system of the departing phenolate. After the phenolate has departed the product complex would now bridge the two metal ions. However, structural evidence for this type of complex has not been obtained. In the final step, the product would be displaced by an incoming solvent molecule prior to the next round of catalysis.

There are a number of issues with regard to this mechanism that remain unresolved. The pK_a value of the bridging solvent molecule is unknown. It has been assumed in the past that it is the pK_a of this group that is reflected in the pK_a values determined by variation of the kinetic parameters with pH. Perhaps this issue can be settled using NMR or EPR spectroscopic methods. It is also unclear if the two metal ions within the binuclear metal center have distinct functions or whether they operate as a true tandem pair. The catalytic properties of the unique Cd/Zn-hybrid would argue for separate functions for each of the two metals. Resolution of this issue may be obtained when the specific binding sites are determine for the Zn^{2+} and Cd^{2+} in the hybrid cluster of PTE by X-ray crystallography.

IV. Substrate Specificity

The substrate specificity for PTE is actually quite broad. For the prototypic reaction illustrated in Scheme 13 the substituent X can either be oxygen or sulfur (Donarski et al., 1989). In a limited investigation it was shown that the thiophosphate esters are hydrolyzed faster than the corresponding phosphate esters when the rate-limiting step was actual bond cleavage

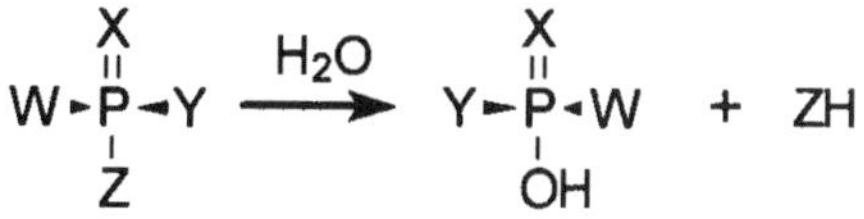

$$W\text{-}\underset{Z}{\overset{X}{P}}\text{-}Y \xrightarrow{H_2O} Y\text{-}\underset{OH}{\overset{X}{P}}\text{-}W \; + \; ZH$$

Scheme 13

(Hong and Raushel, 1996). However, when the rate-limiting step is product release or an associated conformational step, then the turnover of the phosphate derivative was faster than thiophosphate analog.

A variety of substituents have been shown to be viable when attached to the non-leaving group positions (**W** and **Y**). High rates of turnover have been measured for CH_3-, CH_3O-, CH_3CH_2O-, $(CH_3)_2CHO$-, C_6H_5-, C_6H_5O, nC_4H_9O, nC_3H_7O, among others (Donarski et al., 1989; Hong and Raushel, 1999). Fully esterified phosphonates are hydrolyzed as well as organophosphate triesters.

The enzyme exhibits a significant amount of stereoselectivity toward the hydrolysis of chiral organophosphates. The first demonstration of this was reported for the hydrolysis of a racemic mixture of EPN. The enzyme greatly prefers the S_P-isomer over the R_P-isomer (Lewis et al., 1988). More recently, an extensive analysis of the substrate preferences for the chiral forms of a series of organophosphate triesters was initiated. Nearly all possible combinations of the various substituents were analyzed as substrates when **W** and **Y** were changed to CH_3-, CH_3CH_2-, $(CH_3)_2CH$-, or C_6H_5-. In every case examined, the S_P-isomer was hydrolyzed significantly faster then the R_P-isomer. The single exception was with the two smallest substituents, methyl and ethyl. These small alkyl substituents could not be distinguished from one another.

The substituents that can participate as leaving groups are more restricted. A wide range of substituted phenols are substrates but the turnover numbers are very dependent on the pK_a of the phenol (see Brønsted plots). Since the β-value approaches a value of nearly 2, the k_{cat} for substrates with phenol as a leaving group (pK_a ~10) are reduced by a factor of nearly a million relative to those with *p*-nitrophenol as the leaving group. However, the leaving groups are not restricted to phenols. Fluoride and alkylthiols are also tolerated but the turnover numbers are quite slow (Dumas et al., 1989b, 1989c; Chae et al., 1994) . The hydrolysis of extremely toxic substances such as sarin, soman, and VX (Scheme 14) has suggested that this enzyme could be utilized to detect and detoxify a variety of military-type organophosphate nerve agents.

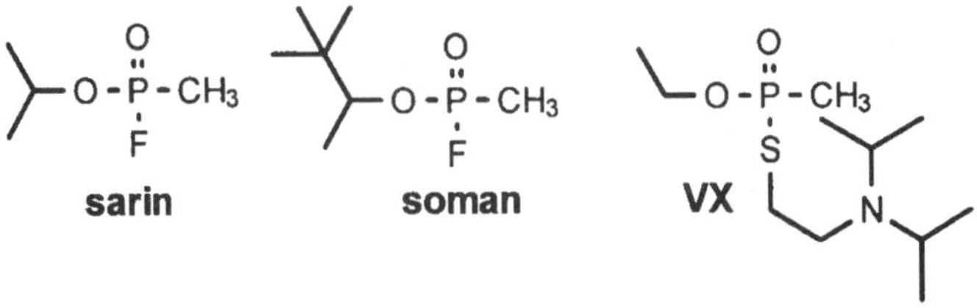

sarin soman VX

Scheme 14

A host of compounds are not hydrolyzed as a significant rate. No reaction was detected with *p*-nitrophenyl acetate and thus the enzyme will not hydrolyze simple carboxylate esters (Donarski et al., 1989). The enzyme will not hydrolyze sulfonate esters or phosphoramides. It was originally thought that organophosphate diesters were not hydrolyzed but it was later shown that diesters could be hydrolyzed but at a rate that was reduced by a factor of 10^5 (Shim et al., 1998).

A. ALTERATIONS TO SUBSTRATE SPECIFICITY

The high catalytic turnover of the wild-type enzyme and the tolerance for modification to the substrate phosphorus core has prompted a number of efforts designed to enhance and expand the substrate specificity. For example, the wild-type enzyme has been shown to hydrolyze phosphorofluoridates such as diisopropyl flurophosphate (DFP) with turnover numbers for the wild-type enzyme that were about 50 s^{-1} when the divalent cation was Zn^{2+} (Dumas et al., 1989b, 1989c). The rate could be enhanced by about an order of magnitude when the native Zn^{2+} was replaced with Co^{2+} (Watkins et al., 1997b). Finally, turnover numbers in excess of 3000 s^{-1} could be achieved when two residues within the substrate binding pocket were mutated (Watkins et al., 1997b). This came about when Phe-132 and Phe-136 were mutated to histidine residues. This substitution provided a more polar environment in the vicinity of the leaving group and also provided potential general acid substituents.

The ability to hydrolyze organophosphate diesters has also been subjected to mutagenesis. If one removes one of the ethyl groups from paraoxon the turnover number decreases by a factor of ~100,000. A significant fraction of the decrease in catalytic activity derives from the fact that organophosphate diesters are inherently less reactive than the corresponding triesters. An examination of the substrate-binding site reveals that the amino acids that constitute the active site are largely hydrophobic and thus not very tolerant of the negative charge of potential diester substrates. Incremental enhancements to the rate of hydrolysis of diester substrates could be obtained by the introduction of cationic residues within the active site (Shim et al., 1998). Enhancement of diester hydrolysis could be obtained by the addition of alkyl amines to the reaction mixture. The amines are thought to enhance the rate of diester hydrolysis by filling the cavity that is left within the active site upon removal of one of the alkyl substituents from a triester substrate and neutralization of the negative charge. The most dramatic effect came about when diethylamine (2.0 M) was added. The V/K value changed from 1.6 $M^{-1} s^{-1}$ to 260 $M^{-1} s^{-1}$. The substrate specificity can also be manipulated. For example, the k_{cat} for dementon-S is increased by a factor of 12 when His-254 and His-257 are mutated to arginine and leucine respectively (diSoudi et al., 1999).

Summary

The bacterial PTE is able to catalyze the hydrolysis of a wide range of organophosphate nerve agents. The active site has been shown to consist of a unique binuclear metal center that has evolved to deliver hydroxide to the site of bond cleavage. The reaction rate for the hydrolysis of activated substrates such as paraoxon is limited by product release or an associated protein conformational change.

Acknowledgments

The work in the author's laboratory on the structure and mechanism of PTE has been supported in part by the NIH (GM33894, GM55513), Office of Naval Research, Robert Welch Foundation (A-840), and the Advanced Technology Program from the State of Texas.

References

Antanaitis BC, Brown RD, Chasteen ND, Freeman JH, Koenig SH, Lilienthal HR, Peisach J, Brewer CF (1987): Biochemistry 26: 7932–7937.

Banzon JA, Kuo JM, Fischer DR, Stang PJ, Raushel FM (1995a): Biochemistry 34: 743–749.

Banzon JA, Kuo JM, Miles BW, Fischer DR, Stang PJ, Raushel FM (1995b): Biochemistry 34: 750–754.

Beese LS, Steitz TA (1991); *EMBO J* 10: 25–33.

Benning MM, Kuo JM, Raushel FM, Holden HM (1994): Biochemistry 33: 15001–15007.

Benning MM, Kuo JM, Raushel FM, Holden HM (1995): Biochemistry 34: 7973–7978.

Blankenship JN, Abu-Soud H, Francisco WA, Raushel FM, Fischer DR, Stang PJ (1991) J Am Chem Soc 113: 8561–8562.

Brouwer AC, Kirsch JF (1982): Biochemistry 25: 1302–1307.

Brown KA (1980): Soil Biol Biochem 12: 105–112.

Buchbinder JL, Stephenson RC, Dresser MJ, Pitera JW, Scanlan TS, Fletterick RJ (1998): Biochemistry 37: 5096–5106.

Caldwell SR, Newcomb JR, Schlecht KA, Raushel FM (1991a): Biochemistry 30: 7438–7444.

Caldwell SR, Raushel FM, Weiss PM, Cleland WW (1991b): Biochemistry 30: 7444–7450.

Caldwell SR, Raushel FM, Weiss PM, Cleland WW (1991c): J Am Chem Soc 113: 730–732.

Chae MY, Omburo GA, Lindahl PA, Raushel FM (1993): J Am Chem Soc 115: 12173–12174.

Chae MY, Postula JF, Raushel FM (1994): Bioorg Med Chem Lett 4: 1473–1478.

Chae MY, Omburo GM, Lindahl PA, Raushel FM (1995): Arch Biochem Biophys 316: 765–772.

diSoudi B, Grimsley JK, Lai K, Wild JR (1999): Biochemistry 38: 6534–6540.

Donarski WJ, Dumas DP, Heitmeyer DP, Lewis VE, Raushel FM (1989): Biochemistry, 28: 4650–4655.

Dumas DP, Caldwell SR, Wild JR, Raushel FM (1989a): J Biol Chem 264: 19659–19665.

Dumas DP, Wild JR, Raushel FM (1989b): Biotech Appl Biochem 11: 235–243.

Dumas DP, Durst HD, Landis WG, Raushel FM, Wild JR (1989c): Arch Biochem Biophys 277: 155–159.

Dumas DP, Raushel FM (1990a): Experientia 46: 729–731.

Dumas DP, Raushel FM (1990b): J Biol Chem 265: 21498–21503.

Farber GK, Petsko GA (1990): Trends Biochem Sci 15: 228–234.

Goldberg J, Huang H-B, Kwon Y-G, Greengard P, Nairn AC, Kuriyan J (1995): *Nature* 376: 745–753.

Holm L, Sander C (1997): *Proteins: Structure Function, and Genetics* 28: 72–82.

Hong S-B, Kuo JM, Mullins LS, Raushel FM (1995): J Am Chem Soc 117: 7580–7581.

Hong S-B, Raushel FM (1996): Biochemistry 35: 10904–10912.

Hong S-B, Mullins LS, Shim H, Raushel FM (1997): Biochemistry 36: 9022–9028.

Hong S-B, Raushel FM (1997): J Enzyme Inhibition 12: 191–203.

Hong S-B, Raushel FM (1999): Biochemistry 38: 1159–1165.

Hough E, Hansen LK, Birknes B, Jynge K, Hansen S, Hordvik A, Little C, Dodson E, Derewenda Z (1989): *Nature* 338: 357–360.

Jabri E, Carr MB, Hausinger RP, Karplus PA (1995): Science 268: 998–1004.

Jencks WP, Carriuollo J (1961): J Am Chem Soc 83: 1743–1750.

Kahn SS, Kirby AJ (1970): J Chem Soc B 1970: 1172–1182.

Ketelaar JAA, Gersmann HR, Beck MM (1956): Nature 77: 1956–1957.

Khangulov SV, Barynin VV, Antonyuk-Barynina SV (1990): Biochim Biophys Acta 1020: 25–33.

Kim EE, Wyckoff HW (1991): J Mol Biol 218: 449–464.

Kuo JM, Raushel FM (1994): Biochemistry 33: 4265–4272.

Kuo JM, Chae MY, and Raushel FM (1997): Biochemistry 36: 1982–1988.

Lai K, Dave KI, Wild JR (1994): J Biol Chem 269: 16579–16584.

Lewis VE, Donarski WJ, Wild JR, Raushel FM (1988): Biochemistry 27: 1591–1597.

McDaniel CS (1985): Ph.D. Dissertation Texas A&M University, College Station, Texas.

Mulbry WW, Karns JS (1989): *J Bacteriol* 171: 6740–6746.

Munneke DM (1976): *Appl Environ Microbiol* 32: 7–13.

Omburo GA, Kuo JM, Mullins LS, Raushel FM (1992): J Biol Chem 267: 13278–13283.

Omburo GA, Mullins LS, and Raushel FM (1993): Biochemistry 32: 9148–9155.

Park I-S, Hausinger RP (1995): Science 267: 1156–1158.

Pollack SJ, Atack JR, Knowles MR, McAllister G, Ragan CI, Baker R, Fletcher SR, Iversen LL, Broughton HB (1994): *Proc Natl Acad Sci USA* 91: 5766–5770.

Poyner RR, Reed GH (1992): Biochemistry 31: 7166–7173.

Reczykowski RS, Ash DE (1992): *J Am Chem Soc* 114: 10992–10994.

Rosenberg S (1977): Ph.D. Dissertation, University of California, Berkeley.

Rowland SS, Speedie MK, Pogell BM (1991): *Appl Environ Microbiol* 57: 440–444.

Scanlan TS, Reid RC (1995): *Chem Biol* 2: 71–75.

Serdar CM, Murdock DC, Rohde MF (1989): Biotechnology 7: 1151–1155.

Shim H, Hong S-B, Raushel FM (1998): *J Biol Chem* 273: 17445–17450.

Stang PJ, Boehshar M, Lin J (1986): *J Am Chem Soc* 108: 7832–7836.

Stowell JK, Widlanski TS (1994): *J Am Chem Soc* 116: 789–790.

Stowell JK, Widlanski TS, Kutateladze TG, Raines RT (1995): J Org Chem 60: 6930–6936.

Sträter N, Lipscomb WN (1995): Biochemistry, 34: 9200–9210.

Sträter N, Klabunde T, Tucker P, Witzel H, Krebs B (1995): *Science* 268: 1489–1492.

Summers MF (1988): Coord Chem Rev 86: 43–134.

Vanhooke JL, Benning MM, Raushel FM, Holden HM (1996): Biochemistry 35: 6020–6025.

Volbeda A, Lahm A, Sakiyama F, Suck D (1991): *EMBO J* 10: 1607–1618.

Watkins LM, Kuo JM, Chen-Goodspeed M, Raushel FM (1997a): *Proteins: Structure. Function, Genetics* 29: 553–561.

Watkins LM, Mahoney HJ, McCulloch JK, Raushel FM (1997b): *J Biol Chem* 272: 25596–25601.

Wilcox DE (1996): *Chem Rev* 96: 2435–2458.

PHOSPHORIBULOKINASE: CURRENT PERSPECTIVES ON THE STRUCTURE/FUNCTION BASIS FOR REGULATION AND CATALYSIS

By HENRY M. MIZIORKO, *Department of Biochemistry, Medical College of Wisconsin, Milwaukee, Wisconsin 53226*

CONTENTS

I. Introduction

The initial observations on the significance of ribulose 1,5–bisphosphate (RuBP) as the CO_2 acceptor in the reductive pentose phosphate cycle prompted efforts to account for production of this metabolite. In spinach leaf extracts, Weissbach et al. (1954) reported the production of

Advances in Enzymology and Related Areas of Molecular Biology, Volume 74: Mechanism of Enzyme Action, Part B, Edited by Daniel L. Purich
ISBN 0-471-34921-6

RuBP from adenosine triphosphate (ATP) and ribose 5-phosphate. The presence of phosphoriboisomerase activity in such extracts prompted the suggestion that the actual substrate that accounts for RuBP production could be ribulose 5-phosphate. Subsequently, Hurwitz and colleagues (1956) were able to produce a spinach leaf preparation of phosphoribulokinase (PRK) that was sufficiently depleted of phosphoriboisomerase to allow the demonstration that ribulose 5-phosphate (Ru5P) is indeed the sugar phosphate substrate from which RuBP is produced (Eq. 1). A divalent cation is required for activity, with Mg^{2+} and Mn^{2+} most effectively supporting activity.

$$\text{M}^{2+}\text{ribulose 5-phosphate} + \text{ATP} \rightarrow \text{ribulose 1,5-bisphosphate} + \text{ADP} \quad (1)$$

The enzyme shows high specificity for the sugar phosphate substrate, Ru5P; ribose 5-phosphate, xylulose 5-phosphate, sedoheptulose 7-phosphate, and fructose 6-phosphate do not support significant activity (Hurwitz, 1962). Nucleotide triphosphate specificity observations have often been compromised due to the common use of a coupled assay that measures adenosine diphosphate (ADP) formation. Using a two-stage assay, Siebert et al. (1981) showed that the *Alcaligenes eutrophus* enzyme could utilize uridine triphosphate (UTP) and guanosine triphosphate (GTP) as alternative substrates.

The two enzymes unique to Calvin's reductive pentose phosphate cycle are PRK and RuBP carboxylase/oxygenase, which produce and utilize RuBP, respectively. Constraints on PRK activity and, consequently, on the production of RuBP can limit metabolism in organisms for which CO_2 assimilation is important (Dietz and Heber, 1984). In work on a *Synechocystis* mutant, Su and Bogorad (1991) documented the lethal consequences of diminished flux through PRK. More recently, in transgenic tobacco plants, Banks et al. (1999) have reported that knockout of the PRK gene can correlate with a diminution of photosynthesis sufficient to limit biomass production under sink-limited conditions. In view of these observations concerning the influence of PRK on metabolism, it is not surprising that, in both prokaryotes and eukaryotes, PRK catalyzes a regulated step in CO_2 assimilation.

Other reviews addressing PRK in the context of gene clusters that encode microbial CO_2 assimilatory enzymes have appeared (Tabita, 1988; Gibson and Tabita, 1996). This account will focus on the enzymology of PRK and the structure/function correlations that are emerging to account for regulation of activity and catalysis.

II. Isolation and Preliminary Characterization of PRK

The early studies that defined PRK's substrate specificity and the stoichiometry of the PRK reaction were not promptly followed by isolation and characterization of highly purified PRK protein. Hart and Gibson (1971) produced a high specific activity preparation of *Chromatium* PRK to support work on enzyme regulation. Another highly purified prokaryotic preparation was reported for *A. eutrophus* PRK by Siebert et al. (1981). Highly purified eukaryotic PRKs have been prepared from spinach (Lavergne and Bismuth, 1973), tobacco (Kagawa, 1982), and Chlamydomonas (Roesler and Ogren, 1990). For many years, there seemed to be intrinsic differences in catalytic efficiency between the prokaryotic and eukaryotic enzymes, with reported specific activities for the bacterial PRKs being consistently and substantially lower than corresponding values for the eukaryotic proteins (Tabita, 1988). However, when the complications inherent in the useful coupled spectrophotometric assay (Racker, 1957) due to inclusion of both reduced nicotinamide-adenine dinucleotide (NADH) (a positive effector) and phosphoenolpyruvate (PEP) (a negative effector) are considered, it is clear that a radioisotopic approach is more reliable for definitive activity estimate. When such estimates are coupled with protein determination by amino acid analysis, specific activities for the bacterial PRKs approach the value (350–400 U/mg) reported for eukaryotic PRKs (Runquist et al., 1996).

The availability of stable, highly purified PRK and the recognition of diverse regulatory mechanisms for prokaryotic and eukaryotic forms of the enzyme led to renewed interest in PRK enzymology in the 1980s. Using an independently developed affinity isolation protocol based on an ATP-derivatized matrix (Krieger and Miziorko, 1986), our lab purified the spinach enzyme that was used in collaborative work with Eckstein's group to demonstrate that PRK catalyzes a single in-line transfer of ATP's γ-phosphoryl group to produce RuBP. This conclusion is based on the observation of inversion of stereochemistry of the phosphoryl group transferred during the reaction (Miziorko and Eckstein, 1984). Hartman's lab utilized Ashton's observation of PRK sensitivity to modification by reactive red dye (Ashton, 1984) to develop a high-yield isolation protocol based on affinity chromatography on a reactive red matrix (Porter et al., 1986). Those investigators pursued fundamental protein chemistry studies of the spinach enzyme, quantitating reactive sulfhydryl groups, determining protein composition, and elucidating the amino terminal sequence (Omnaas et al., 1985; Porter et al., 1986). In affinity labeling studies with the ATP analog, fluorosulfonylbenzoyladenosine, Krieger and Miziorko (1986) detected a reactive cysteine within the ATP binding site. Subsequent Edman analysis (Krieger et al., 1987) mapped this residue[1] as C16 of the mature chloroplast peptide. This residue was also identified by Porter and Hartman (1986) as the preferential alkylation site that accounts for eukaryotic PRK's high sensitivity to sulfhydryl reagents. Subsequent affinity labeling studies with spinach PRK indicated sensitivity to lysine-directed reagents such as periodate-oxidized ATP and adenosine triphosphopyridoxal (Miziorko et al., 1990). The latter reagent was demonstrated to inactivate PRK by modifying K68.

III. Molecular Characteristics and Sequences of PRK Proteins

Comparison of prokaryotic and eukaryotic PRKs reveals many striking contrasts. For example, early consensus on the native and SDS-PAGE estimated subunit molecular weights (90 kDa and 45 kDa, respectively) of the eukaryotic enzyme developed on the basis of work with the spinach and tobacco enzymes (Kagawa, 1982; Krieger and Miziorko, 1986; Porter et al., 1986). As cDNA-deduced sequences have become available, it has become apparent that the precursor form of eukaryotic PRK is a peptide of ~44 kDa. Processing of the leader peptide during transit into the chloroplast converts the precursor into the mature protein, which has a calculated value of ~39 kDa for subunit molecular weight. Such a consensus did not readily develop for prokaryotic PRKs. On the basis of gel filtration and sucrose density gradient centrifugation experiments, Tabita (1980) proposed that the native *Rhodopseudomonas capsulata* enzyme is a hexamer of 36 kDa subunits. A hexameric oligomer (40 kDa subunit) was also proposed by Marsden and Codd (1984) for PRK from the cyanobacterium *Chlorogloeopsis fritschii*. In contrast, Rippel and Bowien (1984) proposed that Rhodopseudomonas acidophila PRK exists as an octamer of 32 kDa subunits. Cross-linking studies on the *A. eutrophus* enzyme allowed Siebert and Bowien (1984) to demonstrate that this prokaryotic PRK is an octamer. This oligomeric status has

[1]Prokaryotic PRK residue numbering starts with the methionine corresponding to the initiation codon. Unless otherwise specified, residue numbering corresponds to the sequence of the prokaryotic enzymes. When specified for eukaryotic PRK, the residue numbering convention is based on the sequence of the mature form of the peptide, which lacks the leader sequence processed during import into the chloroplast.

been confirmed for the prokaryotic PRKs by X-ray crystallography studies (Roberts et al, 1995; Harrison et al., 1998), which demonstrate that *Rhodobacter sphaeroides* PRK is an octamer (Fig. 1). On the basis of deduced sequences available for a variety of prokaryotic PRKs, a subunit molecular weight of ~33 kDa is calculated. In some bacteria, there are two genes that encode highly homologous PRK isozymes that differ only slightly in molecular weight. In the cases of the chemoautotroph *A. eutrophus*, Klintworth et al. (1985) mapped PRK genes to the genome and to megaplasmid pHG1; these encode PRKs of 33,319 and 33,164 Da, respectively (Kossman et al., 1989). In *R. sphaeroides*, Gibson and Tabita (1987) report distinct genes that encode form I (32 kDa) and form II (34 kDa) of PRK. Form I PRK shows a strong dependence on NADH for activation to high catalytic effi-

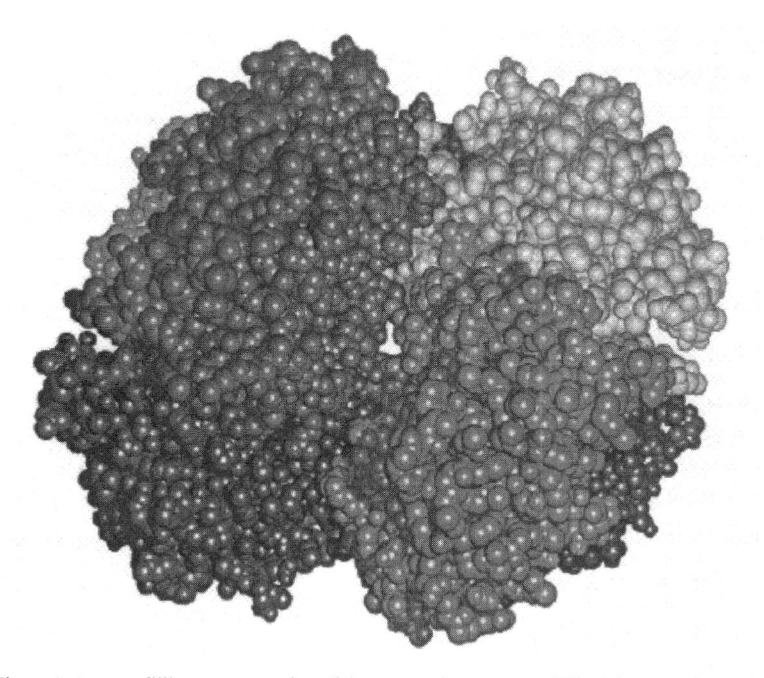

Figure 1. A space filling representation of the octameric structure of *Rhodobacter sphaeroides* phosphoribulokinase (PRK). The octamer consists of a top layer of four subunits displaced by about an eighth of a turn with respect to the bottom layer of four two-fold rotated subunits. The two-fold symmetry that relates the top to the bottom layer creates the impression of a four-bladed pinwheel, with each pinwheel blade (e.g., lightly shaded subunits at right side of figure) corresponding to the dimer believed to represent the eukaryotic form of PRK.

TABLE I
Contrasting Properties of Prokaryotic and Eukaryotic PRKs

	R. sphaeroides PRK	Spinach PRK
Quaternary structure; subunit size	8 × 32 kDa	2 × 40 kDa
Specific activity	340 U/mg	350 U/mg
Regulation	Allosteric	C16/C55
	NADH activation	Thiol/disulfide
	AMP inhibition	Exchange

ciency, while form II exhibits substantial activity in the absence of NADH. A more recent report on PRK from the marine chromophyte *Heterosigma carterae* (Hariharan et al., 1998) suggests that this dithiothreitol (DTT)-stimulated enzyme is a tetramer of 53 kDa subunits. If such an assignment is confirmed, this protein would represent a new PRK quaternary structure.

Within the last decade, PRK-encoding DNA sequences have become available from a variety of prokaryotic and eukaryotic sources. These include spinach (Milanez and Mural, 1988; Roesler and Ogren, 1988); *Arabidopsis* (Horsnell and Raines, 1991); wheat (Raines et al., 1989); ice plant (Michalowski et al., 1992); pea (Wedel et al., 1997); *Chlamydomonas* (Roesler and Ogren, 1990); *Synechocystis* (Su and Bogorad, 1991); *R. sphaeroides* (Gibson et al., 1990, 1991); *A. eutrophus* (Kossman et al., 1989); *R. capsulata* (Paoli et al., 1995); *Nitrobacter vulgaris* (Strecker et al., 1994); and *Xanthobacter flavus* (Meijer et al., 1990). Perhaps none of the contrasts between prokaryotic and eukaryotic PRKs (Table 1) are more dramatic than the differences between amino acid sequences; the alignment of these proteins indicates that only ~13% of total residues are invariant. Some of these sequence differences are explained by regulatory regions of the proteins. As indicated below, the divergent sequences have represented an effective filter that has expedited elucidation of key elements of the catalytic apparatus.

IV. An Overview of PRK Subunit Structure

Recently, Harrison has elucidated the high resolution (2.5 Å) structure of *R. sphaeroides* PRK (Harrison et al., 1998). At present, no report of a structure for eukaryotic PRK has appeared. While, as noted above, there is little homology at the level of amino acid sequence between prokaryotic and eukaryotic PRKs, the bacterial PRK structure appears useful in explaining regulatory and catalytic mechanisms for both prokaryotic and eukaryotic proteins. Figure 2 depicts a schematic diagram of the prokaryotic enzyme's secondary structural elements. This protein is composed of a seven-stranded mixed β-sheet, an auxiliary pair of antiparallel β-strands, and seven α-helices. The PRK subunit exhibits a protein fold (Fig. 3) analogous to the fold that characterizes the family of nucleotide monophosphate (NMP) kinases. As discussed in more detail in the sections below on the structural basis for regulation and catalysis, the C-terminal ends of the five parallel β-strands in the mixed β-sheet define one edge of the active site. Other secondary structure elements that circumscribe the active site include the α-helices comprised of prokaryotic PRK residues 49–62 (helix B), 158–171 (helix E) , and 180–193 (helix F) (Harrison et al., 1998). Many of those few amino acids that are invariant in alignment of the deduced sequences listed above map within these structural elements. For example, prokaryotic PRK helical residues 158–171 include those amino acids designated as a PRK signature sequence in the PROSITE data base. Eukaryotic PRKs contain five insertions and three deletions relative to the prokaryotic amino acid sequence. Speculation concerning

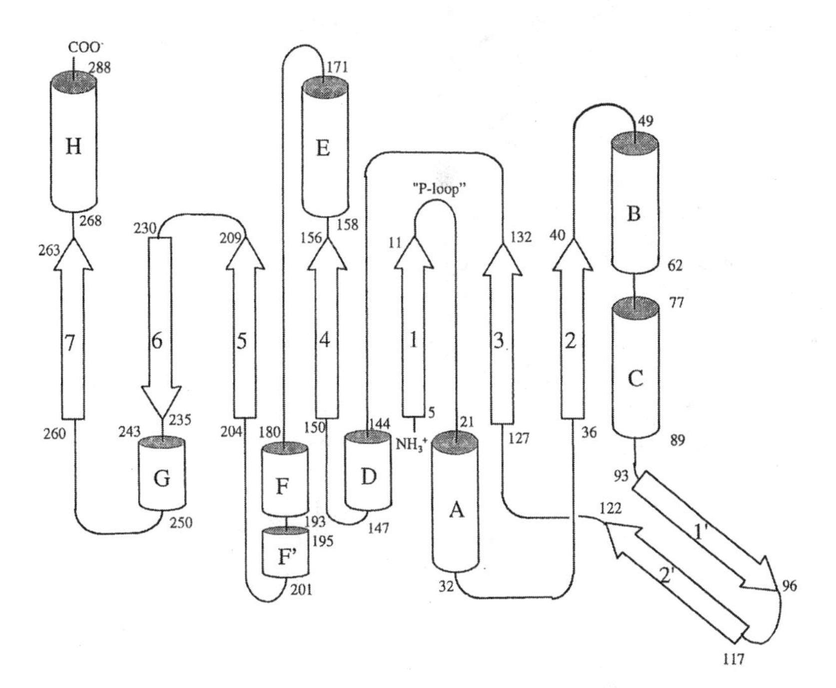

Figure 2. Schematic diagram of the secondary structural elements of *Rhodobacter sphaeroides* phosphoribulokinase (PRK), with numbers indicating the residues that define the N and C termini of α-helices and β-strands. (Adapted from Harrison et al., 1998; 5074–5085.)

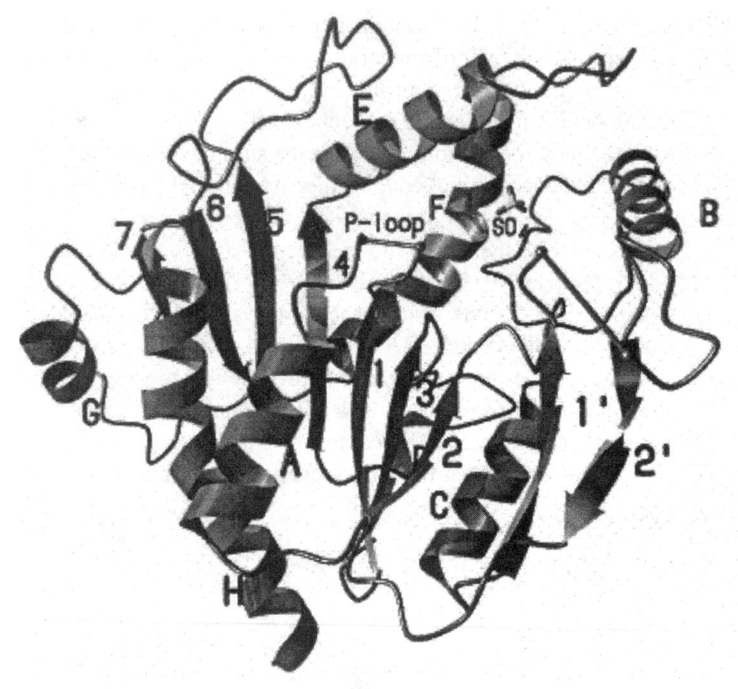

Figure 3. Phosphoribulokinase (PRK) subunit structure. The ribbon diagram illustrates the mixed β-sheet formed by strands 1–7. The C-terminal tips of parallel strands 1–5 form one side of the active site. Helix B as well as the helix E-loop–helix F region represent the "lid" regions believed to approach the "P-loop" and the edge of the β-sheet when substrates ATP and Ru5P bind. Such movement in nucleotide monophosphate kinase fold enzymes effectively closes the active site, sequestering reactive species formed during catalysis.

how these structural differences may account for the contrasting properties of the prokaryotic and eukaryotic enzymes has appeared in the context of discussion of the PRK structure (Harrison et al., 1998).

V. Regulatory Mechanisms

A. REGULATION OF EUKARYOTIC PRK

Diverse regulatory mechanisms also distinguish prokaryotic PRK from eukaryotic PRK. The former class of enzymes is allosterically regulated, while thiol/disulfide exchange is a major determinant of activity for PRKs from higher organisms. Additionally, recent work suggests that eukaryotic PRK's activation state may be influenced by other chloroplast proteins. Early studies indicated that PRK activity could also be effectively inhibited by chloroplast metabolites (Anderson, 1973; Gardemann et al., 1983). The most effective metabolite inhibitor appears to be 6-phosphogluconate, which competes with Ru5P for occupancy of PRK's active site.

The physiological significance of eukaryotic PRK in regulation of carbon assimilation was also demonstrated by the early observation (Latzko et al., 1970) that illumination of chloroplasts resulted in activation of the enzyme. The stimulatory effect could also be produced if DTT was added to broken chloroplasts. Susbequently, Wolosiuk and Buchanan (1978) showed that PRK activation could be attributed to photochemically or chemically reduced thioredoxin. Thioredoxin *f* is the isoform of this physiological reductant that is localized to the chloroplast (Wolosiuk et al., 1979). Thus, as described in detail by Buchanan (1980), the light activation of PRK and other stromal enzymes is attributable to a ferredoxin/thioredoxin mediated pathway.

More recently, work on regulation of eukaryotic PRK has focused on the relative activity of the enzyme in isolated form versus multienzyme complexes. The hypothesis that stromal proteins exist in complexes originated over a decade ago. Sainis and Harris (1986) proposed the association of pea PRK with phosphoriboisomerase and RuBP carboxylase. Nicholson et al. (1987) proposed the existence in *Scenedesmus* of a PRK-glyceraldehyde 3–phosphate dehydrogenase complex that dissociated in the presence of NADPH, DTT, and thioredoxin to release the activated forms of these enzymes. The latter report was followed by a study on spinach leaf enzymes (Clasper et al., 1991) that confirmed the association of PRK with NADPH-linked glyceraldehyde 3–phosphate dehydrogenase. This complex also dissociated when incubated with DTT. In other work on spinach chloroplasts, Gontero et al. (1988) reported isolation of a complex containing five stromal enzymes: phosphoriboisomerase, PRK, RuBP carboxylase, phosphoglycerate kinase, and glyceraldehyde phosphate dehydrogenase. Subsequent investigations (Rault et al., 1991) suggested that, in such a multienzyme complex, PRK exhibited slow kinetics of oxidative inactivation in comparison with isolated enzyme. Conversely, Gontero et al. (1993) showed more rapid thioredoxin-dependent activation of PRK in a multienzyme complex than measured for free PRK. When the stoichiometry of the enzymes in the putative five enzyme complex were addressed (Rault et al., 1993), only two of the component proteins, the PRK dimers and glyceraldehyde 3–phosphate dehydrogenase tetramers, were measured at equivalent levels.

Technically, the demonstration of specificity in the interactions that are crucial to formation of a multienzyme complex is quite difficult. The same group of investigators that reported a five component complex from spinach has most recently focused on a *Chlamydomonas* model, from which they isolate a two component complex that contains equivalent levels of PRK dimer and glyceraldehyde 3-phosphate dehydrogenase tetramer (Avilan et al., 1997a). Directed mutagenesis experiments (Avilan et al., 1997b) indicate that substitutions that eliminate basicity at residue 64 (arginine in wild-type eukaryotic PRK) disrupt the multienzyme complex. Another investigation of this *Chlamydomonas* multienzyme complex (Lebreton et al., 1997a) has prompted speculation that oxidized PRK, which is relatively inactive when enzyme exists as the isolated dimer, can exhibit increased activity when bound to glyceraldehyde 3-phosphate dehydrogenase. Additionally, this report suggests that dilution of the complex results in release of oxidized PRK in a "metastable" state characterized by a higher k_{cat} than measured for oxidized PRK in the complex. This "metastable" species eventually converts to the low-activity species equivalent to oxidized free PRK dimer. A theoretical model has also been developed to accommodate these findings, and a kinetic analysis has been conducted (Lebreton et al., 1997b) and interpreted to suggest that oxidized PRK in the multienzyme complex exhibits a catalytic mechanism different from free PRK. This kinetic analysis relies on the use of ribose 5-phosphate and xylulose 5-phosphate, which are asserted to be alternative substrates. This assertion is not well supported in the PRK literature (Hurwitz, 1962) and the concern that Ru5P may either be present in commercial sources of these metabolites (or, alternatively, be formed by trace phosphoriboisomerase or xylulose 5–phosphate epimerase in the protein complex) remains to be discounted. This issue has not been explicitly addressed in the publication that involves the "alternative" substrates. Such a complication could translate into using a mixture of Ru5P substrate and inhibitor, rather than an alternative substrate, in those kinetic studies. Additionally, in the context of evaluating the model that addresses the changes in activity attributed to multienzyme complex and "metastable" oxidized forms of PRK, it is unfortunate that the oxidation state of PRK cysteinyl sulfhydryls (C16; C55) in these various samples (e.g., complex or metastable form) is never directly tested. Protein chemistry approaches would indicate whether interchain thiol/disulfide exchange has occurred to an extent that would more simply account for the fractional change in the low baseline oxidized PRK activity. If such concerns could be ruled out, the detailed kinetic study and the accompanying modeling work would be elevated in significance.

The suggestion that PRK and glyceraldehyde 3-phosphate dehydrogenase form a multienzyme complex (Nicholson et al., 1987; Avilan et al., 1997a) in *Scenedesmus* and *Chlamydomonas* has been confirmed and extended by Wedel and colleagues on complexes of plant stromal enzymes (Wedel et al., 1997; Wedel and Soll, 1998). Those investigators propose the involvement of a low molecular weight protein, CP12, and suggest that its N-terminal region interacts with a PRK dimer while the C-terminal region interacts with a glyceraldehyde 3-phosphate dehydrogenase tetramer. Two noncovalently linked PRK-glyceraldehyde 3-phosphate dehydrogenase-CP12 species assemble to form a 600 kDa complex. NADPH induces dissociation of the complex; DTT also stimulates dissociation. The observations on spinach proteins are in close agreement with the earlier work on the *Scenedesmus* complex (Nicholson et al., 1987). CP12 is proposed to be widely expressed in wide range of photosynthetic organisms (Wedel and Soll, 1998) and evidence for the existence of a PRK-glyceraldehyde 3–phosphate dehydrogenase-CP12 complex in *Synechocystis* and *Chlamydomonas* has been presented. These observations potentially represent a major new development in the understanding of pentose phosphate cycle regulation and, since CP12 is postulated to be found in a variety of organisms, it will be interesting to learn whether the results of Wedel and colleagues are confirmed in other systems and by other laboratories.

B. THE STRUCTURAL BASIS FOR REGULATION OF EUKARYOTIC PRK

The availability of homogeneous, affinity-isolated spinach PRK facilitated protein chemistry analyses that provided structural information on the regulatory site. The early work of Wolosiuk and Buchanan (1978) indicated that inactive PRK could be reactivated by thioredoxin or DTT, suggesting that a thiol/disulfide exchange process interconverts enzyme between active (reduced) and inactive (oxidized) forms. Porter et al. (1986) measured the free cysteine content of reduced and air-oxidized enzyme, demonstrating that a diminution of two cysteines correlated with oxidative inactivation. Thus, a single disulfide bond is involved. No indication of aggregated PRK was observed upon air oxidation. These experiments were followed by work that demonstrated that ATP protects PRK against oxidative inactivation. Omnaas et al. (1985) had previously shown that the reactive Ru5P analog, bromoacetylethanolamine phosphate, inactivates PRK and that enzyme is protected against modification not by substrate Ru5P but rather by ATP. Because

ATP protects both against oxidative inactivation and alkylation by a substrate analog, Porter and Hartman (1986) speculated that a common modification site might be involved. The observation that oxidatively inactivated PRK does not react with bromacetylethanolamine phosphate supported this hypothesis, as did the pH dependency of oxidation and alkylation. A pK_a of ~7.8 is observed for both processes. Isolation and Edman sequence analysis of the peptide modified by bromacetylethanolamine phosphate indicates that C16 is the alkylation target. This residue has been directly implicated as part of the ATP binding site by the affinity labeling of PRK by the ATP analog fluorosulfonylbenzoyl adenosine (Krieger and Miziorko, 1986; Krieger et al., 1987), accounting for the observation that ATP protects not only against modification by a reactive ATP analog, but also against modification by a reactive Ru5P analog and against air oxidation.

The next task in elucidating the structural basis for PRK regulation by thiol/disulfide exchange involved identification of C16's counterpart in disulfide formation. Porter et al. (1988) took advantage of the fact that spinach PRK contains only four cysteines, allowing these investigators to pursue a differential S-carboxymethylation analysis of free cysteines in active and inactive PRKs. High-performance liquid chromatography (HPLC) analyses of tryptic digests of active and oxidized PRK that had been subject to [^{14}C]-carboxymethylation clearly indicated the peptides that harbor the cysteines involved in disulfide formation. These peptides were isolated and subjected to Edman sequence analysis, which identified the target residues as C16 and C55. Thus, the earlier hypothesis concerning the involvement of C16 in both PRK regulation and modification by a reactive Ru5P analog was confirmed. The identification of C55 and the failure to distinguish any molecular weight difference upon gel filtration of reduced and oxidized PRKs under denaturing conditions prompted the conclusion that formation of a single intrasubunit disulfide between C16 and C55 accounts for inactivation of PRK.

With the identification of the PRK cysteines involved in thiol/disulfide exchange accomplished, the issue of PRK-thioredoxin-*f* interactions was appropriately addressed by Hartman's lab. In experiments focusing on C46 and C49 of thioredoxin-*f* and using fructose bisphosphatase as the enzyme targeted for reduction, Brandes et al., (1993) suggested that C46 is the nucleophile that attacks the disulfide linkage of the oxidized target enzyme. C49 functions in cleavage of the mixed disulfide transiently formed between thioredoxin and the target enzyme, accounting for release of oxidized thioredoxin from a fully reduced enzyme. Having established the thioredoxin nucleophile, Brandes et al. (1996a) addressed whether PRK's C16 or C55 is the site of attack by thioredoxin. Two approaches to this issue were employed. A survey of agents known to catalyze disulfide formation indicated that Cu^{2+} was uniquely well suited to catalyze a PRK-thioredoxin mixed disulfide. Sample mixtures of PRK C16S/thioredoxin C49S, PRK C16S/thioredoxin C46S, PRK C55S/thioredoxin C49S, and PRK C55S/thioredoxin C46S were subjected to Cu^{2+} oxidation. Only PRK C16S/thioredoxin C49S form convincing amounts of the mixed disulfide covalent complex, suggesting that reduced thioredoxin's C46 attacks oxidized PRK's C55. In a second approach, Brandes et al. (1996a) mixed 5-thio-2-nitrobenzoate derivatized PRK C16S or C55S (mimics of the oxidized form of PRK) with thioredoxin C49S. A stable covalent disulfide linked complex only forms with PRK C16S, again suggesting that PRK's C55 is the target of thioredoxin's C46.

Why is oxidized PRK inefficient as a catalyst? Protein modification (Porter and Hartman, 1990) and mutagenesis studies (Milanez et al, 1991; Brandes et al., 1996b) from the Hartman lab indicate that, of the four PRK cysteines, C16, C244, and C250 have little influence on catalysis. C55 plays a only facilitative role since V_{max} of PRK C55S is diminished by only ~10–fold in comparison with wild-type enzyme; an effect of this magnitude does not suggest direct involvement in reaction chemistry. A double mutant (C16S/C55S) of PRK, which lacks both regulatory cysteines, displays minor activity differences in comparison with PRK C55S. Thus loss of these cysteines does not entirely account for the depressed catalytic activity exhibited by oxidized eukaryotic PRK. Conformational constraints have been invoked as attenuating activity in oxidized spinach PRK (Brandes et al., 1992). As discussed below, such an interpretation is strongly supported by the high-resolution X-ray structure that has recently been solved for prokaryotic PRK (Harrison et al., 1998).

In the family of proteins that fold like the nucleotide monophosphate kinases, there commonly is observed a sequence that fits the "Walker A" or "P-loop" motif (Saraste et al., 1990). This sequence invariably loops between an α-helix and one of the β-strands of the β-sheet that defines an edge of the active site. This loop usually contains residues that interact with β– and γ-phosphoryls of the M^{2+}- ATP substrate. In particular, serine or threonine residues in the "P-loop" commonly function as one of the ligands to the divalent cation of the M^{2+}- ATP substrate. In both prokaryotic and eukaryotic PRKs, there are "P-loop" residues that may qualify for such a function. Thus, freedom of motion of this loop would *a priori* be expected to represent a major prerequisite for efficient catalysis of phosphoryl transfer, since binding of

ATP substrate and release of ADP product must occur with each catalytic cycle. The observations of Brandes et al. (1992) in cross-linking studies on spinach PRK indicate considerable conformational flexibility in the region of reduced eukaryotic PRK that harbors the regulatory C55 and C16 of the "P-loop." This region exhibits dynamic motion even in the crystalline state. The bacterial PRK structure exhibits a break in the α-carbon backbone within the "P-loop" (Harrison et al., 1998), probably because this loop exhibits a high degree of motion when a nucleotide is not bound. In eukaryotic PRK, the tethering of C16 to C55 in a disulfide linkage would immobilize the loop and it seems certain that substrate binding and product release would be adversely affected. The issue has been directly addressed in the recent work of Brandes et al. (1998), which indicates that reduced spinach PRK, with a dynamic "P-loop," exhibits a $K_{d\,ATP}$ equaling 37 μM. Oxidized enzyme, with an immobilized "P-loop" exhibits a $K_{d\,ATP}$ equaling 28 mM, demonstrating the predicted impairment in enzyme-substrate interaction.

In addition to the consequences of disulfide formation on "P-loop" mobility, the impact on the region that harbors the other cysteine residue, C55, should be considered. In the prokaryotic sequence, this amino acid would map as residue 40 (which is commonly found as glutamate in bacterial PRKs). Such a residue is situated at the tip of β-strand 2, that is, at the edge of the active site and only two residues from D42, which has the most profound impact on catalysis that has been documented. Constraining a peptide loop by disulfide linkage to this β-strand could easily have steric consequences that would also negatively impact on catalytic efficiency. Any prediction concerning the impact on catalysis of a disulfide linkage to the tip of β-strand 2 is clearly somewhat speculative, since it relies on a bacterial PRK structure that contains no active site ligands. Nonetheless, such a prediction is supported by the observations of Hartman's lab (Milanez et al., 1991; Brandes et al., 1996b), which indicate that replacement of C55 has more impact on reaction efficiency than similar substitution of C16.

For the reasons outlined above, the hypothesis that disulfide linked PRK, with an immobilized "P-loop," can be an efficient catalyst when associated in either a multienzyme complex or when in a "metastable" form immediately after release from a complex raises some concern. In fairness, such proposals were advanced prior to the availability of a PRK structure. As suggested earlier, the direct evaluation of the redox status of the regulatory cysteines of PRK in such multienzyme complexes may resolve this apparent dilemma.

A significant issue to be addressed in judging whether it is legitimate to extrapolate from the bacterial PRK structure and make predictions on eukaryotic

PRK regulation concerns thioredoxin involvement. If the available structural model is relevant, then thioredoxin must be able to approach the C16/C55 disulfide linkage. Such an issue has been addressed by Harrison et al. (1998). Using the structure of *Anabaena* thioredoxin, docking of this protein to a dimer of PRK subunits that is believed to represent the quaternary structure of eukaryotic PRK has been performed. The PRK dimer used in the docking experiment was selected since its subunit interface contains a substantial number of invariant residues and reflects the highest contact area of any of the interfaces in the *R. sphaeroides* PRK structure. The thioredoxin was modeled so that cysteine residues 50 and 53 were aligned with PRK residues 16 and 40. Two orientations of thioredoxin were reported to fit these constraints. Moreover, the width of the active site cavity reportedly matches well with the thickness of the thioredoxin molecule. Clearly, as a structure of thioredoxin-*f* becomes available, more relevant modeling will be possible. However, the speculation outlined above is self-consistent with all available data concerning thiol/disulfide exchange in regulation of isolated eukaryotic PRK. Thus, the prokaryotic PRK structure appears to represent a useful working model.

C. REGULATION OF PROKARYOTIC PRK

Given that microbial autotrophic metabolism (e.g., CO_2 assimilation) is energetically less efficient than heterotrophic metabolism, it seems logical that regulation will be imposed to diminish flux through pathways such as CO_2 assimilation when alternate energy-producing fuels are available. It may also be expected that reactions unique to CO_2 assimilation represent potentially efficient control points. The observation of complex regulation of prokaryotic PRK by several metabolites is in accord with these expectations. For example, using a protein preparation from *R. sphaeroides*, Rindt and Ohmann (1969) showed that PRK activity is stimulated by NADH and inhibited by adenosine monophosphate (AMP). The degree of inhibition by AMP depends on NADH level. Similarly, using extracts of *Hydrogenomonas facilis* (*Pseudomonas facilis*), MacElroy et al. (1969) generated evidence suggesting that NADH activated PRK. Subsequently, using a protein preparation from *P. facilis*, Ballard and MacElroy (1971) demonstrated that PEP could substantially inhibit PRK activity. Shortly thereafter, MacElroy et al. (1972) extended the observation of PEP inhibition to PRK from *Thiobacillus neapolitanus*. Kiesow et al. (1977) reported NADH activation of *Nitrobacter winogradsky* PRK and underscored the requirement for a *reduced* dinucleotide in enzyme activation. That report also indicated that PRK from plant abstracts is *not* sensitive to NADH activation. The observations on NADH activation were subsequently extended to *R. capsulata* and other photosynthetic bacteria by Tabita (1980). Using substantially purified PRK from *Hydrogenomonas eutropha*, Abdelal and Schlegel (1974) reported NADH activation as well as inhibition by both AMP and PEP. Observation of nonhyperbolic ATP saturation curves in several cases (MacElroy et al., 1972; Abdelal and Schlegel, 1974), as well as the structural differences between NADH and ATP, prompted further examination of cooperativity in metabolite binding to PRK. Using homogeneous, affinity purified PRK from *A. eutrophus* (*Ralstonii eutrophus*), Siebert and Bowien (1984) demonstrated the distinct nature of the adenine nucleotide binding sites. In equilibrium binding experiments, stoichiometric binding of substrate ATP (with respect to a 32 kDa subunit) was unaffected by AMP or PEP. Conversely, stoichiometric binding of inhibitor AMP was not influenced by ATP or PEP. While distinct ATP and AMP sites were clearly established, the nature of the binding sites for activator NADH and inhibitor AMP remained to be addressed. Both Rindt and Ohmann (1969) and Abdelal and Schlegel (1974) showed that NADH-stimulated PRK activity was progressively diminished by increasing concentrations of AMP, implying mutually exclusive binding by the dinculeotide and mononucleotide effectors. Runquist et al. (1996) demonstrated a substantial enhancement of NADH fluorescence upon binding to recombinant *R. sphaeroides* PRK. This observation was exploited in competition studies that demonstrated stoichiometric displacement of NADH by AMP, an observation that contrasts with unperturbed ATP substrate site occupancy at comparable AMP levels. Thus, allosteric regulation of prokaryotic PRK and the specificity of substrate versus effector sites has been firmly established. Furthermore, when the fluorescent alternative substrate trinitrophenyl-ATP is used to occupy PRK's active site and NADH binds to the allosteric site, fluorescence energy transfer is observed. This observation allowed Runquist et al. (1996) to estimate a 21 Å distance between these sites on *R. sphaeroides* PRK.

D. THE STRUCTURAL BASIS FOR REGULATION OF PROKARYOTIC PRK

Two distinct approaches have led to recent reports that provide information on the location of the allosteric site in PRK. Novak and Tabita (1999) utilized the genes that encode the two different PRKs (NADH sensitive versus NADH independent) of *R. sphaeroides* to construct a series of chimeric proteins. Working with dialyzed bacterial extracts, these investigators measured enzyme activity and fractional stimulation by NADH to evaluate which regions of PRK are important for NADH activation. Two of the most informative constructs encoded either the N-terminal portion of NADH sensitive PRK I (upstream of E86; remainder of protein from PRK II), producing enzyme that was acutely sensitive to activator NADH or, conversely, encoded the N-terminal portion of PRK II (upstream of E86; remainder from PRK I), producing a chimera that was much less sensitive to NADH activation. The results underscore the importance of PRK's N-terminal region in supporting NADH activation.

An independent approach that mated X-ray diffraction with directed mutagenesis was used by Harrison and colleagues (Kung et al., 1999). Taking advantage of the ability to crystallize PRK bound to the mononucleotide AMP-PCP, a 2.6 Å structure of the binary complex was elucidated. AMP-PCP, normally considered to be an ATP analog, did not bind where the ATP substrate is normally situated in nucleotide monophosphate kinase fold enzymes. Instead, the nucleotide bound at the interface between three subunits

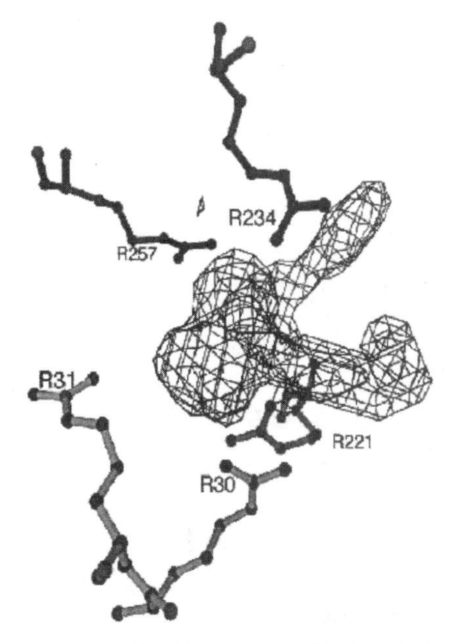

Figure 4. The effector binding site of *Rhodobacter sphaeroides* phosphoribulokinase (PRK). The figures depicts electron density attributable to the nucleotide analog β, γ-methylene ATP, which binds at the interface of three subunits (Kung et al., 1999). Near the bound nucleotide are R221 from one subunit, R234 and R257 from a second subunit, and R30 and R31 from the third subunit. Mutagenesis results indicate that R234 and R257 are crucial to the binding of the allosteric effector, NADH. R31 is key to the transmission of the allosteric signal to the active site.

of the octameric enzyme (Fig. 4). Previously, Koteiche et al. (1995) showed that a spin-labeled ATP analog, *S*-acetamidoproxyl-ATPγS, could bind to PRK's substrate or effector (NADH/AMP) sites. The possibility that the interfacial binding of a mononucleotide identifies the effector site was tested by a series of directed mutagenesis experiments. Arginine residues are frequently implicated in binding of phosphorylated metabolites. The structure of the binary PRK✳AMP-PCP complex implicated five arginines as possible contributors to the allosteric site. R221 is highly conserved in both prokaryotic and eukaryotic PRKs. R234 and R257 are situated on a differ-

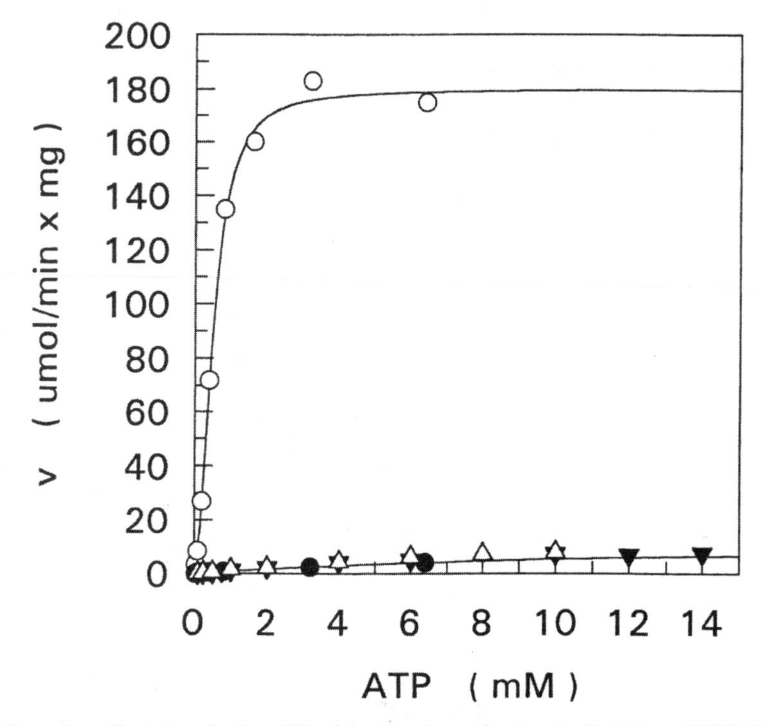

Figure 5. Allosteric activation of *Rhodobacter sphaeroides* phosphoribulokinase (PRK). Enzyme activity as a function of ATP concentration is shown. Wild-type PRK exhibits enhanced activity and a sigmoidal saturation curve (○) in the presence of NADH but low activity and a hyperbolic saturation curve (●) in the absence of NADH. The R31A mutant form of PRK, which binds NADH comparably to wild-type PRK, fails to transmit the allosteric stimulus and exhibits low activity and hyperbolic saturation by adenosine triphosphate (ATP) in the presence (△) or absence (▼) of NADH.

ent subunit and conserved only in prokaryotic (allosterically regulated) PRKs. On the third subunit are located R30 and R31; only the latter is conserved in prokaryotic PRKs. Using site directed mutagenesis, each of the five arginines was replaced with alanine. These mutant PRKs were evaluated by fluorescence techniques, which had been previously developed (Runquist et al., 1996) to detect and characterize binary PRK✻substrate ATP and PRK✻effector nucleotide complexes. The binding experiments indicated that none of the arginines influenced ATP occupancy of the catalytic site. Thus, their tertiary structures are substantially intact. In contrast to these substrate binding data, the effector binding results implicated R234 and R257 in NADH binding, since substitutions that eliminate basicity at these residues disrupt formation of a binary PRK✻NADH complex. Substitutions of the other arginines have a minimal effect on NADH binding. Additionally, the R31A mutant, which effectively binds activator NADH, exhibits hyperbolic saturation by ATP regardless of whether or not NADH is present in the assay (Fig. 5). Wild-type PRK exhibits a diminished V_{max} and hyperbolic saturation by ATP when assayed in the absence of NADH but a high V_{max} and sigmoidal dependence on ATP when NADH is present. The ability of R31A to form a stable complex with NADH coupled with its inability to function as the activated form suggests that R31 is a key link in transmission of the allosteric stimulus.

VI. Catalytic Mechanism and Structure/Function Assignments

The elucidation of a high-resolution structure for *R. sphaeroides* PRK (Harrison et al., 1998) provides several clues to location of the active site. For example, the observation that the PRK fold makes it a member of the nucleotide monophosphate kinase family of proteins can be coupled with the detection of a "P-loop" motif to allow assignment of the region where substrate ATP binds. On the other hand, since the protein crystallizes without bound substrates, analogs, or products, many details would remain to be deduced if complementary solution structure data were not available. From earlier mechanistic work (Miziorko and Eckstein, 1984), it has been established that the transfer of the γ-phosphoryl group from ATP to Ru5P proceeds with inversion of stereochemistry. The most straightforward interpretation of this observation is that the reaction involves a single in-line transfer of the phosphoryl group from donor to acceptor, that is, there is no covalent phosphoryl-enzyme intermediate. Moreover, the data demand that there be simultaneous binding of both substrates in a ternary complex with enzyme. The order of addition of substrates to PRK remains to be unambiguously established. Steady-state kinetic experiments suggested that product RuBP is a competitive inhibitor with respect to Ru5P while mixed inhibition with respect to ATP was assigned. This prompted Lebreton et al. (1997b) to propose that *Chlamydomonas* PRK catalyzes an ordered reaction with Ru5P binding as the first substrate. In contrast, Wadano et al. (1998) argue for ATP as the first substrate on the basis of their kinetic studies on the *Synechococcus* enzyme. Nonequilibrium binding experiments with the *R. sphaeroides* enzyme demonstrate that a stable binary PRK✳ATP complex forms (Koteiche et al., 1995; Runquist et al., 1996); bound nucleotide is released under turnover conditions with concomitant production of RuBP. An alternative substrate, the fluorescent analog trinitrophenyl-ATP will also form a stable binary complex with PRK; competition by the effector AMP does not displace this fluorescent nucleotide, suggesting that it binds at the substrate site rather than the effector site. On the basis of these observations on formation of PRK✳ATP binary complexes, it is unclear whether the binding of substrates, proposed based on kinetic inhibition patterns, must occur in any *strictly* ordered fashion.

Perhaps an equally critical issue to be resolved involves the functional assignment of amino acids that support substrate binding and reaction chemistry. Many enzymes in the nucleotide monophosphate kinase family of proteins primarily catalyze transfer of a phosphoryl group between two nucleotides. While these reactions may not be critically dependent upon a general base catalyst, such assistance appears necessary for production of many phosphorylated metabolites. Several of these, for example, fructose 2,6-bisphosphate, shikimate 3–phosphate, and thymidine 5′-monophosphate, are similar to RuBP in that they are synthesized by enzymes that exhibit a nucleotide monophosphate kinase fold and use ATP as the phosphoryl donor. As discussed below, mechanistic similarities in the reactions that form these metabolites are useful in formulating functional predictions for active site amino acids in PRK.

Several functional assignments have resulted by identification of amino acids that are invariant among the diverse collection of low homology PRK sequences and subsequent investigation of function by directed mutagenesis techniques. In such an approach, amino acids in the longest continuous stretch of invariant sequence (residues 131–135) might be targeted for attention. Prioritization of targets for such a mutagenesis approach might also be influenced by scrutinizing the PRK "signature sequence" (Kossman et al., 1989), which includes amino acids in a "lid" helix that defines an active site boundary. Indeed tryptophan-155/162, which maps in this region, has been scrutinized in mutagenesis work (Brandes et al., 1998) that supports its involvement in ATP binding. Other examples of the utility of this homology or "phylogenetic" strategy are discussed below. In some respects the most productive studies were directed toward a region of relatively *low* homology. The rationale for selecting this region of PRK derives from affinity labeling work. Before launching a detailed discussion on the strategies in target selection and the interpretation of the data on characterization of mutant PRKs, it seems worthwhile to note that much of this work was executed prior to availability of a high-resolution structure, which has largely validated the active site assignments that were made. In order to make these assignments with any degree of surety, our laboratory developed technology to facilitate evaluation of the structural integrity of recombinant PRKs. These experimental approaches, which may have utility in the study of a variety of phosphoryl transfer enzymes, are summarized below.

A. OVEREXPRESSION OF *R. SPHAEROIDES* PRK

With the objective of facilitating the identification of catalytic residues, our lab developed a pET-3d based system (Charlier et al., 1994) to express in *E. coli* the PRK I-encoding DNA that had been provided by Hallenbeck and Kaplan (1987). Anticipating that large diminutions in activity would be encountered if we successfully replaced amino acids crucial to reaction chemistry, the expression system would, at a minimum, be required to produce large amounts of PRK protein to support assays. The expression construct met this requirement, producing PRK as over 20% of total *E. coli* protein, a significant improvement in yield over that obtained for *E. coli* expression of spinach PRK (Milanez et al., 1991; Hudson et al., 1992). Isolation includes a preliminary anion exchange step as a prelude to affinity chromatography (Gibson and Tabita, 1987), which employs a reactive green-19 dye matrix. From a 1.5 l bacterial culture, 15 mg of homogeneous PRK I is routinely isolated, providing sufficient protein for biophysical characterization as well as activity estimates of catalytically deficient PRK mutants.

B. TOOLS FOR EVALUATION OF ACTIVE SITE INTEGRITY

A common theme in protein engineering work involves development of tests to evaluate the integrity of recombinant enzymes, regardless of whether they correspond to wild-type or mutant proteins. In evaluating PRKs that have been expressed in *E. coli*, our lab has used biophysical approaches to characterize the ability of PRK to form stable binary complexes with substrate, effector, or spectroscopically active analogs of these metabolites (Koteiche et al., 1995; Runquist et al., 1996). Wild-type *R. sphaeroides* PRK will stoichiometrically bind ATP, the fluorescent alternative substrate trinitrophenyl ATP, or the spin-labeled analog *S*-acetamidoproxyl-ATPγS to form stable binary complexes. Such complex formation has been used to verify the integrity of catalytic mutants (Runquist et al., 1996, 1998, 1999) as well as effector site mutants (Kung et al., 1999). The test is based on the rationale that, if stoichiometric binding of analog reflects a structurally intact active site, the overall tertiary structure must exhibit a high level of integrity. Similarly, the effector site can be titrated with *S*-acetamidoproxyl-ATPγS when the catalytic site is preloaded with ATP. Additionally a stable binary PRK-*NADH complex can be isolated and has been useful in directly demonstrating competitive displacement of NADH from the effector site by AMP. The stoichiometry of NADH binding has also been used to evaluate the structural integrity of catalytic and substrate binding mutants (Runquist et al., 1996, 1998, 1999). It has also allowed us to discriminate between mutants deficient in effector binding or deficient in transmission of the allosteric signal (Kung et al., 1999). Thus, the approach of mating spectroscopically active metabolite analogs with biophysical characterization of engineered proteins has demonstrated considerable utility and can be potentially applied to a wide variety of enzymes that bind nucleotide triphosphates (Potter and Miziorko, 1997) or other metabolites and their analogs (Narasimhan and Miziorko, 1997).

C. SELECTION OF MUTAGENESIS TARGETS AND INTERPRETATION OF MUTAGENESIS RESULTS

In the initial selection of PRK mutagenesis targets that were likely to yield functional information, affinity labeling results provided some guidance. Mutagenesis of spinach PRK C16 (Milanez et al., 1991), the target of the ATP analog fluorosulfonylbenzoyladenosine (Krieger and Miziorko, 1986; Krieger et al., 1987), demonstrated that this residue's thiol is not crucial to reaction chemistry. This result confirmed expectations based on protein modification work (Porter and Hartman, 1988) as well as on the observation that this residue, involved in regulation of eukaryotic PRKs, is replaced by alanine in prokaryotic PRKs.

The mapping of spinach PRK K68 (prokaryotic PRK K53) to the active site on the basis of its modification by the reactive ATP analog, adenosine triphosphopyridoxal (Miziorko et al., 1990), prompted evaluation of the function of this and other basic residues in the flanking low homology region of PRK. Using an expression construct for *R. sphaeroides* PRK, Sandbaken et al. (1992) produced in *E. coli* the wild-type enzyme and the PRK mutants K53M, H45N, and R49Q. Expectations concerning a catalytic function for lysine had been diminished by the report of an arginine at this position in the *X. flavus* enzyme (Meijer et al., 1990), but the retention of positive charge was potentially compatible with a role in substrate binding. In fact, PRK K53M largely retained wild-type PRK properties. This observation has been confirmed for spinach PRK (Mural et al., 1993). In contrast, H45N and R49Q mutants showed a substantial inflation in $K_{m\ Ru5P}$ (40 and 200–fold, respectively), which seemed significant since no substantial changes in k_{cat} or $K_{m\ ATP}$ were apparent. Prokaryotic PRK R49 corresponds to R64 in the eukaryotic enzyme. The influence of this R64 on $K_{m\ Ru5P}$ was demonstrated for the *Chlamydomonas* enzyme by Roesler et al. (1992) and has subsequently been confirmed by Avilan et al. (1997b). The assumption that inflation of K_m correlates with a function for H45 and R49/64 in Ru5P binding (an issue to be discussed below) raises a question over why this region identified by a reactive ATP analog should exhibit such a function. The answer comes from a consideration of reaction stereochemistry (Miziorko and Eckstein, 1984), which demands formation of a ternary complex with ATP and Ru5P juxtaposed at PRK's active site (Fig. 6). It follows that any reactive group attached to ATP's γ-phosphoryl will be favorably positioned, in the absence of cosubstrate, to modify the sugar phosphate substrate binding site.

While the tryptophan mutagenesis (W155 F/A) results of Brandes et al. (1998) illustrate the contribution of stacking interactions to nucleotide binding by spinach PRK, there is ample precedent for the anchoring of nucleotides and other phosphorylated metabolites by positively charged active site lysines and arginines. These residues may also support transition state stabilization (Mueller-Dieckmann and Schulz, 1994). With different available algorithms and multiple stringency choices (e.g., gap creation and extension penalties) possible, the legitimacy of any particular sequence alignment may raise controversy, as demonstrated by the criticism that R49/64 is not conserved in *R. sphaeroides* or *A. eutrophus* PRKs (Gibson et al., 1990). Despite these legitimate concerns, our lab proposed that H135, K165, R168, R173, and R187 (prokaryotic PRK numbering) should be included with H45 and R49 on the roster of absolutely invariant basic residues in PRK. Previous work (Charlier et al., 1994) had suggested that H135, located within PRKs "Walker B" motif, was not crucial to substrate binding or

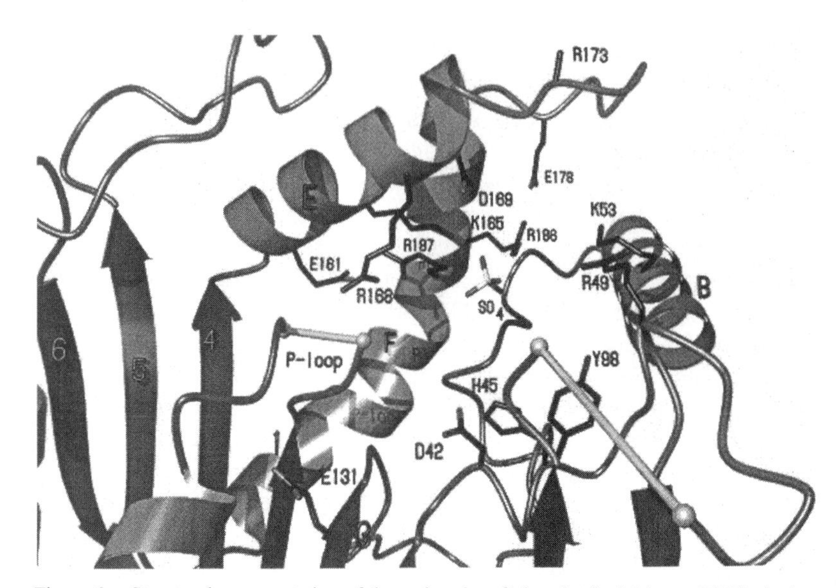

Figure 6. Structural representation of the active site of phosphoribulokinase (PRK). At the base of the figure, the "P-loop," which is highly disordered in this "open" form of the enzyme that lacks any bound substrate, is depicted with a break in the α-carbon backbone. To the right of the β-strand that precedes the "P-loop" is a parallel β-strand that contains as its C-terminal residue the putative Walker B carboxyl group of E131, a candidate for binding cation of the M^{2+}-ATP substrate. Further right is the tip of the edge strand of the parallel β-sheet, which leads to a short loop that contains invariant D42, proposed to function as the general base that deprotonates the C1 hydroxyl of Ru5P. This loop also contains H45, which has been implicated in Ru5P binding. Opposite the β-sheet, at the top and upper right of the figure, are the "lid" helices that contain catalytically essential residues including D169 and several invariant basic residues involved in Ru5P binding (e.g., R49, K165) or catalysis.

catalysis. With the elucidation of a high-resolution structure for PRK (Harrison et al., 1998), it became clear that the remaining invariant basic residues (K165, R168, R173, and R187) are situated in the "lid" region of the nucleotide monophosphate kinase fold (Schlauderer et al., 1996), a region repeatedly demonstrated to form part of the active site. Moreover, K165 is targeted by pyridoxal phosphate upon inactivation of PRK by this reagent (Runquist et al., 1999). This observation directly demonstrates that the "lid" of PRK is part of the active site.

On the basis of the structural rationale and the protein modification data, mutagenesis work was initiated and basicity of each of these residues was eliminated (Runquist et al., 1998, 1999) by the conservative substitutions:

K165M, R168Q, R173Q, R187Q. Additionally, R186, conserved in all prokaryotic PRKs, has also been replaced. Titration of the affinity-isolated mutant proteins with the substrate and effector site probes, trinitrophenyl-ATP and NADH, respectively, indicated that the conservative substitution strategy produced mutant PRKs that were structurally intact, retaining a full complement of substrate and effector sites. Kinetic characterization of these proteins validated the strategy of targeting this "lid" region. While the results indicated that R186 and R187 influence cooperativity of the bacterial enzyme in binding substrate, K165, R168, and R173 were more directly implicated as active site residues. Replacement of K165 depresses k_{cat} by more than 10^3 -fold while K_m values for ATP and Ru5P increase by 10 and 100–fold, respectively. Replacement of R168 diminishes k_{cat} by more than 300-fold; $K_{m\ Ru5P}$ increases by 50–fold. Replacement of R173 diminishes k_{cat} by only 15-fold while $K_{m\ Ru5P}$ increases by 100-fold. Thus, the basic residues in the "lid" of PRK certainly influence both catalytic efficiency and apparent binding of substrate. However, in contrast with the mutagenesis results for H45 and R49, which indicated perturbation of a single kinetic parameter (in these cases, $K_{m\ Ru5P}$), multiple effects were observed upon replacement of K165, R168, and R173. In each case, a significant effect on $K_{m\ Ru5P}$ is detected. Since it seemed unlikely that all five of these basic residues are crucial to Ru5P binding, it became necessary to dissect the K_m perturbations and sort out the contributions of binding parameters from the effects attributable to catalytic constants. The approach employed involved measuring the K_i value of these mutants for a competitive inhibitor, since an inhibitor constant is comprised of only binding parameters. The metabolite selected for these studies is phosphogluconate, which had previously been reported as a competitive inhibitor of eukaryotic PRK (Anderson, 1973; Gardemann et al., 1983) with respect to Ru5P. Elimination of the basic side chain of K165, R49, and H45 results in increases in $K_{i\ phosphogluconate}$, which correlate well with the magnitude of increases in $K_{m\ Ru5P}$ (Runquist et al., 1999). In contrast, mutations at R168 and R173 produce increases in $K_{i\ phosphogluconate}$ that are small in comparison with the significant increases in $K_{m\ Ru5P}$ measured for the mutant PRKs. The k_{cat} perturbations of these mutants may contribute to the changes in their K_m values and, thus, a major role for these arginines in Ru5P binding should not be assumed. In contrast, the strong correlation between K_m and K_i observed for K165, R49, and H45 argues that these three residues are major contributors to Ru5P binding.

The results discussed above implicate as active site residues amino acids situated in two helices and a loop that may be regarded as "lid" elements that are expected to draw closer to the opposing active site boundary comprised of the "P-loop" and β-sheet (Harrison et al., 1998). Are there other active site residues more crucial to catalysis of reaction chemistry? Where do they map? An answer to these questions was provided when, using the precedent generated from structural and mutagenesis investigations of other phosphoryl transfer enzymes (Anderson et al., 1978; Hellinga and Evans, 1987; Berger and Evans, 1992; Green et al., 1993; Hurley et al., 1993), our lab tested the function of PRK's invariant acidic residues (Table 2). Reasoning that such residues could function as a general base catalyst (deprotonating the C1 hydroxyl group of Ru5P) or as a ligand to cation of M^{2+}-ATP, the carboxyl groups of invariant D42, E131, D169, and E178 were conservatively replaced with alanine (Charlier et al., 1994). Studies on the affinity-purified mutants indicate that the kinetic parameters of PRK E178A are not significantly altered. E131A exhibits a 100–fold diminution in k_{cat} and was insensitive to NADH stimulation. In sharp contrast, D42A and D169A exhibit 10^5- and 10^4-fold diminutions, respectively, in k_{cat}. The magnitude of these changes in catalytic efficiency contrasts with the small (less than 10–fold) changes in K_m values observed for these mutants. The issue of the structural integrity of D42A and D169A becomes important for enzymes with such a drop in catalytic efficiency. Initially, this concern was diminished by the observation that these mutants' K_m values are not strongly perturbed (Charlier et al., 1994). Additionally, D42A and D169A have been evaluated by titration with the spin-labeled ATP analog, *S*-acetamidoproxyl-ATPγS or with the fluorescent alternative substrate, trinitrophenyl-ATP (Runquist et al.,

TABLE 2
PRK's Catalytically Essential Invariant Acidic Amino Acids

	57	125	160
Ice plant prk	CL**D**DYHSLDRTGRK	ILVI**E**GLH	KIQR**D**MAERGHSL
Pea prk	CL**D**DYHSLDRTGRK	ILVI**E**GLH	KIQR**D**MAERGHSL
Arabidopsis prk	CL**D**DYHSLDRYGRK	ILVI**E**GLH	KIQR**D**MAERGHSL
Wheat prk	CL**D**DYHSLDRTGRK	IFVI**E**GLH	KIQR**D**MAERGHSL
Spinach prk	CL**D**DFHSLDRNGRK	ILVI**E**GLH	KIQR**D**MKERGHSL
Chlamydomonas prk	CL**D**DYHCLDRNGRK	ILVI**E**GLH	KIQR**D**MAERGHSL
Synechocystis prk	CL**D**DYHSLDRQGRK	VVVI**E**GLH	KIQR**D**MAERGHTY
R. capsulata prk	..**D**AFHRFNRADMK	LLFY**E**GLH	KIHR**D**RAQRGYTT
R. sphaeroides prk	..**D**AFHRFNRADMK	LLFY**E**GLH	KIHR**D**RATRGYTT
A. eutrophus prk	..**D**SFHRYDRAEMK	LLFY**E**GLH	KLWR**D**KKQRGYST
N. vulgaris prk	..**D**AFHRYNRAEMR	LLFY**E**GLH	KLHR**D**RNARGYST
X. flavus prk	..**D**SFHRYDRYEMR	ILFY**E**GLH	KIHR**D**KATRGYTT
	42	131	169

1996). The results indicate that these mutants retain the ability to tightly and stoichiometrically bind these analogs at the ATP site, arguing convincingly for their overall structural integrity. Thus, the kinetic characterization data for D42 and D169 can be straightforwardly interpreted and these residues can be assigned a major function in reaction chemistry.

The two orders of magnitude effect on k_{cat} measured for E131 also requires explanation. While there is currently no high resolution active site structure of PRK containing bound substrates or analogs, a cautious interpretation of the kinetic analyses in the context of the available structural information as well as recognized structural motifs is useful in producing specific functional hypotheses. Phosphotransferases typically use two amino acid side chains to bind cation of the M^{2+}-ATP substrate (Smith and Rayment, 1996). At least one of these is commonly the carboxyl group of an acidic residue. When a Walker B motif is present in a phosphotransferase, it typically contains an acidic amino acid that contributes such a carboxyl group. In PRK, immediately upstream of the invariant EGLH sequence (residues 131–135) appear four hydrophobic amino acids. While these residues are not invariant, they are invariably hydrophobic. The appearance of an invariant acidic residue after the hydrophobic sequence qualifies this region (Table 3) for consideration as a Walker B motif and E131 for assignment as a ligand to M^{2+}-ATP. The second cation ligand remains to be assigned, but work in progress will determine whether an alcohol side chain of a “P-loop” residue, well-precedented to function as a cation ligand, does indeed fulfill such a function in PRK.

The five orders of magnitude k_{cat} effect observed for substitution of D42 commends this residue for consideration as general base catalyst. The posi-

TABLE 3
Identification of PRK’s Walker B Motif:
Implication of E131 in Cation Binding

Arabidopsis prk	ILVI	**E**GLH
Wheat prk	IFVI	**E**GLH
Ice plant prk	ILVI	**E**GLH
Spinach prk	ILVI	**E**GLH
Pea prk	ILVI	**E**GLH
Chlamydomonas prk	ILVI	**E**GLH
Synechocystis prk	VVVI	**E**GLH
R. sphaeroides prk	LLFY	**E**GLH
A. eutrophus prk	LLFY	**E**GLH
N. vulgaris prk	LLFY	**E**GLH
X. flavus prk	ILFY	**E**GLH
R. capsulata prk	LLFY	**E**GLH

tioning of this residue near the tip of strand that forms the edge of the β-sheet supports this functional assignment, which requires the carboxyl to deprotonate Ru5P's C1 hydroxyl and facilitate attack on ATP's γ-phosphoryl. The tip of the adjacent β-strand harbors E131, postulated to anchor the γ-phosphoryl via M^{2+} ligation. Not only is ATP's phosphoryl chain likely to be positioned adjacent to D42, but appropriate positioning of Ru5P is also likely. D42 is situated in a short flexible stretch that links the last strand of the β-sheet with the α-helix that begins with R49. This basic residue, as well as histidine-45 (which also maps within the flexible linker), anchor Ru5P binding (Sandbaken et al., 1992; Runquist et al., 1999). Since R49 is likely to interact with Ru5P's C5 phosphoryl, this substrate's C1 can easily be juxtaposed to D42.

In the absence of any structure depicting an occupied PRK active site, a hypothesis concerning the function of D169 becomes tenuous. This residue has a major impact on catalysis and, prior to recognition of a Walker B motif, was tentatively proposed as contibuting a ligand to M^{2+}-ATP. While such a function for D169 cannot be ruled out, the magnitude of the diminution in k_{cat} upon its replacement is quite large in comparison with effects commonly observed upon elimination of such cation ligating residues in related enzymes (Berger and Evans, 1992; Rider et al., 1994; Hasemann et al., 1996). The most plausible speculation concerning D169's function derives from the observation that it forms a salt bridge to K165. Replacement of either residue diminishes k_{cat} by three to four orders of magnitude. This might be consistent with the need for D169 to precisely orient K165 as that residue's ε-amino group guides transfer of ATP's γ-phosphoryl to Ru5P's C1 hydroxyl. Such a function appears likely for the ε-amino group of a comparable "lid helix" lysine of fructose 6–phosphate 2–kinase (Hasemann et al., 1996), which adopts a fold similar to PRK. Future structural work on substrate*PRK complexes will be required to test this speculation concerning D169 and K165, as well as to elucidate the roles of other active site residues.

D. PRK FUNCTIONAL ASSIGNMENTS: POSSIBLE ANALOGIES IN HOMOLOGOUS ENZYME REACTIONS

The recent elucidation of high-resolution structures for a series of enzymes that are structurally homologous to PRK and catalyze, in some respects, homologous reactions reinforces some of the functional assignments or speculation listed above. Furthermore, this collected body of information prompts the formulation of a consensus model (Fig. 7) based on

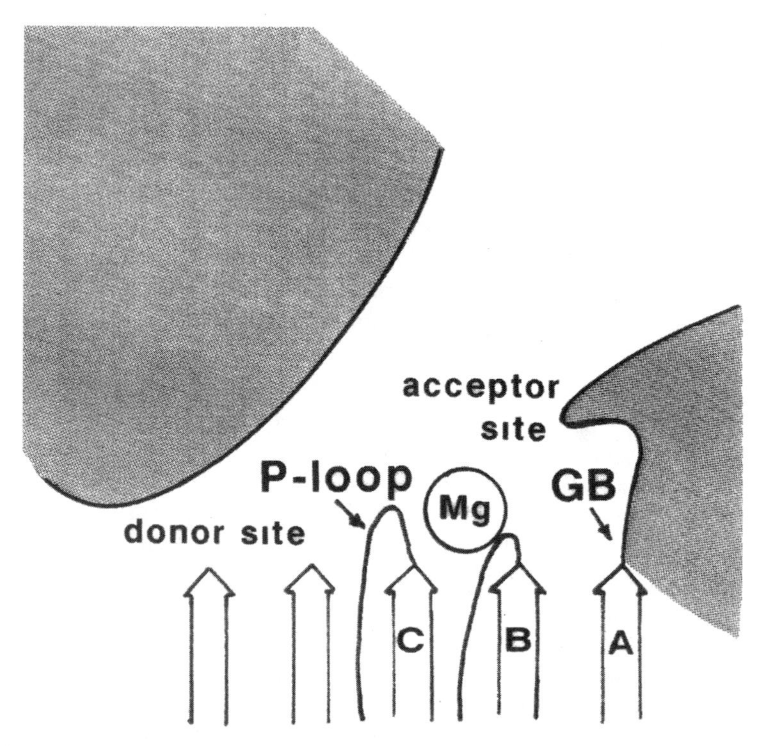

Figure 7. Consensus model based on structure/function correlations observed for a group of nucleoside monophosphate (NMP) kinase fold enzymes that catalyze phospho- or sulfo-transferase reactions. Shaded regions correspond to "lid" elements of the active site, which constitute the active site boundary opposite the boundary edge formed by the C-terminal ends of the strands of the parallel β-sheet. These boundaries are well separated in the "open" form of these enzymes in which no substrates occupy the active site. In the "closed" form of the enzyme, produced upon occupancy of the adenosine triphposphate (ATP) phosphoryl donor and a metabolite phosphoryl acceptor, the "lid" and C-terminal edge of the parallel β-stand become closely juxtaposed. The Walker A or "P-loop" motif is shown situated off the tip of β-strand "C." The Walker B motif, containing an acidic residue that functions as a ligand to M^{++}-ATP, is typically found around the tip of adjacent β-strand "B" in this group of enzymes. Where identified, the general base (GB) catalyst has mapped near the tip of β-strand "A."

structure/function correlations observed across this group of enzymes. PRK, estrogen sulfotransferase, fructose 6-phosphate 2-kinase, thymidine kinase, and shikimate kinase catalyze metabolically unrelated but mechanistically similar phospho- or sulfo-transferase reactions involving a variety of alcohol substrates. All of these enzymes exhibit the NMP kinase fold. In all cases, a "P-loop" or Walker A motif is found at the C-terminal end of the third parallel strand of the β-sheet (Hasemann, et al., 1996; Wild et al., 1997; Harrison et al., 1998; Krell et al., 1998). Additionally, a "Walker B" acidic residue at the C-terminal tip of the second parallel strand has been assigned for several of these enzymes (Hasemann et al., 1996; Wild et al., 1997; Harrison et al., 1998) and a candidate residue for this assignment could be identified in other cases (Krell et al., 1998). Finally, while NMP kinases may require no general base catalysis of their reactions, the transferases listed above are likely to require such assistance to facilitate deprotonation of the alcohol groups of their acceptor substrates. In the sequence following the C-terminus of the first parallel strand of these enzymes, a general base catalyst candidate commonly appears. A clear functional assignment is most obvious in the case of PRK (Charlier et al., 1994; Harrison et al., 1998) but persuasive assignments, based on limited published mutagenesis results, can also easily be made for thymidine kinase (Wild et al., 1997) and estrogen sulfotransferase (Kakuta et al., 1997). Candidate residues are also suitably located in shikimate kinase (Krell et al., 1998). While no assignment of a general base has been offered for fructose 6–phosphate 2–kinase (Hasemann et al., 1996; Uyeda et al., 1997), the general base consensus emerging from these mechanistically related enzymes argues that this issue merits further scrutiny.

In summary, a combination of structural work and mechanistic investigation allows functional assignments and predictions for both prokaryotic and eukaryotic PRKs. Moreover, when the PRK results are interpreted in the context of the database available for other enzymes that exhibit the NMP kinase fold, a consensus emerges that suggests functional assignments not only for PRK but also for a wider array of enzymes that catalyze diverse transferase reactions.

Acknowledgments

PRK studies from the author's laboratory have been supported by the U.S. Department of Agriculture's National Research Initiative Competitive Research Grants Program (Photosynthesis/Photorespiration). The author acknowledges Dr. Jennifer Runquist's extensive contributions to the prokaryotic PRK work that our laboratory has produced. My collaborator, Dr. David H.T. Harrison, generously supported my preparation of this review by producing the figures that are based on his high resolution structures of *R. sphaeroides* PRK.

References

Abdelal ATH, Schlegel HG (1974): *Biochem J* 139: 481–489.

Anderson L (1973): *Biochim Biophys Acta* 321: 484–488.

Anderson CM, Stenkamp RE, McDonald RC, Steitz TA (1978): *J Mol Biol* 123: 207–219.

Ashton AR (1984): *Biochem J* 217: 79–84.

Avilan L, Gontero B, Lebreton S, Ricard J (1997a): *Eur J Biochem* 246: 78–84.

Avilan L, Gontero B, Lebreton S, Ricard J (1997b): *Eur J Biochem* 250: 296–302.

Ballard RW, MacElroy RD (1971): *Biochem Biophys Res Comm* 44: 614–618.

Banks FM, Driscoll SP, Parry MAJ, Lawlor DW, Knight JS, Gray JC, Paul MJ (1999): *Plant Physiol* 119: 1125–1136.

Berger SA, Evans PR (1992): *Biochemistry* 31: 9237–9242.

Brandes HK, Stringer CD, Hartman FC (1992): *Biochemistry* 31: 12833–12838.

Brandes HK, Larimer FW, Geck MK, Stringer CD, Schurmann P, Hartman FC (1993): *J Biol Chem* 268: 18411–18414.

Brandes HK, Larimer FW, Hartman FC (1996a): *J Biol Chem* 271: 3333–3335.

Brandes HK, Hartman FC, Lu TYS, Larimer FW (1996b): *J Biol Chem* 271: 6490–6496.

Brandes HK, Larimer FW, Lu TYS, Dey J, Hartman FC (1998): *Arch Biochem Biophys* 352: 130–136.

Buchanan BB (1980): *Ann Rev Plant Physiol* 31: 341–374.

Charlier HA, Runquist JA, Miziorko HM (1994): *Biochemistry* 33:9343–9350.

Clasper S, Easterby JS, and Powls R (1991): *Eur J Biochem* 202: 1239–1246.

Dietz KJ, Heber U (1984): *Biochim Biophys Acta* 767: 432–443.

Gardemann A, Stitt M, Heldt HW (1983): *Biochim Biophys Acta,* 722: 51–60.

Gibson JL, Tabita FR (1987): *J Bacteriol* 169: 3685–3690.

Gibson JL, Tabita FR (1996): Arch Microbiol 166: 141–150.

Gibson JL, Chen JH, Tower PA, Tabita FR (1990): *Biochemistry* 29: 8085–8093.

Gibson JL, Falcone DL, Tabita FR (1991): *J Biol Chem* 266: 14646–14653.

Gontero B, Cardenas ML, Ricard J (1988): *Eur J Biochem* 173: 437–443.

Gontero B, Mulliert G, Rault M, Giudici-Orticoni MT, RIcard J (1993): *Eur J Biochem* 217: 1075–1082.

Green PC, Tripathi RL, Kemp RG (1993): *J Biol Chem* 268: 5085–5088.

Hallenbeck PL, Kaplan S (1987): *J Bacteriol* 169: 3669–3678.

Hariharan T, Johnson PJ, Cattolico RA (1998): *Plant Physiol* 117: 321–329.

Harrison DHT, Runquist JA, Holub A, Miziorko HM (1998): *Biochemistry* 37: 5074–5085.

Hart BA, Gibson J (1971): *Arch Biochem Biophys* 144: 308–321.

Hasemann CA, Istvan ES, Uyeda K, Deisenhofer J (1996): *Structure* 4: 1017–1029.

Hellinga HW, Evans PR (1987): *Nature* 327: 437–439.

Horsnell PR, Raines CA (1991): *Plant Mol Bio* 17: 183–184.

Hudson GS, Morell MK, Arvidsson YBC, Andrews TJ (1992): *Aust J Plant Physiol* 19: 213–221.

Hurley JH, Faber HR, Worthylake D, Meadow ND, Roseman S, Pettigrew DW, Remington SJ (1993): *Science* 259: 673–677.

Hurwitz J, Weissbach A, Horecker BL, Smyrniotis PZ (1956): *J Biol Chem* 218: 769–783.

Hurwitz J (1962): *Methods Enzymol* 5: 258–261.

Kagawa T (1982): In "*Methods in Chloroplast Biology*." (Edelman E, Hallick RB, Chua NH (eds). New York: Elsevier, pp. 695–705.

Kakuta Y, Pedersen LG, Carter CW, Negishi M, Pedersen LC (1997): *Nat Struct Biol* 4: 904–908.

Kiesow L, Lindsley BF, Bless JW (1977): *J Bacteriol* 130: 20–25.

Klintworth R, Husemann J, Salnikow J, Bowien B (1985): *J Bacteriol* 164: 954–956.

Kossmann J, Klintworth R, Bowien B (1989): *Gene* 85: 247–252.

Koteiche HA, Narasimhan C, Runquist JA, Miziorko HM (1995): *Biochemistry* 34: 15068–15074.

Krell T, Coggins JR, Lapthorn AJ (1998): *J Mol Biol* 278: 983–997.

Krieger TJ, Miziorko HM (1986): *Biochemistry* 25: 3496–3501.

Krieger TJ, Mende-Mueller L, Miziorko HM (1987): *Biochim Biophys Acta* 915: 112–119.

Kung G, Runquist JA, Miziorko HM, Harrison DHT (1999): Biochemistry (in press).

Latzko E, Garnier RV, Gibbs M (1970): *Biochem Biophys Res Comm* 39: 1140–1144.

Lavergne D, Bismuth E (1973): *Plant Sci Lett* 1: 229–236.

Lebreton S, Gontero B, Avilan L, Ricard J (1997a): *Eur J Biochem* 246: 85–91.

Lebreton S, Gontero B, Avilan L, Ricard J (1997b): *Eur J Biochem* 250: 286–295.

MacElroy RD, Johnson EJ, Johnson MK (1969): *Arch Biochem Biophys* 131: 272–275.

MacElroy RD, Mack HM, Johnson EJ (1972): *J Bacteriol* 112: 532–538.

Marsden WJN, Codd GA (1984): *J Gen Microbiol* 130: 999–1006.

Meijer WG, Enequist HG, Terpstra P, Dijkhuizen L (1990): *J Gen Microbiol* 136: 2225–2230.

Michalowski CB, Derocher EJ, Bohnert HJ, Salvucci ME (1992): *Photosyn Res* 31: 127–138.

Milanez S, Mural RJ (1988): *Gene* 66: 55–63.

Milanez S, Mural RJ, Hartman FC (1991): *J Biol Chem* 266: 10694–10699.

Miziorko HM, Eckstein F (1984): *J Biol Chem* 259: 13037–13040.

Miziorko HM, Brodt CA, Krieger TJ (1990): *J Biol Chem* 265: 3642–3647.

Mueller-Dieckmann HJ, Schulz GE (1994): *J Mol Biol* 236: 361–367.

Mural RJ, Lu TY, Hartman FC (1993): *J Protein Sci* 12: 207–213.

Narasimhan C, Miziorko HM (1997): *Adv Biophys Chem* 6: 67–110.

Nicholson S, Easterby JS, Powls R (1987): *Eur J Biochem* 162: 423–431.

Novak JS, Tabita FR (1999): Arch. Biochem. Biophys. 363: 273–282.

Omnaas J, Porter MA, Hartman FC (1985): *Arch Biochem Biophys* 236: 646–653.

Paoli GC, Vichivanives P, Tabita FR (1995): *J Bacteriol* 180: 4258–4269.

Porter MA, Hartman FC (1986): *Biochemistry* 25: 7314–7318.

Porter MA, Hartman FC (1988): *J Biol Chem* 263:14846–14849.

Porter MA, Milanez S, Stringer CD, Hartman FC (1986): *Arch Biochem Biophys* 245: 14–23.

Porter MA, Stringer CD, Hartman FC (1988): *J Biol Chem* 263: 123–129.

Porter MA, Hartman FC (1990): *Arch Biochem Biophys* 281: 330–334.

Potter D, Miziorko HM (1997): *J Biol Chem* 272: 25449–25454.

Racker E (1957): *Arch Biochem Biophys* 69: 300–310.

Raines CA, Longstaff M, Lloyd JC, Dyer TA (1989): *Mol Gen Genet* 220: 43–48.

Rault M, Gontero B, Ricard J (1991): *Eur J Biochem* 197: 791–797.

Rault M, Giudici-Orticoni MT, Gontero B, Ricard J (1993): *Eur J Biochem* 217: 1065–1073.

Rider MH, Crepin KM, DeCloedt M, Bertrand L, Hue L (1994): *Biochem J* 300: 111–115.

Rindt KP, Ohmann E (1969): *Biochem Biophys Res Comm* 36: 357–364.

Rippel S, Bowien B (1984): *Arch Microbiol* 139: 207–212.

Roberts DL, Runquist JA, Miziorko HM, Kim JJP (1995): *Protein Sci* 4: 2442–2443.

Roesler KR, Ogren WL (1988): *Nucleic Acids Res* 16: 7192.

Roesler KR, Ogren WL (1990): *Plant Physiol* 93: 188–193.

Roesler KR, Marcotte BL, Ogren WL (1992): *Plant Physiol* 98: 1285–1289.

Runquist JA, Narasimhan C, Wolff CE, Koteiche HA, Miziorko HM (1996): *Biochemistry* 35: 15049–15056.

Runquist JA, Harrison DHT, Miziorko HM (1998): *Biochemistry* 37: 1221–1226.

Runquist JA, Harrison DHT, Miziorko HM (1999): Biochemistry 38: (in press).

Sainis JK, Harris CG (1986): *Biochem Biophys Res Comm* 139: 947–954.

Sandbaken MG, Runquist JA, Barbieri JT, Miziorko HM (1992): *Biochemistry* 31: 3715–3719.

Saraste M, Sibbald PR, Wittinghofer A (1990): *TIBS* 15: 430–434.

Schlauderer GJ, Proba JA, Schulz GE (1996): *J Mol Biol* 256: 223–227.

Siebert K, Bowien B (1984): *Biochim Biophys Acta* 787: 208–214.

Siebert K, Schobert P, Bowien B (1981): *Biochim Biophys Acta* 658: 35–44.

Smith C, Rayment I (1996): *Biophys J* 70: 1590–1602.

Strecker M, Sickinger E, English RS, Shively JM, Bock E (1994): *FEMS Microbiol Lett* 120: 45–50.

Su X, Bogarad L (1991): *J Biol Chem* 266: 23698–23705.

Tabita FR (1980): *J Bacteriol* 143: 1275–1280.

Tabita FR (1988): *Microbiol Rev* 52: 155–189.

Uyeda K, Wang XL, Mizuguchi H, Li Y, Nguyen C, Hasemann CA (1997): *J Biol Chem* 272: 7867–7872.

Wadano A, Nishikawa K, Hirahashi T, Satoh R, Iwaki T (1998): *Photosyn Res* 56: 27–33.

Wedel N, Soll J (1998): *Proc Natl Acad Sci USA* 95: 9699–9704.

Wedel N, Soll J, Paap BK (1997): *Proc Natl Acad Sci USA* 94: 10479–10484.

Weissbach A, Smyrniotis PZ, Horecker BL (1954): *J Am Chem Soc* 76: 5572–5573.

Wild K, Bohner T, Folkers G, Schulz GE (1997): *Protein Sci* 6: 2097–2106.

Wolosiuk RA, Buchanan BB (1978): *Arch Biochem Biophys* 189: 97–101.

Wolosiuk RA, Crawford NA, Yee BC, Buchanan BB (1979): *J Biol Chem* 254: 1627–1632.

THE MOLECULAR EVOLUTION OF PYRIDOXAL-5′-PHOSPHATE-DEPENDENT ENZYMES

By PERDEEP K. MEHTA and PHILIPP CHRISTEN,
Biochemisches Institut der Universität Zürich, CH-8057 Zürich, Switzerland

CONTENTS

Advances in Enzymology and Related Areas of Molecular Biology, Volume 74: Mechanism of Enzyme Action, Part B, Edited by Daniel L. Purich
ISBN 0-471-34921-6

I. Introduction

The large and diverse group of pyridoxal-5′-phosphate(PLP)-dependent enzymes offers a unique opportunity to reconstruct the at least 3000-million-year-long course of evolution of enzyme mechanisms starting from the primordial protein catalysts in the first cells and leading up to the modern enzymes with defined reaction and substrate specificities. This review focuses on PLP-dependent enzymes that catalyze reactions in the metabolism of amino acids. Such enzymes are found in all organisms. In a multitude of reactions they synthesize, degrade, and interconvert amino acids. B_6 enzymes are essential in linking the carbon and nitrogen metabolism, replenishing the pool of one-carbon units, and forming biogenic amines. PLP is easily the most versatile coenzyme (Jencks, 1969). The diversity of the B_6 enzymes acting on amino acid substrates is evident from the fact that they belong to no fewer than five of the total six enzyme classes as defined by the Enzyme Nomenclature Committee of the International Union of Biochemistry and Molecular Biology (Webb, 1992). PLP is a derivative of vitamin B_6; henceforth, "B_6 enzymes" is used as a convenient abbreviation for PLP-dependent enzymes.

All B_6 enzymes acting on amino acids as substrates share important mechanistic features. Invariably, PLP is bound covalently via an imine bond to the ε-amino group of an active-site lysine residue, forming the so-called "internal" aldimine. In the first covalency change, the amino group of the incoming substrate replaces the ε-amino group forming a coenzyme-substrate imine. This "external" aldimine intermediate is common to all enzymic and nonenzymic reactions of PLP with amino acids. In the next step, the different reaction pathways diverge, depending on which bond at Cα is cleaved and on the following covalency changes, one of the diverse reactions catalyzed by B_6 enzymes becomes realized (Fig. 1). In all reactions, the coenzyme acts as an electron sink, storing electrons from cleaved substrate bonds and dispensing them for the formation of new linkages with incoming protons or second substrates. All reactions catalyzed by B_6 enzymes are assumed to take place also very slowly with amino acid substrates and PLP alone. It is the protein moiety of a given B_6 enzyme that determines which of the many potential pathways is adopted by the coenzyme-substrate adduct. PLP and pyridoxamine-5′-phosphate serve also as cofactors in a particular group of bacterial enzymes that participate in the biosynthesis of dideoxy and deoxyamine sugars (Rubinstein and Strominger, 1974) and are evolutionarily related with the B_6 enzymes in amino acid metabolism. The role of PLP in glucan phosphorylases (Palm et al., 1990) is entirely different; these enzymes constitute another independent evolutionary lineage that will not be discussed here in detail.

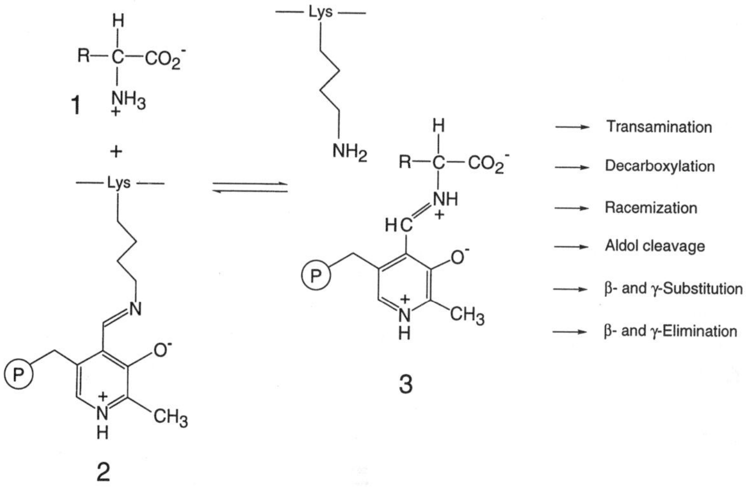

Figure 1. Transformations of amino acids by PLP-dependent enzymes. **(1)** Amino acid substrate. **(2)** "Internal" aldimine intermediate in which PLP has formed a Schiff base with the ε-amino group of an active-site lysine residue. **(3)** "External" aldimine intermediate. In nonenzymic reactions, the aldimine intermediate is produced by *de novo* formation from PLP and the amino acid. After the aldimine intermediate, the pathways of the various PLP-dependent reactions diverge. Formation of the planar pyridine-aldimine Schiff base **3** is a prerequisite for the catalytic efficacy of PLP, which is due to the electron-withdrawing effect exerted on Cα by the positively charged imine group and pyridine nitrogen atom and is mediated through the extensive resonance system of the pyridine ring and the imine double bond.

A competent brief introduction to the metabolic and mechanistic aspects of B_6 enzymes has been provided by Metzler (1977). For a comprehensive review on B_6 enzymes covering their physiological, structural, and mechanistic aspects, see the two-volume treatise edited by Dolphin et al. (1986); limited to the aminotransferases is the treatise edited by Christen and Metzler (1985). Newer reviews on B_6 enzymes are those by Hayashi et al. (1990), John (1995), and most recently by Jansonius (1998). The stereochemical aspects of B_6 enzymes are discussed in a review by Vederas and Floss (1980).

This review presents the current knowledge on the evolutionary emergence and development of the apoenzyme moieties of B_6 enzymes. A considerable increase in pertinent data on B_6 enzymes, that is, in the number of available amino acid sequences and three-dimensional (3-D) structures, as well as progress in the methodology of homology searches have contributed to make a comprehensive, though not quite definitive, account on the molecular evolution of B_6 enzymes feasible. The study summarizes earlier reports and complements them with results obtained on the basis of recently published structural data and with improved computational procedures (Mehta et al., 1999). The B_6 enzymes clearly are of multiple evolutionary origin. All B_6 enzymes of which the amino acid sequences are known to date belong to one of four mutually unrelated families of homologous proteins, the α family comprising by far the largest number of member enzymes. The ancestor proteins of these families apparently were regio-specific proto-enzymes catalyzing reactions with covalency changes limited to Cα or extending to Cβ. From regio-specific protein catalysts reaction-specific ancestor enzymes (e.g., aminotransferases, amino acid decarboxylases, etc.) developed, which then diverged further into the modern substrate-specific enzymes (e.g., aspartate aminotransferase, ornithine aminotransferase, etc.). For most B_6 enzymes, this development appears to have been already completed in the last universal ancestor cell before the three biological kingdoms of archebacteria, eubacteria, and eukaryotes diverged; in general, the evolution of B_6 enzymes may thus be explored independently of phylogenetic divergence. The course of evolution of a given B_6 enzyme family will thus be traceable by comparing the homologous, or more precisely paralogous, sequences of enzymes with different functions, e.g., serine hydroxymethyltransferases with 1-aminocyclopropane-1-carboxylate synthases, regardless of the species that they come from. In evolutionary studies, the generic term "homologous" is often replaced either by "paralogous" to indicate homologous nucleotide or amino acid sequences of different functions or by "orthologous" to denote homologous sequences with the same or similar function in different species (Fitch, 1970). Comparison of paralogous sequences elucidates the evolutionary relationship between related genes or proteins whereas comparison of orthologous sequences traces the phylogenetic relationship of the species concerned.

II. Databases and Methodology

A. SEQUENCE AND 3-D STRUCTURE DATABASES

The available database of the structures of B_6 enzymes is growing rapidly; 23 spatial structures and more than 600 amino acid sequences of more than 60 B_6 enzymes are now known. For our analyses of sequences, the SWISS-PROT database (Release 37 of April 1999 with a total of 70,000 sequences; Bairoch and Boeckmann, 1991), supplemented with sequences of B_6 enzymes available from TrEMBL, was used. All enzymes that were included in the present study are listed in Table 1. The structural data are, however, dis-

TABLE 1
Structural Data on Families of PLP-Dependent Enzymes that Act on Amino Acid Substrates

EC Number	Enzyme	3-D Structure PDB Code[a]	Number of Sequences
	α Family		
1.4.4.2	Glycine dehydrogenase, decarboxylating		14
2.1.2.1	Glycine (serine) hydroxymethyltransferase	1BJ4[b]	37
2.3.1.29	2-Amino-3-ketobutyrate CoA ligase		1
2.3.1.37	5-Aminolevulinate synthase		18
2.3.1.47	8-Amino-7-oxononanoate synthase	–[c]	9
2.3.1.50	Serine palmitoyltransferase		4
2.6.1.1	Aspartate aminotransferase	1AAW[d]	43
2.6.1.2	Alanine aminotransferase		7
2.6.1.5	Tyrosine aminotransferase		4
2.6.1.9	Histidinol-phosphate aminotransferase		18
2.6.1.11	Acetylornithine aminotransferase		13
2.6.1.13	Ornithine aminotransferase	1OAT[e]	9
2.6.1.18	β-Alanine-pyruvate aminotransferase	–[f]	1
2.6.1.19	4-Aminobutyrate aminotransferase	1GTX[g]	15
2.6.1.36	Lysine-ε-aminotransferase		2
2.6.1.44	Alanine glyoxylate aminotransferase		1
2.6.1.45	Serine-glyoxylate aminotransferase		2
2.6.1.46	2,4-Diaminobutanoate-pyruvate aminotransferase		2
2.6.1.51	Serine aminotransferase		5
2.6.1.52	Phosphoserine aminotransferase	1BJN[h]	18
2.6.1.57	Aromatic-amino-acid aminotransferase	1AY4[i]	3
2.6.1.62	AdoMet 8-amino-7-oxononanoate aminotransferase		12
2.6.1.64	Glutamine-phenylpyruvate aminotransferase		1
2.6.1.66	Valine-pyruvate aminotransferase		1
2.6.1.–	Succinylornithine aminotransferase		1
2.6.1.–	DNTP-hexose aminotransferase		1
2.6.1.–	*malY* gene product		2
2.9.1.1	L-Seryl-tRNA(Sec) selenium transferase		4
3.7.1.3	Kynureninase		4
4.1.1.15	Glutamate decarboxylase		25
4.1.1.17	Ornithine decarboxylase (prokaryotic)	1ORD[j]	21
4.1.1.18	Lysine decarboxylase		5
4.1.1.19	Arginine decarboxylase (biodegradative)		1
4.1.1.22	Histidine decarboxylase		8
4.1.1.28	Tyrosine/DOPA decarboxylase		17
4.1.1.64	2,2-Dialkylglycine decarboxylase	1DGD[k]	1
4.1.2.5	Threonine aldolase		4
4.1.99.1	Tryptophan indole-lyase (tryptophanase)	1AX4[l]	5
4.1.99.2	Tyrosine phenol-lyase	1TPL[m]	4

(*Continued*)

TABLE 1 (*continued*)
Structural Data on Families of PLP-Dependent Enzymes that Act on Amino Acid Substrates

EC Number	Enzyme	3-D Structure PDB Code[a]	Number of Sequences
4.2.99.9	Cystathionine γ-synthase	1CS1[n]	9
4.2.99.–	*O*-Succinylhomoserine sulfhydrylase		2
4.4.1.1	Cystathionine γ-lyase		4
4.4.1.4	Alliinase		4
4.4.1.8	Cystathionine β-lyase	1CL1[o]	6
4.4.1.11	Methionine γ-lyase		2
4.4.1.14	1-Aminocyclopropane-1-carboxylate synthase	1B8G[p]	44
5.4.3.8	Glutamate-1-semialdehyde 2,1-aminomutase	2GSA[q]	19
5.–.–.–	Isopenicillin *N* epimerase		2
–.–.–.–	*cobC* gene product		2
–.–.–.–	Cysteine desulfurase (*nifS* gene product)		15
–.–.–.–	tRNA splicing protein SPL1		4
–.–.–.–	Rhizopine catabolism regulatory protein		1
	β Family		
4.1.99.4	1-Aminocyclopropane-1-carboxylate deaminase		5
4.2.1.13	L-Serine dehydratase		4
4.2.1.14	D-Serine dehydratase		2
4.2.1.16	Threonine dehydratase	1TDJ[r]	10
4.2.1.20	Tryptophan synthase β chain	1BKS[s]	32
4.2.1.22	Cystathionine β-synthase		4
4.2.99.2	Threonine synthase		9
4.2.99.8	Cysteine synthase (*O*-acetylserine sulfhydrylase)	1OAS[t]	23
4.3.1.15	Diaminopropionate ammonia-lyase		1
	D-Alanine aminotransferase family		
2.6.1.21	D-Alanine aminotransferase	1DAA[u]	5
2.6.1.42	Branched-chain amino acid aminotransferase	1A3G[v]	22
4.–.–.–	4-Amino-4-deoxychorismate lyase		3
	Alanine racemase family (β/α-barrel proteins)		
4.1.1.17	Ornithine decarboxylase (eukaryotic)	7ODC[w]	20
4.1.1.19	Arginine decarboxylase (eukaryotic and *E. coli*)		9
4.1.1.20	Diaminopimelate decarboxylase		11
5.1.1.1	Alanine racemase	1BDO[x]	16

[a]Bernstein et al., 1977; [b]Renwick et al., 1998; Scarsdale et al., 1999; the recommended name for this enzyme is glycine hydroxymethyltransferase (Webb, 1992) and will be used henceforth; [c]Alexeev et al., 1998; [d]Ford et al., 1980; McPhalen et al., 1992; Okamoto et al., 1994; [e]Shen et al., 1998; [f]Watanabe et al., 1989; [g]Storici et al., 1999; [h]Hester et al., 1999; [i]Okamoto et al., 1998; [j]Momany et al., 1995a; [k]Toney et al., 1993; [l]Isupov et al., 1998; [m]Antson et al., 1993; [n]Clausen et al., 1998; [o]Clausen et al., 1996; [p]Capitani et al.; [q]Hennig et al., 1997; [r]Gallagher et al., 1998; [s]Hyde et al., 1988; [t]Burkhard et al., 1998; [u]Sugio et al., 1995; [v]Okada et al., 1997; [w]Kern et al., 1999; [x]Shaw et al., 1997.

tributed somewhat uneven among the known B_6 enzymes: Of the total 114 B_6 enzymes listed with EC numbers (Webb, 1992) about 50 enzymes have as yet not been structurally characterized at all. In contrast, for some enzymes, for example, aspartate aminotransferase, the sequences of many source variants have been reported. Determination of sequences and 3-D structures in the as yet uncharted areas of B_6 enzymes seems highly desirable.

B. PROFILE ANALYSIS

In view of the often quite low degrees of identity between the paralogous sequences that were compared, profile analysis of sequences (Gribskov et al., 1990) was an important tool in our study. A set of aligned homologous amino acid sequences, the so-called probe sequences, and an amino acid substitution matrix serve—with the help of the program PROFILEMAKE in the UWGCG package (Devereux et al., 1984)—to construct the profile that is a position-specific scoring table. The profile is then used to screen the database with the program PROFILESEARCH for sequences that are related with the probe sequences. The profile will attribute a certain score value to every position of the target sequence depending on what amino acid residue is found there. For deletions and insertions in the target sequences variable penalties are given. The overall score of the target sequence corresponds to the algebraic sum of the score values attributed to all its residues, gaps, and insertions. The so-called Zscore corresponds to the difference between the score of the target sequence and the mean score of the unrelated sequences in the database expressed in standard deviations. The mean Zscore of the sequences unrelated to the profile thus per definition is zero and the standard deviation is 1. A Zscore value higher than 3 standard deviations may indicate a relationship, a value higher than 6 corresponds to a definitive relationship between the target sequence and the probe sequences (Gribskov, 1992). Profile analysis is uniquely suited to detect distant relationships because the profile incorporates position-specific information on invariant and variant regions in the probe sequences. It can detect and unequivocally verify homology between proteins even if their relationship is not evident from the low degree of sequence identity. However, the discriminatory power of profile analysis does not reach as far as that of a comparison of the folding patterns of the polypeptide chains.

C. FAMILY PROFILE ANALYSIS

Family profile analysis (FPA), a newly developed algorithm (Mehta et al., 1999), was used for the verification of very distant relationships between protein families as well as for estimating the relative distance between different branches of the evolutionary pedigrees of B_6 enzymes. FPA compares all available homologous amino acid sequences of a target family with the profile of a probe family while conventional sequence profile analysis (see above) considers only a single target sequence in comparison with the probe family (Fig. 2). The increased input of sequence information in FPA expands the range for sequence-based recognition of structural relationship. In the FPA algorithm, Zscores of each of the target sequences, obtained from a probe profile search over all known amino acid sequences of the target family, are averaged and then compared with the mean scores for sequences of 100 reference families in the same probe family search. The resulting F-Zscore of the target family, expressed in "effective standard deviations" of the mean Zscores of the reference families, with value above a threshold of 3.5 indicates a statistically significant evolutionary relationship between the target and probe families. In comparison to other methods, the FPA technique is significantly more sensitive and is useful to test a suspected distant relationship between probe and target families. For details of both sequence profile analysis and FPA the reader is referred to the original reports (Gribskov et al., 1990; Mehta et al., 1999).

Evolutionary trees of the B_6 families were constructed by comparison of the mean F-Zscores of all enzymes in each family. All enzymes were treated independently without consideration of the degree of mutual sequence relatedness, that is, sequences of different enzymes were not multiply aligned. The standard procedure of profile analysis (see above) was used to construct the profiles, the position-specific scoring matrices, for all sequences of each enzyme. Profile searches were conducted against the SWISS-PROT database. For enzymes with less than four sequences, of which a reliable probe family profile could not be made, results of BLAST programs (Altschul et al., 1997) and profile analysis under stringent conditions were combined. Once all profile search results had been obtained, FPA was run on each profile search output file and the F-Zscores of all enzymes were computed with the FPA algorithm. The arithmetic mean of the F-Zscores of each pair of enzymes (probe enzyme to target enzyme and vice versa) was taken as an estimate of relative distance between two enzymes. Though F-Zscores are asymmetric, it seems useful to apply a symmetric function to results from an asymmetric procedure to obtain a symmetric answer.

The GrowTree program of UWGCG was used to build the evolutionary trees of the B_6 families. The program requires the distances among all the member enzymes of a given family. Since the mean F-Zscores are similarity

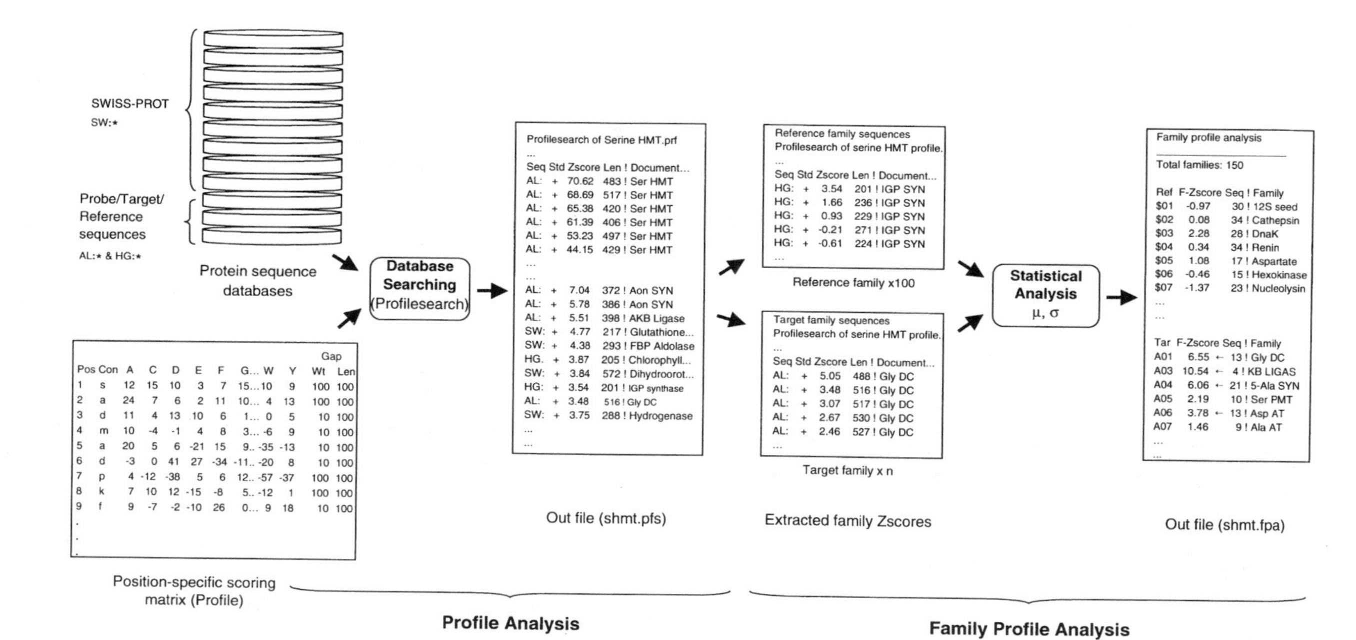

Figure 2. Flow chart of family profile analysis (FPA). The profile analysis part includes construction of a probe profile and searching the SWISS-PROT database with the ProfileSearch program (see text). The symbols SW:*, AL:*, and HG:* along the databases denote sequences that were included from the SWISS-PROT, probe/target, and reference databases, respectively. In the profile table, amino acid residues are shown in standard single letter code (A to Y). Pos and Con are the position number and consensus sequence of the family, respectively. Gap Wt and Len indicate the position-specific penalties applied for opening a gap and its extension. The FPA part includes the Zscores for each of the 100 reference families as well as for the probe and target families. The arrows in the output file of the FPA indicate the families obtaining F-Zscores above the statistically significant threshold of 3.5 (see text). FPA will be available at http://www.unizh.ch/biochem/fpa/index.html

scores, the reciprocal of each mean F-Zscore for a given pair of enzymes was computed to serve as measure of evolutionary distance. A routine was written in standard ANSI C to compute automatically the reciprocal mean F-Zscores from the F-Zscores of the whole family. It was especially useful in the case of the α family, which comprises over 50 enzymes that are evolutionarily related with each other. The reciprocal values of mean F-Zscores <0.1 generated extremely large distance measures. Hence, all values of mean F-Zscores <0.1 were replaced with a value of 0.1. Replacing the mean F-Zscores values <0.1 did not show significant effects on the tree topology. A table of distances (reciprocal mean F-Zscores) was used to construct evolutionary pedigrees with the help of the GrowTree program, using UPGMA (unweighted pair group method using arithmetic averages) with the default parameters. For clustering, the method of UPGMA was preferred over neighbor-joining methods, since UPGMA expects a constant rate of amino acid substitution and thus a linear relationship between distance measure and evolutionary time.

III. Emergence and Evolution of B_6 Enzymes

The indispensable mechanistic role of PLP, the multiple evolutionary origin of the enzymes that depend on it, and the uniqueness of PLP as cofactor for the transformation of amino acids all argue for PLP to have arrived on the evolutionary scene before the emergence of the apoenzymes. This notion probably holds for all organic and, of course, inorganic cofactors of enzymes. Conceivably, in a prebiotic world a primitive metabolism catalyzed by cofactors provided building blocks and polypeptides that in their turn assisted the cofactors in their catalytic functions. Such a self-complicating system can evolve without genetic transfer of information. Nucleotides and oligonucleotides might have acted as cofactors in roles similar to some of those of modern rRNA and tRNA. The advent of polynucleotides with mRNA-like function and self-replicating capacity initiated the transition from prebiotic to biotic evolution. To the best of our knowledge, no pathway for the prebiotic synthesis of PLP has been proposed to date. However, other pyridine derivatives such as nicotinamide have been found to be formed under what are assumed to be prebiotic conditions, and it thus seems possible that PLP might also have arisen by prebiotic synthesis. Remarkably, the in part still hypothetical biosynthetic pathway leading to PLP (Hill et al., 1996) includes a step that is catalyzed by a B_6 enzyme (Lam and Winkler, 1990).

A. HYPOTHETICAL SCENARIO OF EMERGENCE OF B_6 ENZYMES

The emergence of a B_6 enzyme family might include the following sequence of events. The first step was the reaction of PLP with a lysine residue of the progenitor protein. This reaction was facilitated by noncovalent binding of the cofactor. A further prerequisite for the development toward an effective protein catalyst was the preexistence of a rudimentary substrate-binding site adjacent to the PLP-binding site that allowed the transimination with an amino acid substrate and thus the formation of the planar coenzyme-substrate aldimine adduct. Formation of the external aldimine, the starting point for all PLP-dependent reactions (Fig. 1), is an absolute necessity for the catalytic efficacy of PLP toward amino acid substrates. Subsequent optimization of the noncovalent interactions between the protein and the PLP-substrate adduct may be assumed to have led to improved catalytic efficiency and to specialization for reaction and substrate specificity.

B. EVOLUTIONARY LINEAGES OF B_6 ENZYMES

On the basis of sequence comparison with profile analysis and FPA, the B_6 enzymes, as far as structural information on them is available, can be subdivided into few independent families of paralogous proteins, that is, the α family with aspartate aminotransferase as prototype enzyme, the β family with tryptophan synthase β as prototype enzyme, the D-alanine aminotransferase family, and the alanine racemase family. In each of the four families, the 3-D structures of at least two member enzymes have been determined by X-ray crystallography (Table 1). The four B_6 enzyme folds are indeed unrelated with each other (for a comprehensive review of the folds, see Jansonius, 1998). For clarity's sake, the various names used by different authors for the B_6 enzyme families are compiled in Table 2.

Possibly, additional evolutionarily independent lineages of B_6 enzymes may be found when sequences of new enzymes will be determined; as yet there is still no structural information available on 40% of the B_6 enzymes listed with EC numbers. The possibility of evolutionary convergence (see Section III C) increases the chances of finding yet unknown lineages.

1. α Family

The α family is not only the largest but also the functionally most diverse family. All crystal structures of α enzymes that have been determined to date indeed show a similar folding pattern of the polypeptide chain. The ma-

TABLE 2
Denotations of B_6 Enzyme Families Used by Different Authors

This review	Mehta et al., 1993; Alexander et al., 1994; Mehta and Christen, 1994; Sandmeier et al., 1994	Grishin et al., 1995	Jansonius, 1998
α Family	α Family and γ family	Fold type I	Aspartate aminotransferase family
β Family	β Family	Fold type II	Tryptophan synthase β family
D-Alanine aminotransferase family	Aminotransferase subgroup III	Fold type IV	D-Amino acid aminotransferase family
Alanine racemase family	Amino acid decarboxylase group IV	Fold type III	Alanine racemase family

jority of the α enzymes show a specific common functional feature: In the reactions they catalyze, the covalency changes are limited to Cα (defined as the carbon atom that engages in the imine bond with PLP). There are, however, exceptions to the rule that α enzymes catalyze α reactions (see Section III E for a discussion of the course of functional specialization within the B_6 enzyme families). The γ subfamily, which previously was regarded to be an independent family (Alexander et al., 1994), is, by the criteria of the more sensitive family profile analysis and 3-D structures, an evolutionary sublineage of the large α family. The γ subfamily is both structurally and functionally quite homogenous. The majority of the enzymes of the γ subfamily catalyze reactions in which the covalency changes extend from Cα to Cγ, cystathionine β-lyase being the exception. The α family also comprises gene products that with either PLP or pyridoxamine-5′-phosphate as prosthetic group catalyze reactions in the biosynthesis of deoxyamino and dideoxy sugars (Rubinstein and Strominger, 1974; Pascarella et al., 1993; Pascarella and Bossa, 1994; Huber, 1995; Bruntner and Bormann, 1998). These gene products are not listed in Table 1 as their precise catalytic functions are still unknown.

All three additional B_6 enzyme families have considerably fewer members than the α family, are structurally much more homogenous, and functionally limited in scope. These additional families overlap functionally with the α family, each of them comprises enzymes that catalyze reactions of a type being also catalyzed by enzymes of the α family (Table 1). Conceivably, the α family was first to emerge in evolution and thus occupied the bulk of the biocatalytic space of PLP. An alternative though seemingly less likely explanation is that the folds of the other three families proved less adaptable in the development of new reaction and substrate specificities. At least for the alanine racemase family with its $(\beta/\alpha)_8$ fold, also called TIM-barrel fold, this notion is hardly tenable as the $(\beta/\alpha)_8$ fold is probably the most widely distributed and functionally most diverse protein fold (Brändén, 1991; Reardon and Farber, 1995; Babbitt and Gerlt, 1997; Mehta et al., 1999).

2. *β Family*

The β family is much smaller than the α family. High scores in FPA indicate a structurally homogenous family of as yet total 9 enzymes. As with the α family, this family has been defined by purely structural criteria. Nevertheless, all β enzymes share a conspicuous functional feature as they are lyases catalyzing reactions in which not only Cα but also Cβ participates in the covalency changes. Threonine synthase is an exception as it catalyzes a β,γ-replacement reaction (Table 1). As yet there is no enzyme of the β family known that catalyzes a reaction in which the covalency changes are limited to Cα. Included are two enzymes that act on enantiomers of the same substrate. Cystathionine β-synthase, which catalyzes the condensation of homocysteine and serine to cystathionine, is a unique B_6 enzyme as it appears to depend on both heme and PLP for catalytic activity, the role of heme being unclear (Taoka et al., 1999).

3. *D-Alanine Aminotransferase Family*

The D-alanine aminotransferase family, named after its prototype enzyme, represents a third independent evolutionary lineage. This family comprises—on the basis of the as yet available amino acid sequences—only two additional enzymes, that is, branched-chain amino acid aminotransferase and 4-amino-4-deoxychorismate lyase (formerly known as *pabC* gene product), which both act on L-amino acids. Together with the mutually paralogous L- and D-serine dehydratases of the β family, the example of the D-alanine aminotransferase family testifies that the development of specificity for enantiomers of the substrate starting with the same ancestor enzyme is a problem biological evolution has solved more than once. However, not in all cases the problem of enantiomeric substrates was solved in this way, for instance the lactate dehydrogenases producing D- and L-lactate belong to two evolutionarily unrelated families of 2-hydroxyacid dehydrogenases (Taguchi and Ohta, 1991; Kochhar et al., 1992).

4. Alanine Racemase Family

The alanine racemase family is an equally small family. Its other three members are without exception basic amino acid decarboxylases formerly denoted as amino acid decarboxylase group IV (Table 2). The available sequences of arginine decarboxylase are from eukaryotic species and also from *Escherichia coli*, the sequences of ornithine decarboxylase are from eukaryotic species, and the sequences of diaminopimelate decarboxylase are as yet only from eubacterial and archebacterial species. The enzymes of this family are $(\beta/\alpha)_8$ proteins (Farber and Petsko, 1990). The enzymes of the alanine racemase family are the only B_6 enzymes the fold of which is related to that of non-B_6 proteins. FPA shows a distant relationship between alanine racemase and tryptophan synthase α (Mehta et al., 1999), indicating that in these cases the $(\beta/\alpha)_8$ fold is the result of divergent rather than convergent evolution. The 3-D structures of alanine racemase (Shaw et al., 1997) and mammalian ornithine decarboxylase (Kern et al., 1999) show the active site of this B_6 enzyme to be at the same location as in other (non-B_6) enzymes with $(\beta/\alpha)_8$ barrel structure. Both structural comparison, in particular of the phosphate-binding site, and FPA indicate that also many other $(\beta/\alpha)_8$ barrel proteins might be evolutionarily related with each other (Wilmanns et al., 1991; Mehta et al., 1999).

C. EXAMPLES OF EVOLUTIONARY CONVERGENCE

There are two conspicuous examples of functional evolutionary convergence among the B_6 enzymes (Table 3). Basic amino acid decarboxylases are found in both the α family and the alanine racemase family. Similarly, aminotransferases, though not of the same substrate specificity, occur in the α family as well as in the D-alanine aminotransferase family (Table 1). Both the mode of cofactor binding and the catalytic mechanism deduced from the crystal structure of D-alanine aminotransferase is completely analogous to that of aspartate aminotransferase of the α family (Peisach et al., 1998). These findings are reminiscent of the well-known example of evolutionary convergence of the bacterial protease subtilisin, the trypsin protease family, and serine carboxypeptidase (Liao et al., 1992); again, proteins have been developed that catalyze the same or a very similar reaction by the same mechanism, that is, by exploiting the catalytic potential of

TABLE 3
Examples of Evolutionary Convergence Among B_6 Enzymes[a]

Enzyme	Occurrence	Family Affiliation
Ornithine decarboxylase 1 (EC 4.1.1.17)	Prokaryotes	α Family
Ornithine decarboxylase 2 (EC 4.1.1.17)	Eukaryotes	Alanine racemase family
Arginine decarboxylase 1 (EC 4.1.1.19)	*E. coli* (biodegradative isoenzyme)	α Family
Arginine decarboxylase 2 (EC 4.1.1.19)	Eukaryotes and *E. coli* (biosynthetic isoenzyme)	Alanine racemase family

[a]The number of primary and tertiary structures available is indicated in Table 1.

aldimine formation with PLP, without having any sequence or conformational homology.

D. EVOLUTIONARY CONSTRAINTS DUE TO PYRIDOXAL-5'-PHOSPHATE

The mechanistic uniformity of B_6 enzymes (see Introduction) is in apparent contrast to their multiple evolutionary origin. The inevitable conclusion to be drawn is that the common features are not historical traits acquired by chance and passed on from a common ancestor protein, but were necessitated by the chemical properties of the coenzyme and provided an advantage in evolutionary selection for mechanistic or other reasons. An important basis for the reaction specificity of B_6 enzymes seems to be the orientation of the bonds at Cα of the substrate moiety in the external aldimine adduct. The bond to be broken is supposed to lie in a plane orthogonal to the plane of the coenzyme-imine π system. This conformation minimizes the energy of the transition state for bond breaking as it allows maximum σ-π overlap between the bond to be broken and the π system of the pyridine ring-imine system (Dunathan, 1966, 1971). This hypothesis has been confirmed by the determination of high-resolution structures of enzyme-substrate intermediates of aspartate aminotransferase (Kirsch et al., 1984) and all other B_6 enzymes of which the structures of complexes with analogs of the covalent coenzyme-substrate adducts have been determined (Table 1). The structures of the external aldimine intermediates **2a–2c** in the mechanistic scheme on the reactions catalyzed by B_6 enzymes (Fig. 3) take this stereoelectronic control of the reaction pathways into consideration. The

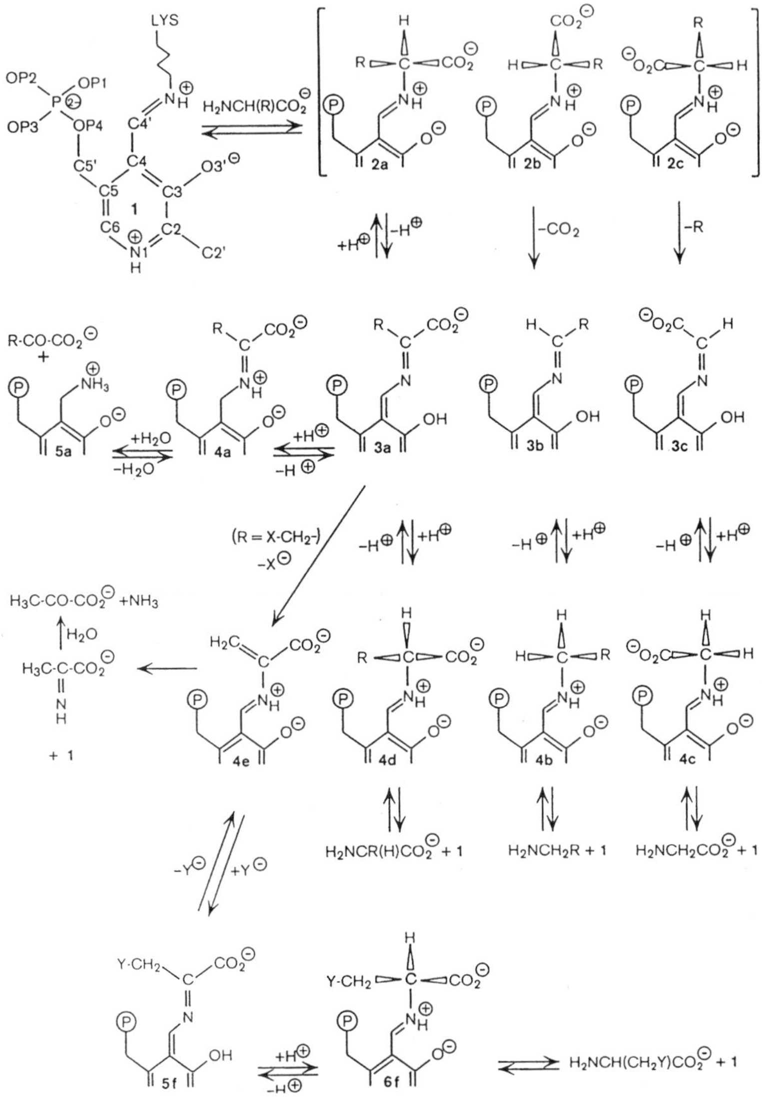

Figure 3. Pathways of reactions catalyzed by B_6 enzymes. Structure **1** depicts the protonated internal aldimine formed by PLP and the active-site lysine residue with *cisoid* aldimine double bond, protonated pyridine N1, and deprotonated 3-hydroxy group. Transimination with the uncharged amino group of the in-coming amino acid substrate produces the external aldimine **2.** According to Dunathan (1966), the pathway of transamination will lead from conformation **2a** through deprotonation at Cα to the quinonoid intermediate **3a,** by reprotonation at C4′ to the

following paragraphs discuss the chemical features of PLP and its reactions that underlie the common properties of B_6 enzymes.

1. Internal Aldimine

The covalent linkage of PLP to an active-site lysine residue (Fig. 1) is thought by many authors to reflect a kinetic advantage because the existence of the internal aldimine **1** replaces the *de novo* formation of the external aldimine (**2a–2c**) by a supposedly faster transimination reaction. However, the ubiquitous internal aldimine might not be a mechanistic necessity but rather reflect the fact that the very first step in the evolution of any B_6 enzyme was the covalent binding of the coenzyme to the primordial apoenzyme. This notion is supported by the following arguments: (1) The same PLP-lysine linkage exists also in glucan phosphorylases in which the carbonyl group of PLP does not participate at all in the catalytic process (Palm et al., 1990). (2) In the reverse half reaction of enzymic transamination (Fig. 3), enzyme-bound pyridoxamine-5′-phosphate forms an imine with the oxo acid substrate. This reaction in all known cases occurs fast enough for not being rate-limiting, though the ketimine intermediate is formed *de novo* and not by transimination. (3) The experimental evidence for a kinetic advantage of the internal aldimine linkage in the PLP-catalyzed reactions of amino acids is ambiguous both in nonenzymic model systems (Schonbeck et al., 1975, and references cited therein) and in B_6 enzymes of which the active-site lysine residues have been replaced by another residue. Most experiments of the latter kind have been conducted with aspartate aminotransferase. In this en-

---

letimine intermediate **4a** and by its hydrolysis to the oxo acid product and PMP **5a.** Reversal of this half-reaction with another oxo acid as amino group acceptor completes the transamination cycle. Alternatively, along the pathway of racemization **3a** may be reprotonated from the opposite side to give the external aldimine **4d,** which by transimination produces the D-amino acid product and **1.** Starting from **3a** is also the β or γ elimination of an $X^{\ominus}$ group from the side chain R. In the example, $X^{\ominus}$ is eliminated from a side chain $X{-}CH_2{-}$ producing the aminoacrylate intermediate **4e.** Reprotonation and transimination give **1** and the secondary imine product that hydrolyzes to pyruvate and ammonia. In β replacement reactions, a group $Y^{\ominus}$ is added to Cβ, resulting in the quinonoid intermediate **5f,** which is reprotonated at Cα to the external aldimine **6f.** Transimination releases the β replacement product and **1.** The external aldimine **2b** leads to α decarboxylation, producing the quinonoid intermediate **3b.** Reprotonation at Cα gives the external aldimine **4b** and then the amine product and **1.** Cleavage of the Cα-Cβ bond in **2c** produces, with tetrahydrofolate as acceptor, 5,10-methylenetetrahydrofolate and, via **3c** and the external aldimine **4c,** glycine and **1.** This type of reaction is catalyzed by glycine hydroxymethyltransferase and related Cα-Cβ-bond-cleaving enzymes. The scheme is taken from the review of Jansonius (1998).

zyme, the ε-amino group is thought to serve as proton acceptor and donor in the tautomerization of the aldimine **2a** to the ketimine intermediate **4a** (Kirsch et al., 1984). Replacement of the active-site lysine residue (K258) by alanine (Malcolm and Kirsch, 1985; Kirsch et al., 1987) or arginine (Toney and Kirsch, 1991) had resulted in virtually inactive enzyme variants. However, if the coenzyme-binding Lys258 of aspartate aminotransferase was exchanged for a histidine residue, the enzyme retained partial catalytic competence (Ziak et al., 1990). The transamination cycle with glutamate and oxalacetate proceeded only three orders of magnitude more slowly than the overall reaction of the wild-type enzyme. Reconstitution of the mutant apoenzyme with [4′-^{3}H]pyridoxamine 5′-phosphate resulted in rapid release of ^{3}H with a first-order rate constant similar to that of the wild-type enzyme (Tobler et al., 1987). Apparently, in aspartate aminotransferase, histidine can to some extent substitute for the active-site lysine residue. The imidazole ring of H258, however, seems too distant from Cα and C4′ to act as an efficient proton donor/acceptor group in the aldimine-ketimine tautomerization, suggesting that the prototropic shift might be mediated by an intervening water molecule (Ziak et al., 1993). Partial catalytic activity is also maintained by D-amino acid aminotransferase on replacement of active-site Lys145 by glutamine (Futaki et al., 1990). The data suggest but do not prove that transimination between the internal and the external aldimine could be replaced by *de novo* formation of the latter and by its hydrolysis in the reverse direction. (4) In pyruvoyl-dependent amino acid decarboxylases the carbonyl group is not engaged in an "internal" imine and the imine with the substrate is formed *de novo* and not by transimination (Recsei and Snell, 1970). Taken together, there is evidence for the notion that the internal aldimine linkage was indispensable for the evolutionary emergence of B_6 enzymes rather than for their catalytic efficiency. The collision and interaction between PLP and a candidate apoenzyme, that is, a protein with rudimentary coenzyme and substrate binding sites, must have been a rare event due to the very likely quite low concentrations of the two reactants. Covalent binding of the coenzyme preventing its rapid loss presumably was a decisive advantage in the biological selection process.

2. *Cisoid Conformation of Aldimines*

Yet another common feature is the conformation of the internal and external aldimine. In all crystal structures of B_6 enzymes determined to date the C4′-N bond assumes the *cisoid* conformation (Fig. 4). Calculation of the

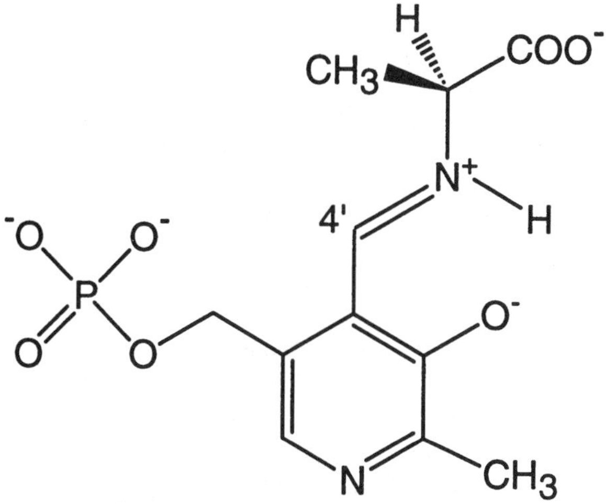

Figure 4. Aldimine adduct of PLP and L-alanine. The dihedral angle (C4′, N, Cα, C_{COO}) in this conformation of the adduct is 180°. Model studies and calculations have shown that the *cisoid* conformation of the aldimine (imine N on the same side of the C4-C4′ bond as O3′) is energetically favored (Tumanyan et al., 1974; Tsai et al., 1978).

minimum-energy conformation of the coenzyme-L-alanine aldimine adduct yields a conformer in which the α-carboxylate group with its negative charge is at maximum distance from the phosphate group of the cofactor if a dielectric constant of 20 or less is used in the calculations (Christen et al., 1996). Such a value of the dielectric constant within the molecule, that is, the coenzyme-substrate adduct, seems an appropriate assumption. This minimum-energy conformation of the coenzyme-substrate adduct may be assumed to contribute to stronger binding to the apoenzyme.

3. *Stereochemistry of C4′ Protonation*

Apparently related to the *cisoid* conformation of the aldimines is the invariable stereochemistry of the protonation at C4′ (Christen et al., 1996), another feature that has previously been interpreted as to indicate a common evolutionary origin of B_6 enzymes. Numerous B_6 enzymes are converted, either in their main reaction (aminotransferases and certain amino acid decarboxylases) or in a side reaction with the amino acid substrate, from the PLP form to the pyridoxamine-5′-phosphate form **5a** (Fig. 3). The transamination reaction requires the tautomerization of the aldimine **2a–2c** to the corresponding ketimine intermediate (e.g., **4a** in the case of aldimine **2a**) by release of a substituent from Cα and addition of a proton to C4′. The protonation at C4′ of **3a–3c** was found to occur on its *si* face in all seven enzymes that were examined (Dunathan and Voet, 1974). This invariance in absolute stereochemistry was interpreted as evidence for a common ancestor of the B_6 enzymes. For five of the seven enzymes, this supposition holds true in the light of the present knowledge on the molecular evolution of B_6 enzymes. Alanine aminotransferase, aspartate aminotransferase, 2,2-dialkylglycine decarboxylase, glutamate decarboxylase, and serine hydroxymethyltransferase indeed belong to the large α family of homologous B_6 enzymes (Table 1). However, the PLP-dependent β subunit of tryptophan synthase, which shows the same stereochemistry, is a member of the β family. (On the seventh enzyme, pyridoxamine pyruvate aminotransferase, no information on primary or tertiary structure and thus on evolutionary relationship is available.) The invariant absolute stereochemistry of enzymes that are evolutionarily unrelated with each other and the inverse stereochemistry in the case of also unrelated D-alanine aminotransferase (Yoshimura et al., 1993) might reflect stereochemical constraints in the emergence of these enzymes rather than an accidental historical trait passed on from a common ancestor enzyme. Conceivably, the coenzyme and substrate binding sites of primordial PLP-dependent enzymes had to fulfill the following prerequisites in order to allow their development toward effective catalysts: (i) The negatively charged α-carboxylate group of the enzyme-bound amino acid substrate had to be positioned at maximum distance from the negatively charged phosphate group of the cofactor (see above), and (ii) the Cα-H bond together with the imine N atom had to lie in a plane close to orthogonal to the plane of the coenzyme-imine π system (Dunathan, 1966; Kirsch et al., 1984) and had to be oriented toward the surface of the protein in order to allow protein-assisted deprotonation at Cα, an integral step of transamination and many other reactions catalyzed by B_6 enzymes (Fig. 3). If a coenzyme-substrate-binding site meets these two criteria, the *si* face with respect to C4′ of the coenzyme-L-amino acid adduct (and in the case of serine hydroxymethyltransferase the *si* face with respect to the adduct with D-alanine that is the substrate undergoing transamination (Voet et al., 1973)) will be facing the protein and its *re* side will be exposed to the solvent. Thus, the protonation at C4′ will occur from the *si* side if protonation is assumed to be assisted by a side chain group of the protein as it has been found to be the case in aspartate aminotransferase (Kirsch et al., 1984). Indeed, in all enzymes for which the pertinent structural information is available, for example, aspartate aminotransferase (Kirsch et al., 1984), tyrosine phenol-lyase (Antson et al., 1993), cystathionine γ synthase (Clausen et al., 1998), dialkylglycine decarboxylase (Toney et al., 1995), and tryptophan synthase β (Hyde et al., 1988), the structures of the covalent coenzyme-substrate adducts either determined directly by X-ray crystallography or deduced from the position of the putative substrate-binding sites invariably show the coenzyme and substrate moieties to be arranged as displayed in Fig. 4.

As a corollary of the above hypothesis, the inverse stereochemistry of protonation at C4′ is expected in enzymes acting on D-amino acids. Indeed, in D-alanine aminotransferase (and in homologous branched-chain L-amino acid aminotransferase) C4′ has been found to be protonated from the *re* side (Yoshimura et al., 1993). The two enzymes belong to the evolutionarily independent D-alanine aminotransferase family (Table 1). In agreement with C4′ protonation from the *re* side, the crystal structure of D-alanine aminotransferase (Sugio et al., 1995) shows the active-site lysine residue to extend toward the *re* face of the coenzyme.

In summary, the multiple evolutionary origin precludes to explain the manifold common structural and mechanistic features of B_6 enzymes by a common ancestor protein. The cofactor PLP is the evolutionary origin that is common to all B_6 enzymes. It seems that specific properties of the cofactor resulted in specific features of the apoproteins that provided evolutionary advantages in binding of the cofactor and in exploiting its catalytic potential.

E. FAMILY PEDIGREES AND COURSE OF FUNCTIONAL SPECIALIZATION

Because of its many diverse member enzymes, the evolutionary pedigree of the α enzyme family is most informative for deducing how functional specialization proceeded. Scrutiny of the pedigree (Fig. 5) reveals a clear pattern in the temporal sequence of events that led to the functional specialization of these enzymes. Apparently, the ancestor enzyme of the α family was specific for covalency changes limited to Cα. The regio-specific ancestor enzyme then diverged into reaction-specific enzymes, such as the three subfamilies of aminotransferases, the three subfamilies of amino acid decarboxylases, the CoA-dependent acyltransferases like 5-aminolevulinate synthase, etc. The subdivision of the enzymes into subfamilies of aminotransferases (subfamilies AT I, II, and IV in the α family and subfamily AT III in the D-alanine aminotransferase family; Mehta et al., 1993),

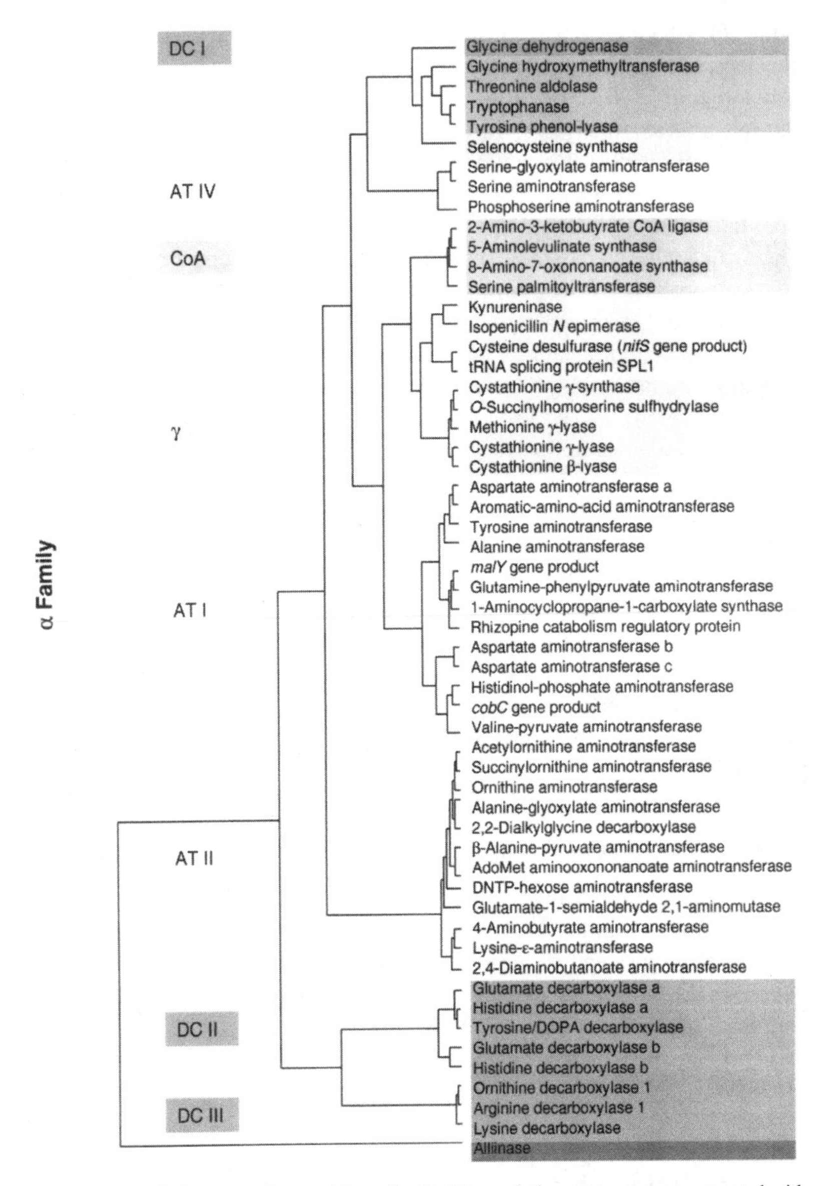

Figure 5. Evolutionary pedigree of the α family. The evolutionary tree was constructed with the GrowTree program on the basis of FPA data (see Databases and Methodology). The branch lengths represent the relative evolutionary distances as measured in reciprocal values of the mean F-Zscores between the branching points. The color code in the left column defines

subfamilies of amino acid decarboxylases (subfamilies DC I–III in the α family and subfamily DC IV in the alanine racemase family; Sandmeier et al., 1994), a γ subfamily (Alexander et al., 1994) and a CoA subfamily (Huber, 1995) agrees with earlier published work. The enzymes that catalyze reactions extending to Cβ and Cγ, for example, the glycine (serine) hydroxymethyltransferase/tyrosine phenol-lyase group or the γ subfamily seem to have diverged from sublineages of the α family just prior to the specialization for substrate specificity. The evolutionary tree shows clearly that the last and also shortest phase was the specialization for substrate specificity. For some enzymes, a further subdivision of subfamilies into different groups seems appropriate on the basis of their relative degree of relationship among themselves and with other enzymes. The subdivision of subfamily ATI into groups *a–c* of aspartate aminotransferase and of subfamily DC II into groups *a* and *b* of glutamate decarboxylase and groups *a* and *b* of histidine decarboxylase seems by and large to reflect the course of phylogenesis (Table 4). Conceivably, these enzymes were already fully specialized for their particular reaction and their substrate very early in phylogenetic evolution (see also Section III G).

The pedigree of the α family suggests that the as yet unidentified gene product *cobC* and perhaps also *malY* as well as the rhizopine catabolism regulatory protein in the aminotransferase subfamily AT I are aminotransferases. 1-Aminocyclopropane-1-carboxylate synthase is as yet the only enzyme in the three AT subfamilies of the α family that is not an aminotransferase but rather catalyzes an α,γ-elimination reaction. The other enzymes of the α family (Fig. 5) that do not catalyze an α reaction, include the γ-family (with one

the subfamilies. The subfamilies are numbered as previously (Mehta et al., 1993; Sandmeier et al., 1994; Alexander et al., 1994). The aminotransferase subfamilies (AT I, II, and IV) are shown in *yellow* and the amino acid decarboxylase subfamilies (DC I–III) in *magenta*. Within subfamilies, some enzymes have been subdivided into different groups, that is, aspartate aminotransferase groups *a* to *c* in subfamily AT I, glutamate decarboxylase groups *a* and *b*, as well as histidine decarboxylase groups *a* and *b* in decarboxylase subfamily DC II, because the sequences of the different groups could not be aligned with each other due to a distant relationship. A different situation is found with ornithine and arginine decarboxylases. The members of ornithine decarboxylase group 1 have an α family fold and are all of prokaryotic origin. The single known amino acid sequence of arginine decarboxylase group 1 is from *Escherichia coli* (biodegradative isoenzyme). However, other forms of these enzymes belong to the alanine racemase family with a $(\alpha/\beta)_8$-barrel fold (ornithine decarboxylase 2 and arginine decarboxylase 2; Fig. 6). The position of selenocysteine synthase [L-seryl-tRNA(Sec) selenium transferase, EC 2.9.1.1] in the evolutionary tree, tentatively reported in an earlier report (Tormay et al., 1998), has been slightly charged by the new, more comprehensive procedure. (See also Color Plates.)

TABLE 4
Species Representation in the Different Groups of Subfamily AT I of Aspartate Aminotransferases and Subfamily DC II of Glutamate Decarboxylases and Histidine Decarboxylases[a]

		Occurrence in		
Enzyme	Group	Eukaryotes	Eubacteria	Archebacteria
Aspartate amino-	*a*	+	+	
transferase in	*b*		+	+
subfamily AT I	*c*			+
Glutamate decarboxylase	*a*	Animals		
in subfamily DC II	*b*	Plants	+	+
Histidine decarboxylase	*a*	Animals		
in subfamily DC II	*b*	Plants	+	

[a]All enzymes belong to the α family (Table 1 and Fig. 5). The subdivision of aspartate aminotransferase subfamily AT I into group a and b was first proposed by Okamoto et al. (1996) and Nakai et al. (1999).

member enzyme, cystathionine β-lyase, catalyzing a β reaction), an evolutionary branch with glycine hydroxymethyltransferase, threonine aldolase, tryptophanase, tyrosine phenol-lyase, selenocysteine synthase, an evolutionary branch with kynureninase, isopenicillin *N* epimerase, cysteine desulfurase, as well as the very distantly related alliinase.

The pedigrees of the β family, the D-alanine aminotransferase family, and the alanine racemase family are small and do not allow general conclusions (Fig. 6). The two enzymes of the β family that act on enantiomeric substrates, that is, D- and L-serine dehydratase fall into two different branches of the tree.

What were the functional and structural properties of the hypothetical intermediates of the evolutionary process delineated by the pedigrees of the B_6 enzymes? The primordial α enzyme with covalently bound PLP probably collected L-amino acids of any type and facilitated the formation of PLP-amino acid aldimine adducts thus accelerating all types of PLP-dependent transformations of amino acids. In a later evolutionary phase, the reaction-specific ancestor enzymes must have favored one particular reaction over all others without differentiating between the various amino acids. Why came specialization for reaction specificity before specialization for substrate specificity? Two explanations may be offered, one mechanistic and one metabolic.

Mechanistically, specialization of the catalytic apparatus for reaction specificity may be assumed to require more extensive structural adaptations than the modifications of the substrate-binding site necessary to make it interact with a specific substrate. This notion is supported by the relatively short evolutionary time that was used for developing substrate specificity (Fig. 5). A substrate-unspecific enzyme is not necessarily tantamount to an enzyme with low affinity for the substrates. The relative facility to interact with different substrates is illustrated by the relatively broad substrate specificity of many B_6 enzymes, in particular by the aminotransferases, many of which accept a particular amino acid/oxo acid substrate pair, for example, aromatic or basic, in one half reaction and act on glutamate and 2-oxoglutarate in the second half reaction with reverse direction. The recommended names of the aminotransferases (Figs. 5 and 6) are derived from the amino acid substrates for which the given enzyme is considered to be specific. However, it should be borne in mind that for many enzymes the second substrate pair, glutamate and 2-oxoglutarate, provides the link between the metabolism of the amino acid carbon skeleton and nitrogen metabolism.

For the organization of metabolism in the uncompartmented progenote cell, the development of catalysts that accelerate one particular reaction of diverse substrates seems more important than the development of catalysts that act only on one substrate but do not accelerate any of the diverse reactions to an extent exceeding that reached already by the primordial ancestor B_6 enzyme. Evolutionary precedence of reaction specificity over substrate specificity is also observed in numerous other instances, for example, in families of proteinases or sugar kinases (Perona and Craik, 1997).

As a corollary of the course of functional specialization, it should be easier to change experimentally the substrate than the reaction specificity of a B_6 enzyme. This observation is consonant with the conclusion drawn from experiments of protein engineering (Hedstrom, 1994) that it might be easier first to select an enzyme with the desired catalytic activity and then to tailor its substrate binding site to a new substrate instead of trying to adapt preexisting binding sites for catalysis of a new reaction.

In the specialization for substrate specificity there is a remarkable similarity between the aminotransferases and the amino acid decarboxylases (Fig. 5). In both cases, the enzymes acting on basic amino acids and their derivatives and the enzymes with acidic and aromatic substrates form separate subfamilies. Apparently, acidic and neutral amino acids may be accommodated in similar substrate-binding sites whereas binding of basic sub-

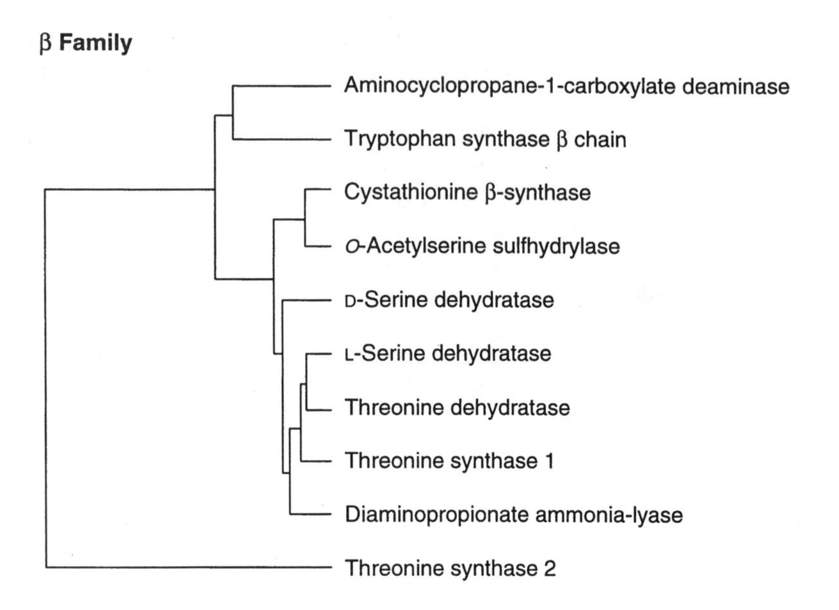
β Family
Aminocyclopropane-1-carboxylate deaminase
Tryptophan synthase β chain
Cystathionine β-synthase
O-Acetylserine sulfhydrylase
D-Serine dehydratase
L-Serine dehydratase
Threonine dehydratase
Threonine synthase 1
Diaminopropionate ammonia-lyase
Threonine synthase 2

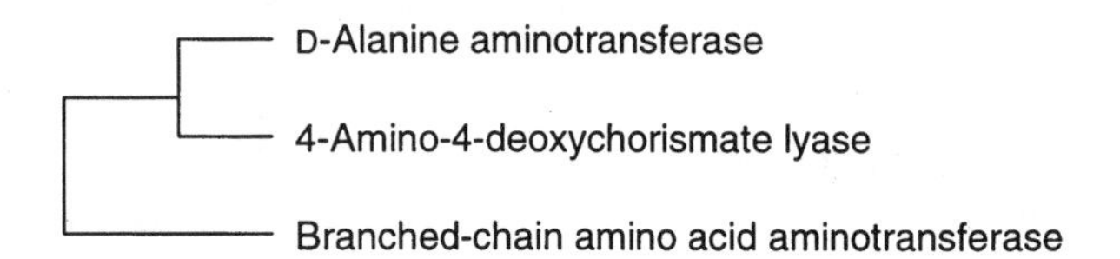
D-Alanine aminotransferase family
D-Alanine aminotransferase
4-Amino-4-deoxychorismate lyase
Branched-chain amino acid aminotransferase

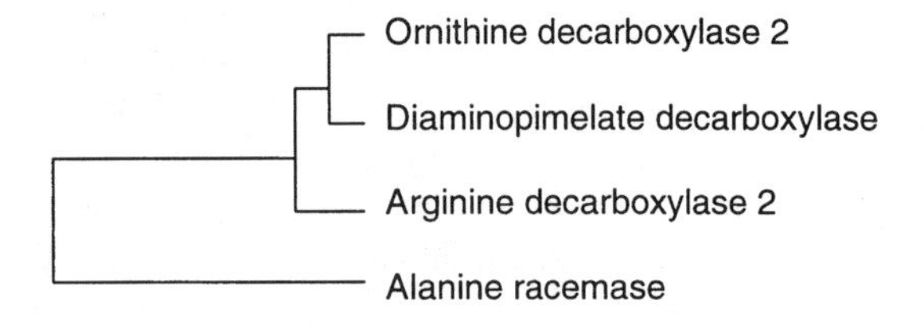
Alanine racemase family
Ornithine decarboxylase 2
Diaminopimelate decarboxylase
Arginine decarboxylase 2
Alanine racemase

strates requires more specific features. Experimental results from site-directed mutagenesis of aspartate aminotransferase point into the same direction (Cronin and Kirsch, 1988; Onuffer and Kirsch, 1995; Malashkevich et al., 1995).

F. DIVERGENCE IN CATALYTIC MECHANISMS

In view of the numerous constraints apparently due to the particular chemical properties of PLP and the mechanistic similarity of its interactions with the amino acid substrates, it seems remarkable that the following mechanistic features vary rather widely (Jansonius, 1998, and references cited therein): (1) The noncovalent interactions between the phosphate group of the coenzyme and the apoprotein are quite different both in type and number, resulting in different degrees of local electrostatic compensation of the negative charge of the phosphate group. (2) The group interacting with *N*1 of the pyridine ring is the distal carboxylate group of an aspartate or glutamate residue in all known enzyme structures of the α family, the D-alanine aminotransferase family and ornithine decarboxylase of the alanine racemase family; in the β family and in alanine racemase, however, it is a serine and an arginine residue, respectively. Apparently, the stabilization of

◄

Figure 6. Evolutionary pedigrees of the β family, D-alanine aminotransferase family, and alanine racemase family. For details on the construction of the evolutionary trees, see legend to Fig. 5. Within a given family, the length of the branches is a measure of the relative evolutionary time. In the **β family**, the sequences of threonine synthase fall into two groups (1 and 2). Threonine synthase 1 includes gram-positive bacterial and archebacterial sequences. Threonine synthase 2 includes sequences from mainly gram-negative bacteria and sequences from yeast. This group proved still to be closely related to threonine synthase 1 (F-Zscore > 13.0) and to some other members of the β family, for example, L-serine dehydratase and cystathionine β-synthase. The **D-alanine aminotransferase family** includes branched-chain amino acid aminotransferase acting on L-valine and L-isoleucine. It is to be noted that this enzyme (EC 2.6.1.42) is distinct from the valine-pyruvate aminotransferase (EC 2.6.1.66) in the α family, the two enzymes use glutamate/2-oxoglutarate and alanine/pyruvate, respectively, as the second substrate pair in the transamination reaction. The **alanine racemase family** also includes three decarboxylases acting with basic amino acids as substrates. Ornithine decarboxylase 2 (OrnDC 2), which appears to be exclusively eukaryotic, and arginine decarboxylase 2 (ArgDC 2), which has been found to occur in both eukaryotes and eubacteria, represent the second group of the respective enzymes. Groups OrnDC 1 and ArgDC 1 are part of the α family (see Fig. 5). The polypeptide chain of alanine racemase is much shorter than those of the other member enzymes. Since decarboxylases are multidomain proteins (Kern et al., 1999), the sequences of OrnDC 2 and ArgDC 2 were truncated before family profile analysis at the unaligned regions by approximately 100 residues at both the amino and carboxy termini.

a positive charge at *N*1 is not a prerequisite for PLP to become catalytically effective. The effect of the group interacting with *N*1 on the electron distribution within the cofactor has been experimentally probed by site-directed mutagenesis experiments with aspartate aminotransferase of the α family (Yano et al., 1992; Onuffer and Kirsch, 1994) and the β-subunit of tryptophan synthase of the β family (Jhee et al., 1998). (3) K^+ ions have been found to be essential for catalytic activity of particular B_6 enzymes such as 2,2-dialkylglycine decarboxylase (Toney et al., 1993, 1995), tryptophan synthase β (Rhee et al., 1996), tyrosine phenol-lyase (Antson et al., 1994; Sundararaju et al., 1997), and tryptophan phenol-lyase (tryptophanase; Isupov et al., 1998). In most B_6 enzymes, however, monovalent cations play neither a mechanistic nor structural role. (4) Virtually all B_6 enzymes seem to be oligomeric proteins, the degree of oligomerization ranging from dimers such as aspartate aminotransferase to dodecameric ornithine decarboxylase of prokaryotes (Momany et al., 1995b). The active sites in all B_6 enzymes with known crystal structure consist of residues of two adjacent subunits. The only B_6 enzyme that has been reported to be a monomer is D-serine dehydratase of the β family (Dowhan and Snell, 1970).

On the basis of the present knowledge on the structure-function relationship of B_6 enzymes, it does not seem possible to attribute a functional significance to the above divergent features. They might have arisen by chance in the different evolutionary lineages, quite in contrast to the features that are common to all B_6 enzymes (see Section III D).

G. TIME OF EMERGENCE AND RATE OF EVOLUTION

In three B_6 enzyme families, enzymes are found of which sequences from all biological kingdoms are available (Table 5). For each enzyme, its sequences in archebacteria, eubacteria, and eukaryotes are all homologous with each other and are more closely related among each other than with the sequences of any other B_6 enzyme in the three biological kingdoms. We may thus conclude that these enzymes were already completely developed, that is, endowed with reaction and substrate specificity, in the universal ancestor cell. These enzymes must have arisen and specialized before the three biological kingdoms branched off from each other 1500 to 1000 million years ago. An emergence so early in biological evolution probably applies to numerous other B_6 enzymes, e.g. other aminotransferases and amino acid synthases, as they catalyze reactions that link different areas of metabolism and

TABLE 5
Enzymes with Homologous Sequences in All Three Biological Kingdoms

Family	Enzyme (EC Number)
α	Glycine hydroxymethyltransferase (2.1.2.1)
	8-Amino-7-oxononanoate synthase (2.3.1.47)
	Histidinol-phosphate aminotransferase (2.6.1.9)
	Acetylornithine aminotransferase (2.6.1.11)
	Phosphoserine aminotransferase (2.6.1.52)
	AdoMet 8-amino-7-oxononanoate aminotransferase (2.6.1.62)
β	1-Aminocyclopropane-1-carboxylate deaminase (4.1.99.4)
	Tryptophan synthase β chain (4.2.1.20)
	Threonine synthase (4.2.99.2)
D-Alanine aminotransferase	Branched-chain amino acid aminotransferase (2.6.1.42)

must have been essential in organizing the metabolic pathways in the progenote. This notion of an early organization of amino acid metabolism is consonant with a general study on the phylogenetic distribution of protein families into the three biological kingdoms. The analysis showed that in the last universal ancestor of contemporary cells the major metabolic pathways were already established (Ouzounis and Kyrpides, 1996).

1. *Functional Specialization During Phylogenesis*

In some instances, however, specialization for (or change of) substrate specificity seems to have overlapped with phylogenetic evolution. Within the aminotransferase subfamily AT I, the aspartate aminotransferase sequences are subdivided into three groups *a–c* (Fig. 5). Aspartate aminotransferase group *a* includes eukaryotic sequences, that is of higher vertebrates and plants, and certain eubacterial sequences. Aspartate aminotransferase group *b* consists exclusively of prokaryotic sequences, some of them of hyperthermophilic and archebacterial species. Aspartate aminotransferase group *c* is exclusively archebacterial (Table 4).

The sequences of group *a* prove to be more closely related to those of aromatic-amino-acid aminotransferase, tyrosine aminotransferase, and alanine aminotransferase in the same evolutionary branch than to the aspartate aminotransferases in groups *b* and *c*. The latter groups, though distantly related with group *a,* clearly constitute evolutionary branches on their own.

TABLE 6
Rate of Evolution of B_6 Enzymes

Enzyme (EC Number)	UEP (Myr)[a]
Serine pyruvate aminotransferase (2.6.1.51)[b]	4.6
Aromatic 1-amino acid decarboxylase (4.1.1.28)[c]	9.7
Aspartate aminotransferase, mitochondrial (2.6.1.1)[d]	10.3/19.4
Phosphoserine aminotransferase (2.6.1.52)[e]	10.9
Alanine aminotransferase, cytosolic (2.6.1.2)[f]	11.6
Aspartate aminotransferase, cytosolic (2.6.1.1)[g]	12.9/13.6
4-Aminobutyrate aminotransferase (2.6.1.19)[h]	12.9
5-Aminolevulinic acid synthase, erythroid specific, mitochondrial (2.3.1.37)[i]	13.8
5-Aminolevulinic acid synthase, nonspecific, mitochondrial (2.3.1.37)[j]	14.5
Ornithine aminotransferase, mitochondrial (2.6.1.13)[k]	15.6
Glycine dehydrogenase, decarboxylating (1.4.4.2)[l]	18.7
Ornithine decarboxylase (4.1.1.17)[m]	19.6
Glutamate decarboxylase isoenzyme 1 (4.1.1.15)[n]	45.1

[a]The unit evolutionary period (UEP) of a given protein is defined as the time required for its amino acid sequence to change by 1%. The values are expressed in millions of years (Myr) and are corrected for multiple amino acid substitutions at identical positions. The footnotes indicate the species from which the sequences considered in determining the evolutionary rate were taken. The data are from Salzmann et al., 2000.

[b]*Callithrix jacchus, Felis silvestris catus, Homo sapiens, Oryctolagus cuniculus, Rattus norvegicus.*

[c]*Bos taurus, Catharanthus roseus, Cavia porcellus, Drosophila melanogaster* isoenzymes 1 and 2, *Homo sapiens, Manduca sexta, Mus musculus, Papaver somniferum* isoenzymes 1, 2 and 5, *Petroselinum crispum* isoenzymes 1, 2, 3 and 4, *Rattus norvegicus, Sus scrofa.*

[d]*Arabidopsis thaliana, Bos taurus, Equus caballus, Gallus gallus, Homo sapiens, Mus musculus, Rattus norvegicus, Saccharomyces cerevisiae, Sus scrofa.* The second value of the UEP was determined by considering only the sequences from vertebrate species (see Fig. 7).

[e]*Arabidopsis thaliana, Oryctolagus cuniculus, Saccharomyces cerevisiae, Schizosaccharomyces pombe, Spinacia oleracea.*

[f]*Chlamydomonas reinhardtii, Homo sapiens, Hordeum vulgare, Panicum miliaceum, Rattus norvegicus, Saccharomyces cerevisiae, Schizosaccharomyces pombe, Zea mays.*

[g]*Saccharomyces cerevisiae, Arabidopsis thaliana* isoenzymes 1 and 2, *Bos taurus, Daucus carota, Equus caballus, Gallus gallus, Homo sapiens, Medicago sativa, Mus musculus, Oryza satina, Rattus norvegicus, Sus scrofa.* The second value of the UEP was determined by considering only the sequences from vertebrate species (see Fig. 7).

[h]*Emericella (aspergillus) nidulans, Homo sapiens, Rattus norvegicus, Saccharomyces cerevisiae, Schizosaccharomyces pombe, Sus scrofa, Ustilago maydis.*

[i]*Drosophila melanogaster, Homo sapiens, Mus musculus, Opsanus tau.*

[j]*Agaricus bisporus, Emericella (aspergillus) nidulans, Gallus gallus, Homo sapiens, Kluyveromyces lactis, Opsanus tau, Rattus norvegicus, Saccharomyces cerevisiae.*

(*Continued*)

TABLE 6 (*continued*)
Rate of Evolution of B_6 Enzymes

[k]*Drosophila ananassae, Emericella (aspergillus) nidulans, Homo sapiens, Mus musculus, Rattus norvegicus, Saccharomyces cerevisiae, Vigna aconitifolia.*

[l]*Flaveria anomala, Flaveria pringlei* isoenzymes a and b, *Flaveria trinervia, Gallus gallus, Homo sapiens, Pisum sativum, Saccharomyces cerevisiae, Schizosaccharomyces pombe, Solanum tuberosum.*

[m]*Bos taurus, Candida albicans, Cricetulus griseus, Datura stramonium, Drosophila melanogaster* isoenzymes 1 and 2, *Gallus gallus, Haemonchus contortus, Homo sapiens, Lycopersicon esculentum, Mus musculus, Mus pahari, Neurospora crassa, Panagrellus redivivus, Rattus norvegicus, Saccharomyces cerevisiae, Xenopus laevis.*

[n]*Homo sapiens, Mus musculus, Rattus norvegicus, Sus scrofa, Felis silvestris.*

Apparently, in the case of these aminotransferases the specialization for substrate specificity occurred in parallel with early phases of phylogenetic evolution when the three kingdoms diverged. Another example of molecular evolution concomitant with phylogenesis and perhaps due to functional constraints specific for a particular phylogenetic branch, is also observed in the amino acid decarboxylase subfamily DC II. Glutamate decarboxylase group *a* is more closely related to histidine decarboxylase group *a* than to glutamate decarboxylase group *b*. Glutamate decarboxylase group *a* occurs exclusively in animals; apparently it developed after the divergence of animals from plants and lower eukaryotes and might perhaps relate to the newly acquired function of the decarboxylation product 4-aminobutyrate (GABA) as a neurotransmitter. A third example of specialization during phylogenetic evolution are the two threonine synthase groups in the β family (Fig. 6).

2. *Rates of Evolution*

How fast did B_6 enzymes—after their specialization for reaction and substrate specificity—change their amino acid sequence? Two prerequisites have to be met in order to explore this question: enough sequences that can be reliably aligned and a reliable time scale of phylogenetic divergence. Screening of the B_6 enzyme sequence database with these two criteria allowed estimation of the rate of evolution of a number of B_6 enzymes (Table 6). The values cover a relatively broad range from the fast-changing serine pyruvate aminotransferase with a unit evolutionary period of only 4.6 million years, which is slightly longer than the 3.7 million years for the α chain of hemoglobin, to the slowly changing glutamate decarboxylase with 45 million years, which compares with other relatively slowly changing intracellular enzymes such as triosephosphate isomerase and glutamate dehydrogenase with unit evolutionary periods of 20 and 55 million years, respectively (Wilson et al., 1977; Salzmann et al., 2000). The glutamate decarboxylase sequences that were considered are all of mammalian species in which this enzyme plays a delicate role as producer of the neurotransmitter 4-aminobutyrate (GABA, the most important inhibitory transmitter in the central nervous system). Glutamate decarboxylase is found only in neurons that use 4-aminobutyrate as transmitter.

Rate of evolution and family affiliation do not seem to correlate. Fast-changing serine pyruvate aminotransferase and slowly changing glutamate decarboxylase belong both to the α family and the evolutionary rates of the enzymes of the other families seem to intermingle at random with those of the α enzymes. The differential rates of neutral evolution of the B_6 enzymes do not correlate with their different folds. Apparently, the rates of evolution are largely determined by functional constraints.

3. Evolution of Homologous Heterotopic Isoenzymes: The Example of Aspartate Aminotransferase

Mitochondrial and cytosolic aspartate aminotransferase are homologous isoproteins. In vertebrate species, the sequence identity between the two isoenzymes is close to 50%. A similar situation may apply to other homologous heterotopic isoenzymes (Hartmann et al., 1991a). Sequence comparisons date the gene duplication that gave rise to the divergence of the two heterotopic isoenzymes of aspartate aminotransferase at the time of the emergence of eukaryotic cells about 1000 million years ago (Graf-Hausner et al., 1983). Their genes appear as single copies in the genomes of man and other mammalian species (Sonderegger et al., 1985, and references cited therein). According to the endosymbiotic hypothesis on the origin of mitochondria (Bogorad, 1975), the genes of the two isoenzymes of aspartate aminotransferase have a rather intricate history. A common ancestral gene diverged into the aspartate aminotransferase gene of ancestor eukaryotic cells and the aspartate aminotransferase gene of respiring microorganisms that were later to become mitochondria. After independent development of the two aspartate aminotransferases and their genes in these two different lineages, endosymbiosis joined them as constituents of the eukaryotic cell. On transfer of the greatest part of the original mitochondrial genome to the nucleus (Nomiyama et al., 1985) their genes became part of the same genome.

Determination of the structure of the genes of cytosolic and mitochondrial aspartate aminotransferase in mouse (Tsuzuki et al., 1987; Obaru et al., 1988) and chicken (Juretić et al., 1990) showed that the genes are very similar. In both species, there are a total of eight and nine introns in the genes of the cytosolic and mitochondrial isoenzymes, respectively. All exon-intron boundaries are exactly at the same positions in the two species. In the genes of the two isoenzymes, five exon-intron boundaries are at the same position. Apparently, these introns had been present before the two genes diverged into the two lineages. This finding corroborates the notion (Gilbert et al., 1986) that introns existed at an early stage of biological evolution, certainly before the advent of eukaryotes. The intermediary development of the two genes in two separate lineages of cells seems to be reflected in a difference at their 3′ splice sites (Juretić et al., 1987).

Another difference between cytosolic and mitochondrial aspartate aminotransferase is the slightly slower evolutionary change of the mitochondrial isoenzyme during vertebrate evolution (Fig. 7). The present data confirm previous conclusions that the mitochondrial isoenzyme was undergoing sequence changes at a slower rate than its homologous cytosolic counterpart (Gehring et al., 1975; Sonderegger et al., 1977). An even more marked difference in the evolution rate of the two isoenzymes was found in an immunological comparison. Quantitative microcomplement fixation, a method that may be assumed to probe preferentially the protein surface, yielded an immunological distance between the mitochondrial aspartate aminotransferases of chicken and pig that was only about half of that of the cytosolic isoenzymes (Sonderegger and Christen, 1978). The difference between the two isoenzymes indicates that in the evolution of the mitochondrial isoenzyme during vertebrate speciation some specific evolutionary constraints were operative, perhaps related to the translocation of the protein from its site of synthesis into the mitochondria and the heterotopic localization of the two isoenzymes in the cell. The following observations indicate the existence of special structural features of the mitochondrial isoenzyme. Upon expression in yeast, the authentic precursor of mitochondrial aspartate aminotransferase of chicken was four times faster imported into the mitochondria than a chimeric protein with the presequence of the mitochondrial isoenzyme attached to the NH_2-terminus of cytosolic aspartate aminotransferase (Hartmann et al., 1991b). What structural features of the mature moiety of the mitochondrial isoenzyme might promote its import? Possibly, the precursor more readily maintains or achieves a translocation-competent conformation than the chimeric protein. Indeed, mature mitochondrial aspartate amino-

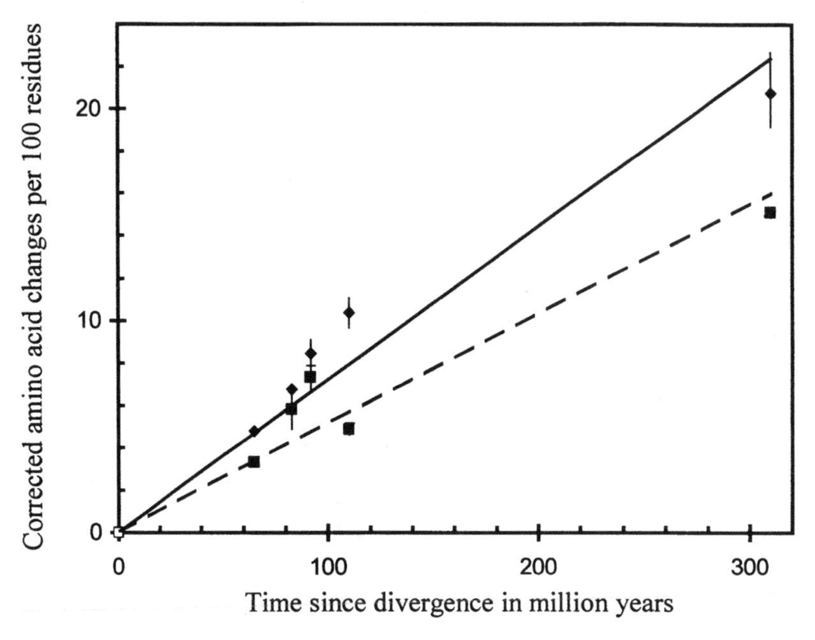

Figure 7. Comparison of the rates of evolution of the mitochondrial (■) and cytosolic (◆) isoenzymes of aspartate aminotransferase. Only sequences from vertebrate species were considered (see Table 5). The vertical bars indicate the range of the amino acid changes (PAM values) if more than two sequences were compared at a given branching point. The unit evolutionary periods during vertebrate evolution are 13.6 and 19.4 million years for the cytosolic and the mitochondrial isoenzyme, respectively.

transferase is faster heat-inactivated than the mature cytosolic isoenzyme (Hartmann et al., 1991b). Specific structural features of the mitochondrial isoenzyme might also facilitate the interaction with molecular chaperones and other components of the translocation machinery before, during, and after translocation through the mitochondrial envelope (Hartl and Neupert, 1990; Schatz and Dobberstein, 1996). A further notable difference between the cytosolic isoenzyme and the mature moiety of its mitochondrial counterpart is their electric charge. A comparison of the isoelectric points of pairs of homologous mitochondrial and cytosolic isoenzymes of vertebrate species has shown that their average pI values are 7.2 and 5.8, respectively (Hartmann et al., 1991a; Jaussi, 1995). The force moving the precursor proteins through the mitochondrial envelope is thought to derive at least in part from the membrane potential (Neupert, 1997). The evolutionary selection for high pI values in mitochondrial isoproteins thus might relate either to facilitated importation into mitochondria or to an adaptation to conditions inside the organelles such as higher pH or higher calcium-ion concentration. It is to be noted that a differential rate of evolution of the two heterotopic isoenzymes is only found during vertebrate evolution (cf. Salzmann et al., 2000). Possibly, the mechanism of importation into mitochondria was particularly optimized in vertebrates.

IV. Experimental Simulation of Steps of B_6 Enzyme Evolution

This section reports on attempts to re-enact experimentally some evolutionary steps, such as the specialization for reaction and substrate specificity, by engineering of B_6 enzymes. PLP-dependent catalytic antibodies were generated as analogs of primordial B_6 enzymes.

A. ENGINEERING OF B_6 ENZYMES

All four lineages of B_6 enzymes known to date, foremost the α family, show that with similar protein scaffolds, for example, the fold of the α family, quite diverse reactions of amino acids can be catalyzed. Thus, alteration of the reaction and substrate specificity of a given enzyme by substitution of a limited number of critical amino acid residues seems feasible. On the basis of the course of the molecular evolution of B_6 enzymes (see Fig. 5) it is to be expected that a change in substrate specificity is more readily achieved than a change in reaction specificity. Examples of such attempts follow.

1. Conversion of Tyrosine Phenol-Lyase to Dicarboxylic Amino Acid β-Lyase

An example for changing the substrate specificity is the conversion of tyrosine phenol-lyase to dicarboxylic amino acid β-lyase (Mouratou et al., 1999). Tyrosine β-lyase is a member of the α family. It catalyzes the β-elimination of L-tyrosine to produce phenol, pyruvate, and ammonium:

$$\text{L-Tyrosine} + H_2O \rightleftharpoons \text{phenol} + \text{pyruvate} + NH_4^+$$

X-ray crystallographic structure analysis has shown the folding pattern of the polypeptide chain of tetrameric tyrosine phenol-lyase from *Citrobacter*

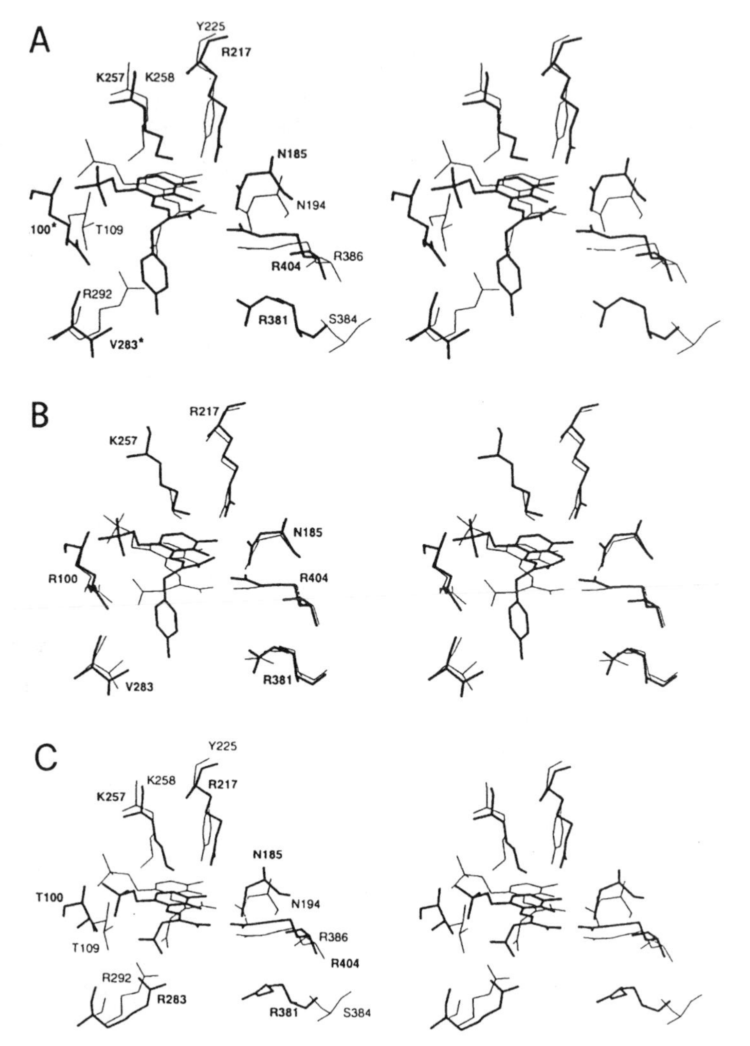

Figure 8. Comparison of the active sites of tyrosine phenol-lyase (TPL) and aspartate aminotransferase (AspAT). **A,** The structures of the quinonoid intermediate of TPL with L-tyrosine (*thick lines*) and of AspAT with L-aspartate (*thin lines*) were obtained by molecular modeling and dynamics simulations on the basis of the crystal structures of the enzymes (for details, see Mouratou et al., 1999). The amino acid residues in TPL corresponding to Arg292 and Thr109 in AspAT are marked with asterisks. **B,** The modeled quinonoid intermediates of wild-type TPL with L-tyrosine (*thick lines*) and with L-aspartate as substrates (*thin lines*). **C,** The modeled quinonoid intermediates with L-aspartate of TPL R100T/V283R (*thick lines*) and wild-type AspAT (*thin lines*). The figures are from Mouratou et al., 1999.

freundii to be similar to that of dimeric aspartate aminotransferase (Antson et al., 1992), which is also a member of the α-family. However, the two enzymes show low sequence identity, for example, 23% between tyrosine phenol-lyase of *C. freundii* and aspartate aminotransferase of *E. coli.* Aspartate aminotransferase catalyzes the reversible transamination reaction of the dicarboxylic amino acids L-aspartate and L-glutamate with the cognate 2-oxo acids—2-oxoglutarate and oxalacetate.

The structures of the active sites of tyrosine phenol-lyase and aspartate aminotransferase are similar; most of the residues that participate in the binding of the coenzyme and the α-carboxylate group of the substrate in aspartate aminotransferase (Kirsch et al., 1984) are conserved in the structure of tyrosine phenol-lyase (Antson et al., 1993) (Fig. 8). To change the substrate specificity of tyrosine phenol-lyase in favor of dicarboxylic amino acids, tyrosine phenol-lyase was compared with aspartate aminotransferase using homology modeling and molecular dynamic simulations. The specificity of aspartate aminotransferase for dicarboxylic amino acids and oxo acids seems to be based primarily on the salt bridge-hydrogen bond interaction of the side chain of Arg292 (of the adjacent subunit) with the distal carboxylate group of these substrates (Kirsch et al., 1984). In agreement with this notion, Arg292 is conserved among all aspartate aminotransferases (Mehta et al., 1989). Since the sequence identity between aspartate aminotransferase and tyrosine phenol-lyase is too low to allow the use of standard alignment algorithms, comparison of their 3-D structures (Antson et al., 1993; Jäger et al., 1994) by superposition was used to identify Val283 of tyrosine phenol-lyase as the residue corresponding to Arg292 in aspartate aminotransferase. Another significant difference in the active sites of the two enzymes is the replacement of a residue interacting with the phosphate group of the coenzyme. Arg100 in tyrosine phenol-lyase apparently corresponds to Thr109 in aspartate aminotransferase, which is also conserved among all aspartate aminotransferases (Fig. 8A).

Molecular modeling showed that the quinonoid adduct of L-aspartate and PLP can be sterically accommodated in the active site of wild-type tyrosine phenol-lyase (Mouratou et al., 1999; Fig. 8B). However, the orientation of the leaving group of the substrate relative to the planar coenzyme-substrate adduct did not appear to be optimum for a β-elimination reaction. Positively charged Arg100 in the hydrophobic part of the active site of tyrosine phenol-lyase interacted with the distal carboxylate group of dicarboxylic substrates and thus did not allow the required orthogonal orientation of the plane defined by Cα, Cβ, and Cγ of the amino acid substrate relative to the plane defined by the π system of the coenzyme-substrate adduct including Cβ (Dunathan, 1966, 1971). Thus, the introduction of an arginine residue into position 283 of tyrosine phenol-lyase together with the substitution of Arg100 with an uncharged residue, that is, the double mutation R100T/V283R, was thought to mimic the binding site for dicarboxylic substrates of aspartate aminotransferase and thus to bring the β-carboxylate group of the substrate into an orientation favorable for reaction, resulting in a corresponding alteration in the substrate specificity of tyrosine phenol-lyase. Both expectations were met: In tyrosine phenol-lyase R100T/V283R, the β-carboxylate group of the substrate assumed in the modeled structure a similar position as in wild-type aspartate aminotransferase and as the phenyl ring of tyrosine in wild-type tyrosine phenol-lyase (Fig. 8C). Indeed, the double-mutant tyrosine phenol-lyase catalyzed the following β-elimination reactions with the dicarboxylic substrates L-aspartate and L-glutamate:

$$\text{L-Aspartate} + H_2O \rightleftharpoons \text{formate} + \text{pyruvate} + NH_4^+$$

$$\text{L-Glutamate} + H_2O \rightleftharpoons \text{acetate} + \text{pyruvate} + NH_4^+$$

These new reactions do not occur with the wild-type enzyme and are only one order of magnitude slower than the reaction of L-tyrosine with the wild-type enzyme (Table 7). The k_{cat} value of tyrosine phenol-lyase R100T/V283R toward L-tyrosine was decreased 30-fold as compared with wild-type tyrosine phenol-lyase without significant change in the K_m value. Apparently, dicarboxylic substrates adopt in the active site of tyrosine phenol-lyase R100T/V283R a similar configuration as L-tyrosine in wild-type tyrosine

TABLE 7
Kinetic Parameters for the β-Elimination Reactions Catalyzed by Wild-type and Mutant Tyrosine Phenol-lyase

	Wild-type Enzyme		R100T/V283R	
Substrate	k_{cat} (s^{-1})	K_m (mM)	k_{cat} (s^{-1})	K_m (mM)
L-Tyrosine	3.7	0.2	0.11	0.32
L-Aspartate	Below detection		0.21	54
L-Glutamate	Below detection		0.10	5.3

Source: Data from Mouratou et al. (1999).

phenol-lyase and interact in a similar way with the critical residues of tyrosine phenol-lyase that control reaction specificity (Fig. 8C). Dicarboxylic amino acid β-lyase is an enzyme that is not found in nature.

2. *Conversion of Aspartate Aminotransferase into Aspartate β-Decarboxylase*

An important feature of the R100T/V283R double mutation in tyrosine phenol-lyase (see above) is the shift of an arginine residue from the wild-type position to another position. This displacement of a positive charge may be expected to influence two important determinants of the reaction pathways taken by the coenzyme-substrate adduct—the bond angles and the electron redistribution in the coenzyme-substrate adduct. Introduction of a similar arginine-shift mutation converted aspartate aminotransferase into a L-aspartate-β-carboxylase (Garber et al., 1995). In aspartate aminotransferases, Arg386 binds the α-carboxylate group of the substrate. Arg386 is one of the four residues invariant in aminotransferases (Mehta et al., 1993) and appears to be absent in some enzymes that are homologous with aminotransferases but catalyze reactions other than transamination. The R386A/Y225R double mutation as well as the R386A/Y225R/R292K triple mutation only slightly change the topography of the active site (Fig. 9) but drastically alter the reaction specificity of the enzyme (Table 8). The double-mutant enzyme possesses β-decarboxylase activity toward L-aspartate, cleaving this substrate into L-alanine plus CO_2. Consonant with an arginine-shift effect, the single mutations Y225R or R386A do not elicit β-decarboxylase activity. R292K as a third mutation reduces the aminotransferase activity without reducing the β-decarboxylase activity (Graber et al., 1999). Thus, aspartate aminotransferase is converted into a true aspartate β-decarboxylase, which catalyzes transamination as a side reaction. Molecular dynamics simulation on the basis of the high-resolution crystal structures of the double- and triple-mutant enzymes suggest that a new hydrogen bond formed between the newly introduced Arg225 and the imine *N* of the coenzyme adduct reinforces the electron sink capacity of the π system of the imine and the cofactor pyridine ring to such an extent that, even after deprotonation at Cα, it remains effective enough to stabilize the carbanion produced by β-decarboxylation (Fig. 10, intermediate VI).

The data illustrate the feasibility to change the reaction specificity of a B_6 enzyme by active-site mutations. Similar amino acid substitutions might

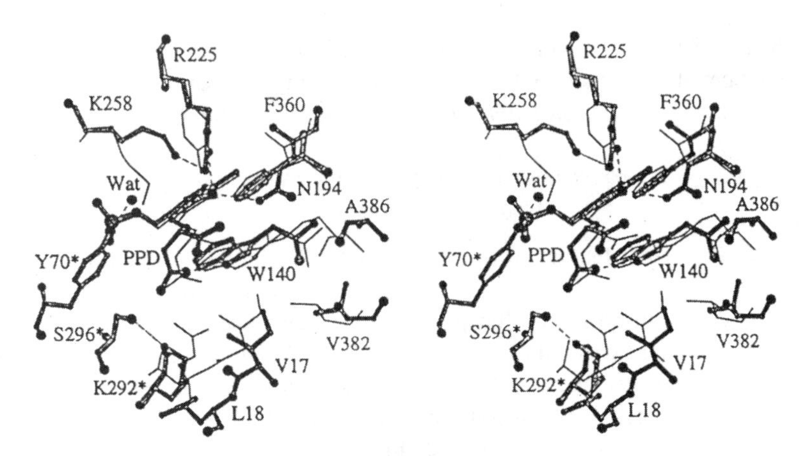

Figure 9. The active site in the crystal structures of the 5′-phosphopyridoxyl-L-aspartate (PPD) complexes of AspAT R386A/Y225R/R292K (*thick lines*) and of wild-type AspAT (*thin lines*). Asterisks indicate residues of the adjacent subunit of the AspAT dimer; Wat, a fixed water molecule. The figure is from Graber et al., 1999.

have occurred in early phases of the molecular evolution of B_6 enzymes when the ancestor enzyme of one of the B_6 enzyme lineages diverged and specialized for different reaction specificities. The molecular activity of the newly generated aspartate β-decarboxylase ($k_{cat} = 5\ min^{-1}$ at 25°C; Table 8) is still about three orders of magnitude lower than that of an average B_6 en-

TABLE 8
Aspartate β-Decarboxylase Activity of Mutant and Wild-type Aspartate Aminotransferase

AspAT	β-Decarboxylation (k_{cat}, s^{-1})	Transamination (k_{cat}, s^{-1})	Ratio β-Decarboxylation/ Transamination
R386A/Y225R/R292K	0.08	0.01[a]	8
R386A/Y225R	0.08	0.19	0.42
R386A	<Wild-type	0.33	
Y225R	<Wild-type	0.45	
R292K	$1.8 \cdot 10^{-3}$	0.5	$3.6 \cdot 10^{-3}$
Wild-type	$6 \cdot 10^{-5}$	180	$3.3 \cdot 10^{-7}$

[a]The K_m values of the triple-mutant enzyme for aspartate and 2-oxoglutarate are increased in comparison to the wild-type enzymes by less than one order of magnitude.

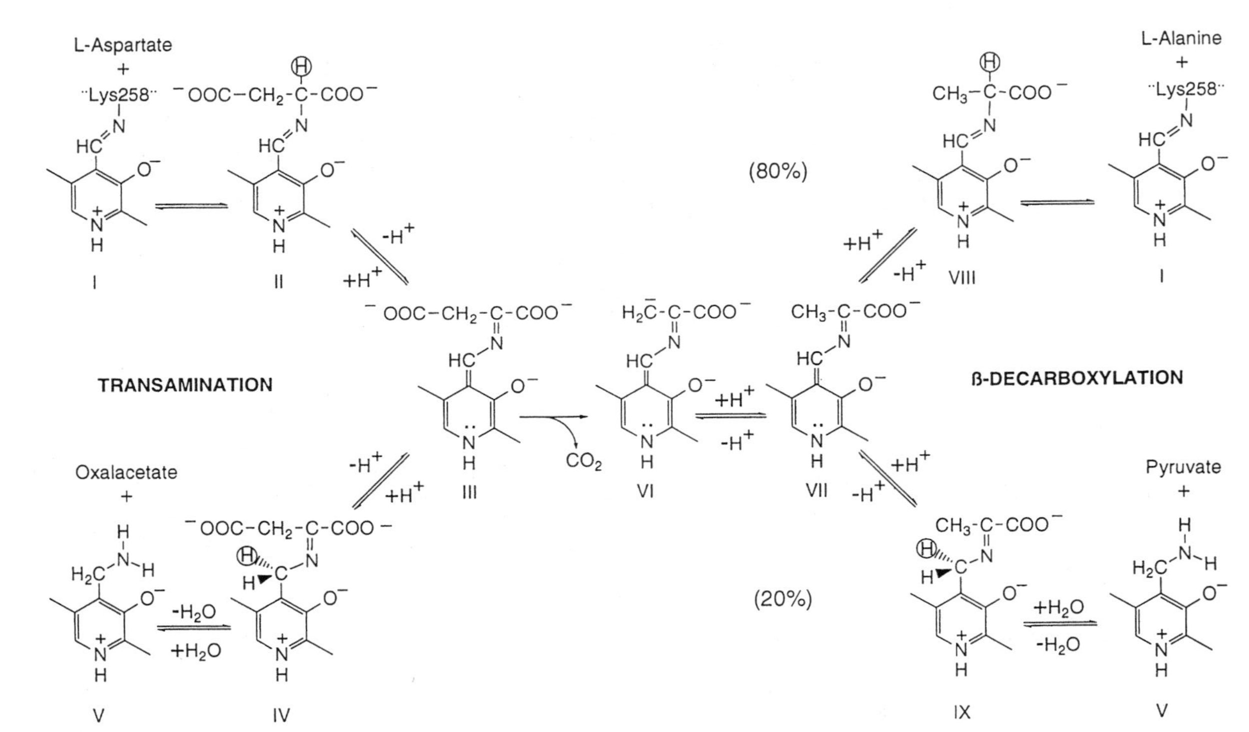

Figure 10. Reaction pathways of enzymic transamination and β-decarboxylation.

zyme. As in evolution, further changes including non-active-site residues would be required to improve its catalytic potency.

B. PYRIDOXAL-5′-PHOSPHATE-DEPENDENT CATALYTIC ANTIBODIES

PLP-dependent catalytic antibodies were generated in an effort to simulate experimentally the molecular evolution of B_6 enzymes. In accordance with the hypothetical scenario of the molecular evolution of B_6 enzymes outlined in Section III A, monoclonal antibodies that are capable of binding a planar PLP-amino acid adduct might be taken as substitutes for a primordial PLP-dependent enzyme. Antibodies elicited against N^{α}-(5′-phosphopyridoxyl)-L-lysine (Fig. 11) may be assumed to comply almost ideally with the most important criterion for an ancestor PLP-enzyme, that is, the existence of a binding site for a covalent coenzyme-amino acid adduct. However, the hapten phosphopyridoxyl lysine is—in contrast to the covalent coenzyme-substrate adducts (Fig. 3)—nonplanar due to the reduction of the C4′-N^{α} double bond. As a consequence, the antibodies elicited with phosphopyridoxyl lysine will not necessarily be able to accommodate the planar coenzyme-substrate adduct, a feature that is essential for the catalytic effect of PLP (Fig. 1). This deficiency of the antigen used for immunization was compensated by the screening protocol. The multitude of possible transformation products of amino acids is a major problem in the design of a screening procedure for PLP-dependent catalytic antibodies. In view of this difficulty, the protocol screened for the occurrence of two successive crucial reaction steps rather than for a final product.

First, the hapten-binding IgG-antibodies were screened for binding of the planar Schiff base formed from PLP and D- or L-norleucine by a competition enzyme-linked immunosorbant assay (ELISA) (for experimental details, see Gramatikova and Christen, 1997). In this ELISA, the antibodies were tested to determine whether their binding to immobilized antigen (with nonplanar 5′-phosphopyridoxyl-L-lysine as hapten) was inhibited by PLP plus D- or L-norleucine. Binders of the planar coenzyme-substrate aldimine adduct were assumed to be inhibited more strongly by the conjoint effect of PLP plus amino acid than by PLP or the amino acid alone. Antibodies 13B10, 8H4, 15A9, 11C2, and 14G1 showed indeed that their binding to the antigen was inhibited more strongly in the presence of PLP plus D- or L-norleucine than in the presence of PLP or the amino acid alone (Fig. 12). The inhibition of antibody-antigen binding by the Schiff base formed from PLP plus glycine indicates the existence of binding sites for the substituents at Cα of the hapten,

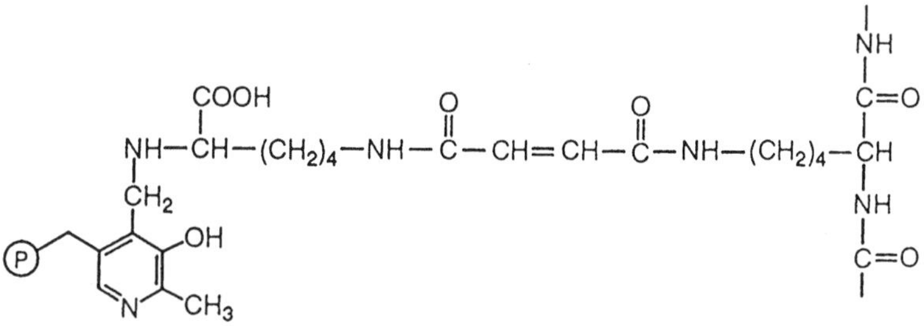

Figure 11. Structure of the antigen. N^{α}-(5′-Phosphopyridoxyl)-L-lysine, produced by reduction of the PLP-lysine aldimine with sodium borohydride, was covalently coupled to a carrier protein.

and the difference in inhibition by PLP plus norleucine and PLP plus glycine reflects the contribution of the amino acid side chain to the binding of the Schiff base. The inhibition profiles of antibodies 5G12 and 6E9 illustrate the binding properties of the great majority of the antibodies that did not show a significant difference in the inhibition by PLP and by PLP plus amino acids. Apparently, these antibodies cannot accommodate the planar aldimine adduct in their binding site. The antibodies differ from PLP enzymes by their lack of a lysine residue at the coenzyme-binding site. Because of the merely noncovalent binding of PLP to the antibodies, the external aldimine intermediate has to be formed *de novo* rather than by transimination.

In the second screening step, the aldimine-binding antibodies were screened for a catalytic effect, that is, the cleavage of the Cα-H bond of the substrate moiety. In the molecular evolution of PLP-dependent enzymes, the analogous step after acquiring the capacity of aldimine binding may be assumed to have been the development of a catalytic apparatus facilitating the cleavage of one of the bonds between Cα and its substituents. The easily measured α,β-elimination of β-chloro-D/L-alanine served to test for Cα-deprotonation, which underlies the majority of PLP-dependent reactions of amino acids (Fig. 3). Due to its good leaving group in the β-position, this substrate analog is decomposed to chloride, ammonia, and readily detectable pyruvate in an α,β-elimination reaction that is initiated by the deprotonation at Cα (Morino et al., 1974). β-Chloroalanine thus allows a convenient and almost universal screening for deprotonation at Cα. Antibody 13B10 was found to catalyze the α,β-elimination of β-chloro-L-alanine, which is consistent with its enantiomeric binding specificity. In contrast, antibody 15A9, which preferably

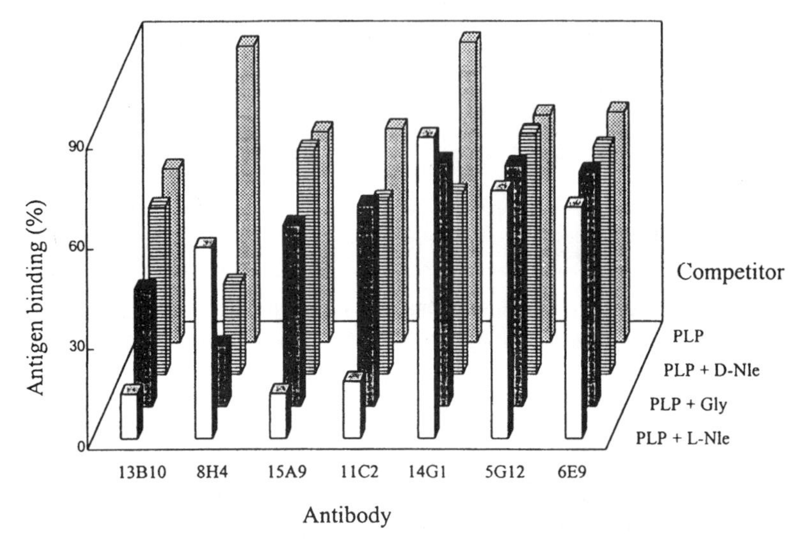

Figure 12. Competition ELISA of antibodies for aldimine binding. The Schiff base **3** was formed from PLP in the presence of D- or L-norleucine or glycine. The assay measures the binding of the antibodies to the antigen (Fig. 11) in the absence and presence of PLP or of PLP plus amino acid, which react nonenzymically to form the Schiff base **3** (aldimine; compound **3** in Fig. 1). At a PLP concentration of 100 μM, an amino acid concentration of 25 mM ensured that at least 80% of the cofactor in the incubation mixture was present as Schiff base. In all cases, inhibition of antibody-antigen binding by the amino acids alone was negligible. The figure is from Gramatikova et al., 1997.

binds the aldimine with L-amino acids (Fig. 12), catalyzed exclusively the reaction of β-chloro-D-alanine. Apparently, the Cα-H bond of the L-amino acid substrate is directed toward an inert surface region of the antibody.

Three reactions were found to be catalyzed by antibody 15A9: formation of aldimine, deprotonation of Cα as reflected by β-elimination of β-chloroalanine, and transamination with hydrophobic D-amino acids (Table 9). Catalysis of aldimine formation might reflect a favorable relative orientation of bound PLP and amino acid. β-Elimination of β-chloroalanine and transamination share one important feature: the crucial reaction steps are proton transfers (Fig. 3). Apparently, in antibodies 13B10 and 15A9 general acid-base groups are positioned in proximity of Cα and Cα/C4′, respectively. Alternatively, water molecules might have access to these atoms and mediate the proton transfers. With antibody 15A9, transamination is two orders of magnitude slower than β-elimination, suggesting that reprotonation at C4′ is

TABLE 9
PLP-Dependent Reactions Catalyzed by Antibody 15A9

Reaction	Substrate(s)	k'_{cat} (min^{-1})	K'_m (mM)
α,β-Elimination	β-Chloro-D-alanine	50	10
Transamination[a]	D-Alanine + PLP	0.42	25
Transamination	Pyruvate + PMP[b]	0.1	19
Transamination	D-Norleucine + PLP	0.07	25

[a]Slower transamination reactions were detected with *N*ε-acetyl-D-lysine, D-lysine and other D-amino acids.

[b]PMP, pyridoxamine-5′-phosphate.

rate-limiting. Antibody 15A9 was also found to catalyze the stereoselective exchange of the α-protons of glycine (Mahon et al., 1998).

Antibody 15A9 is the only antibody catalyzing the transformation of a natural amino acid (Gramatikova and Christen, 1996). The antibody is remarkably reaction-specific, transamination being the only observable reaction. The antibody accelerates the transamination reaction not only of PLP and an amino acid but also in the reverse direction with pyridoxamine-5′-phosphate and an oxo acid as substrates. The orientation of the Cα-substituents relative to the plane of the resonance system of imine and coenzyme together with the presence (and absence) of catalytically effective protein side chains serving as general acid-base groups or modulating the electron repartition in the coenzyme-substrate adduct are thought to determine the reaction specificity in PLP-dependent enzymes (Fig. 3). In contrast to the reaction specificity, the substrate specificity of 15A9 is less strictly defined, apparently hydrophobic amino acids, and oxo acids in the reverse reaction with pyridoxamine-5′-phosphate, are generally accepted as substrates.

As in B_6 enzymes, PLP and antibody effectively complement each other (Table 10). The protein enhances the catalytic efficacy of the cofactor and ensures reaction specificity, including stereospecificity and substrate specificity. The selection criteria of the screening protocol were formation of the planar resonance system of the external aldimine as well as catalysis of deprotonation at Cα. These successive screening steps plausibly simulate the functional selection pressures that probably have been operative in the molecular evolution of protein-assisted pyridoxal catalysis. The analogy is in the interplay of chance and necessity being at work in both cases.

TABLE 10
Rate Acceleration by PLP, Catalytic Antibody 15A9 (Ab), and Aspartate Aminotransferase (AspAT)[a]

Reaction	Reactive Species	Relative Rate Constants
α,β-Elimination	β-Chloro-D-alanine[b]	*1*
	Aldimine [β-chloro-D-alanine-PLP]	10^4
	[β-Chloro-D-alanine-PLP]·Ab	$2 \cdot 10^7$
	[β-Chloro-D-alanine-PLP]·AspAT[c]	$1.2 \cdot 10^8$
Transamination	Aldimine [D-alanine-PLP]	*1*
	[D-Alanine-PLP]·Ab	$5 \cdot 10^3$
	[L-Aspartate-PLP]·AspAT	$2 \cdot 10^8$
	D-Norleucine plus pyruvate[d]	*1*
	D-Alanine plus PLP	10^4

[a]The first-order rate constants of the reactions of the indicated species at 25°C are compared. AspAT is included in the comparison as a prototype PLP-dependent enzyme.

[b]Rate of spontaneous α,β-elimination reaction of β-chloroalanine.

[c]β-Chloro-L-alanine is a mechanism-based inhibitor of AspAT; however, inactivation is at least two orders of magnitude slower than α,β-elimination.

[d]The second-order rate constant of transamination between 20 mM D-norleucine and 40mM pyruvate was determined under anaerobic conditions at 80°C for 10 h. Assuming a temperature coefficient of 2, the rate was compared with the rate of transamination with PLP at 25°C.

V. Alternatives for and to Pyridoxal-5′-phosphate

A. B_6 ENZYMES NOT ACTING ON AMINO ACID SUBSTRATES

The versatility of PLP is not limited to B_6 enzymes acting on amino acids and amino acid derivatives. There are two groups of enzymes that use PLP in an entirely different way. In the glucan phosphorylases (prototype glycogen phosphorylase, EC 2.4.1.1.) the phosphate group of PLP serves as general acid/base group in the phosphorolytic cleavage of glycosidic bonds (Palm et al., 1990). Nevertheless, PLP is covalently bound by an imine linkage to the ε-amino group of a lysine residue. In contrast to the mechanism of B_6 enzymes acting on amino acid substrates, the carbonyl function of PLP is not a mechanistic necessity. The imine double bond may be reduced with sodium borohydride to a stable secondary amine without impairing the catalytic activity. Comparison of folds does not show any similarity with any of the other B_6 enzyme families.

A group of bacterial enzymes exploits the B_6 cofactor in a way reciprocal to that of B_6 enzymes acting on amino acids. As in the reverse half reaction of transamination (Fig. 3), the amino group of pyridoxamine-5′-phosphate reacts with carbonyl groups of their substrates. These enzymes together with PLP-dependent enzymes catalyze reactions in the biosynthesis of dideoxy and deoxyamino sugars, for example, the conversion of 4-oxo-6-deoxy-CDP-glucose to 4-oxo-3,6-dideoxy-CDP-glucose, the reaction requiring NADH or NADPH in addition to pyridoxamine-5′-phosphate (Rubinstein and Strominger, 1974). Reactions of this kind are participating in the synthesis of surface antigens and antibiotics. In most cases, however, only the genes have been identified without examination of the respective protein, the precise function of the gene products still being unknown. Sequence comparison indicates a clear relationship to the α family (Pascarella et al., 1993; Pascarella and Bossa, 1994; Huber, 1995; Bruntner and Bormann, 1998).

B. ALTERNATIVES TO PYRIDOXAL-5′-PHOSPHATE AS COFACTOR

During biological evolution, nature has found in numerous instances on the macroscopic and the molecular scale more than one mechanism to fulfill a functional task. A well-known example in the field of enzymes are the four different types of proteases that have been developed for the hydrolytic cleavage of peptide bonds. Their catalytic activity depends on serine, cysteine, aspartic acid residues, or a zinc ion at their active sites. Another example are two classes of fructrose-1,6-bisphosphate aldolases. Class I aldolases form Schiff bases with their carbonyl substrates and are inactivated by sodium borohydride in the presence of substrate (Horecker et al., 1961) whereas class II aldolases depend on a divalent metal ion, mostly zinc, at their active site and do not form an imine linkage with the substrate (Rutter, 1964). Enzymes of both classes, however, form an intermediary carbanion detectable with tetranitromethane and other electron acceptors (Christen and Riordan, 1968; Healy and Christen, 1972). B_6 enzymes provide yet another example, some reactions of amino acids are catalyzed not only by a B_6 enzyme but also by enzymes operating with an entirely different catalytic machinery (Table 11).

Pyruvoyl-dependent histidine decarboxylases have been found in eubacteria, the PLP-dependent counterparts occur both in eukaryotes and eubacteria. A detailed mechanistic analysis based on the high-resolution crystal structure of the pyruvoyl-dependent histidine decarboxylase of *Lactobacillus*

TABLE 11
Non- B_6 Enzymes Catalyzing B_6 Enzyme Reactions

Enzyme (EC Number)	Cofactor	Reaction
Histidine decarboxylase[a] (4.1.1.22)	Pyruvoyl	Histidine = histamine + CO_2
L-Serine dehydratase[b] (4.2.1.13)	Iron-sulfur cluster	Serine + H_2O = pyruvate + NH_3 + H_2O
Aspartate decarboxylase[c] (4.1.1.11)	Pyruvoyl	Aspartate = β-alanine + CO_2
Glutamate racemase[d] (5.1.1.3)	None	L-Glutamate = D-glutamate
Aspartate racemase[d] (5.1.1.13)	None	L-Aspartate = D-aspartate

[a]Riley and Snell, 1968; Rosenthaler et al., 1965; Recsei and Snell, 1970
[b]Hofmeister et al., 1992; Grabowski et al., 1993
[c]Williamson and Brown, 1979
[d]No PLP or other cofactors such as FAD, NAD^+ and metal ions have been detected in these enzymes (Lamont et al., 1972; Yamauchi et al., 1992; Gallo and Knowles, 1993).

30a has indicated that in analogy to the PLP-dependent amino acid decarboxylases the first reaction step is the formation of a Schiff base between the active-site pyruvoyl group and the α-amino group of the substrate (Gallagher et al., 1989). The examples of Table 11 not only show that at least two different ways exist to catalyze these transformations of amino acids but also that the alternative mechanisms proved, at least in prokaryotes, equivalent in the evolutionary selection.

VI. Conclusions and Perspectives

PLP provides an impressive example of how biological evolution has exploited the catalytic potential of organic cofactors. PLP serves as cofactor not only of enzymes that catalyze manifold transformations of amino acids but also of glucan phosphorylases, the active parts of the cofactor being the carbonyl function and the phosphate group, respectively. This review focuses on the molecular evolution of B_6 enzymes of amino acid metabolism. The α family includes also enzymes that use the amino form of the cofactor (i.e., pyridoxamine-5′-phosphate) to react with carbonyl group-containing substrates. Such a reciprocal function of the cofactor is found in enzymes that reduce dideoxy sugars. The versatility of PLP as prosthetic group of enzymes is perhaps equaled by zinc as inorganic cofactor of all six EC classes of enzymes (Vallee and Falchuk, 1993). The multifunctionality of PLP as organic cofactor is surpassed only by nucleotides and polynucleotides (RNA) in their roles as allosteric effectors, structural components, contributors to catalytic effects, and as function-programming template in enzymes and ribonucleoprotein complexes.

The process of formation of a primordial B_6 enzyme by reaction of PLP with a suitable apoprotein has happened more than once and has led to the emergence of several independent evolutionary lineages. In the progenote, the subsequent divergent evolution of the protoenzymes was first determined by functional specialization for reaction as well as substrate specificity and then by the phylogenetic divergence, that is, by speciation (Fig. 13). The present state of knowledge of the molecular evolution of B_6 enzymes is incomplete; as yet, no structural information is available on perhaps half of all B_6 enzymes existing in nature. Genome research in conjunction with functional genomics may be expected to provide most of the lacking information within the next decades. Possibly, more examples of convergent evolution, such as found in the case of amino acid decarboxylases and aminotransferases (Table 3) might be revealed. More sensitive homology search methods in both sequence and fold comparison might detect non-B_6 enzymes other than $(\beta/\alpha)_8$ barrel proteins to be related with B_6 enzymes.

Knowledge of the evolution of B_6 enzymes provides a basis for the experimental engineering of these enzymes. Site-directed mutagenesis, in part led by evolutionary considerations, has already succeeded in changing both the reaction and substrate specificity of a few B_6 enzymes. A selection protocol simulating the functional constraints that might have been operative in the molecular evolution of B_6 enzymes proved successful in the generation of PLP-dependent catalytic antibodies. However, most engineered B_6 enzymes and the catalytic antibody 15A9 possess catalytic efficiencies that are three to five orders of magnitude lower than that of wild-type B_6 enzymes. The great majority of amino acid substitutions at the active site of B_6 enzymes have two mutually opposite kinetic consequences: a decrease in the rate of the specific reaction and an increase in the rates of side reactions. Clearly, the reaction specificity of PLP-dependent enzymes and possibly of many other enzymes is not only achieved by accelerating the specific reaction but also by preventing potential side reactions (Rétey, 1990; Vacca et al., 1997). Enhancement of the catalytic efficiency of engineered enzymes and catalytic antibodies requires subtle adaptations in the active-site topochemistry as they

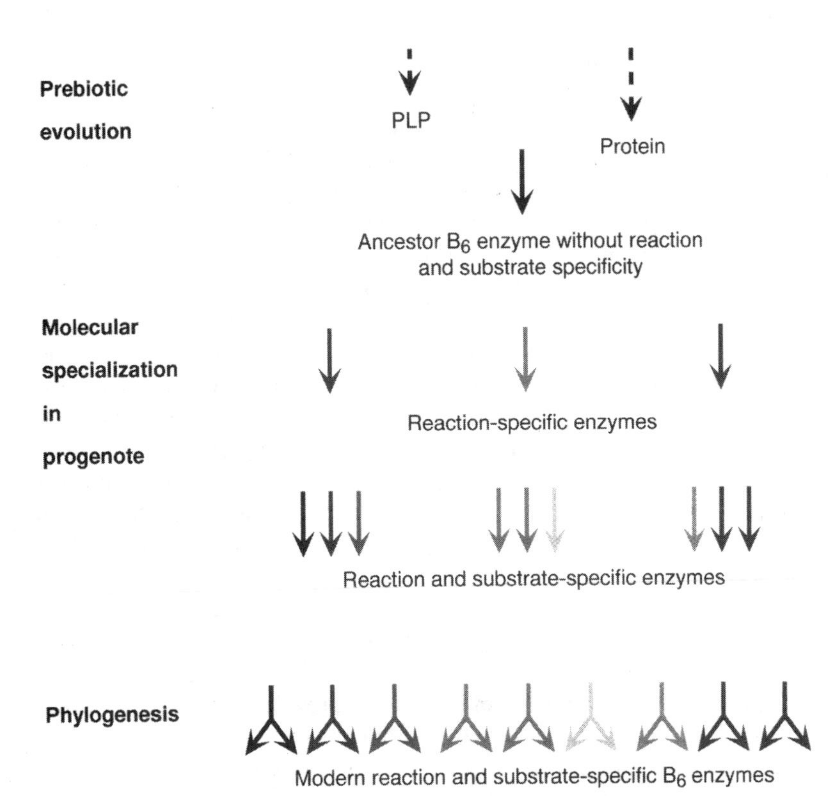

Figure 13. Synopsis of the molecular evolution of B_6 enzymes. The scheme depicts the temporal sequence of the processes that led from the ancestor of a given B_6 enzyme family, which very likely was an allrounder catalyst, to the reaction and substrate-specific enzymes of amino acid metabolism in recent species. The B_6 enzymes are of multiple evolutionary origin, the course of events outlined in the scheme has occurred in parallel a couple of times, each time starting from a different ancestor protein with another fold. For brevity's sake, the development of only one B_6 enzyme family is shown. The main features of molecular evolution are most clearly evident from the data on the large α family (Fig. 5). However, the data on the other much smaller B_6 enzyme families (Fig. 6) (i.e., the β family, as well as the D-alanine aminotransferase and the alanine racemase families) comply with the same scheme. The first cells and the last common ancestor of prokaryotes and eukaryotes are estimated to have been in existence about 3500 and 1500–2000 million years ago, respectively (Doolittle et al., 1996). (See also Color Plates.)

must have occurred in the biological molecular evolution of these enzymes. These optimization processes have to involve also non-active-site residues and have to accelerate the specific reaction with the specific substrate and to suppress potential side reactions. Syncatalytic conformational changes (Gehring and Christen, 1978; Kirsch et al., 1984; Pfister et al., 1985) will have to adjust the topochemistry of the coenzyme-substrate adduct and the active-site groups in a spatially and temporally coordinated manner along the reaction coordinate.

The generation of new B_6 enzymes by forced molecular evolution starting out from PLP-dependent catalytic antibodies or engineered PLP enzymes seems an attainable goal. In all likelihood, the most successful screening protocol will simulate the natural evolution both with respect to the temporal order of the selection criteria to be applied and the constraints that the particular chemical features of PLP will impose on the apoprotein. One experimental possibility is the application of directed evolution. In the first attempt with a B_6 enzyme, metabolic screening of randomly mutagenized aspartate aminotransferase has succeeded in increasing the activity of the enzyme toward branched-chain amino acids by a factor of 10^6 (Yano et al., 1998; Oue et al., 1999). Only one residue out of a total of 17 mutagenized residues appears to interact directly with the substrate. Improvement of the catalytic efficiency of engineered B_6 enzymes by other, more generally applicable, means of forced evolution, for example, by phage display of randomly mutagenized enzyme and direct screening for catalytic activity, is an experimental challenge of topical interest.

Acknowledgments

We are very grateful to Johan N. Jansonius, Heinz Gehring, and Erika Sandmeier for their year-long excellent collaboration, including fruitful discussions of many topics touched upon in this review. Special thanks are due to Margrit Mathys and Silvia Kocher for their help in preparing the manuscript. Part of the work for this review was supported by Swiss National Foundation grants No. 31-36542 and No. 31-45940.

References

Alexander FW, Sandmeier E, Mehta PK, Christen P (1994): Eur J Biochem 219: 953–960.

Alexeev D, Alexeeva M, Baxter RL, Campopiano DJ, Webster SP, Sawyer L (1998): J Mol Biol 284: 401–419.

Altschul SF, Madden TL, Schaffer AA, Zhang JH, Zhang Z, Miller W, Lipman DJ (1997): Nucl Acids Res 25: 3389–3402.

Antson AA, Strokopytov BV, Murshudov GN, Isupov MN, Harutyunyan EH, Demidkina TV, Vassyiyer DG, Dauter Z, Terry H, Wilson KS (1992): FEBS Lett 302: 256–260.

Antson AA, Demidkina TV, Gollnick P, Dauter Z, von Tersch RL, Long J, Berezhnoy SN, Phillips RS, Harutyunyan EH, Wilson KS (1993): Biochemistry 32: 4195–4206.

Antson AA, Dodson GG, Wilson KS, Pletnev SV, Harutyunyan EG, Demidkina TV (1994): In Marino G, Sannio G, Bossa F (eds): "Biochemistry of Vitamin B_6 and PQQ." Basel: Birkhäuser, pp. 187–191.

Babbitt PC, Gerlt JA (1997): J Biol Chem 272: 30591–30594.

Bairoch A, Boeckmann B (1991): Nucl Acids Res 19: 2247–2249.

Bernstein FC, Koetzle TF, Williams GJB, Meyer EF Jr, Brice MD, Rodgers JR, Kennard O, Shimanouchi T, Tasumi M (1977): J Mol Biol 112: 535–542.

Bogorad L (1975): Science 188: 891–898.

Brändén CI (1991): Curr Opinion Struct Biol 1: 978–983.

Bruntner C, Bormann C (1998): Eur J Biochem 254: 347–355.

Burkhard P, Jagannatha Rao GS, Hohenester E, Schnackerz KD, Cook PF, Jansonius JN (1998): J Mol Biol 283: 121–133.

Capitani G, Hohenester E, Feng L, Storici P, Kirsch JF, Jansonius JN, (1999) J Mol Biol 294: 745–756.

Christen P, Metzler DE (eds) (1985): "Transaminases." New York: Wiley-Interscience.

Christen P, Riordan JF (1968): Biochemistry 7: 1531–1538.

Christen P, Kasper P, Gehring H, Sterk M (1996): FEBS Lett 389: 12–14.

Clausen T, Huber R, Laber B, Pohlenz H-D, Messerschmidt A (1996): J Mol Biol 262: 202–224.

Clausen T, Huber R, Prade L, Wahl MC, Messerschmidt A (1998): EMBO J 17: 6827–6838.

Cronin CN, Kirsch JF (1988): Biochemistry 27: 4572–4579.

Devereux J, Haeberli P, Smithies O (1984): Nucl Acids Res 12: 387–395.

Dolphin D, Poulson R, Avramovic O (eds) (1986): "Vitamin B_6 Pyridoxal Phosphate, Parts A and B." New York: Wiley.

Doolittle RF, Feng DF, Tsang S, Cho G, Little E (1996): Science 271: 470–477.

Dowhan W, Snell EE (1970): J Biol Chem 245: 4618–4628.

Dunathan HC (1966): Proc Natl Acad Sci U S A 55: 712–716.

Dunathan HC (1971): Adv Enzymol 35: 79–134.

Dunathan HC, Voet JG (1974): Proc Natl Acad Sci U S A 71: 3888–3891.

Farber GK, Petsko GA (1990): TIBS 15: 228–234.

Fitch WM (1970): Syst Zool 19: 99–113.

Ford GC, Eichele G, Jansonius JN (1980): Proc Natl Acad Sci U S A 77: 2559–2563.

Futaki SF, Ueno H, Martinez del Pozo A, Pospischil MA, Manning JM, Ringe D, Stoddard B, Tanizawa K, Yoshimura T, Soda K (1990): J Biol Chem 265: 22306–22312.

Gallagher DT, Gilliland GL, Xiao G, Zondlo J, Fisher KE, Chinchilla D, Eisenstein E (1998): Structure 6: 465–475.

Gallagher T, Snell EE, Hackert ML (1989): J Biol Chem 264: 12737–12743.

Gallo KA, Knowles JR (1993); Biochemistry 32: 3981–3990.

Gehring H, Christen P (1978): J Biol Chem 253: 3158–3163.

Gehring H, Wilson KJ, Christen P (1975): Biochem Biophys Res Commun 67: 73–78.

Gilbert W, Marchionni M, McKnight G (1986): Cell 46: 151–153.

Graber R, Kasper P, Malashkevich VN, Sandmeier E, Berger P, Gehring H, Jansonius JN, Christen P (1995): Eur J Biochem 232: 686–690.

Graber R, Kasper P, Malashkevich VN, Strop P, Gehring H, Jansonius JN, Christen P (1999): J Biol Chem 274: 31203–31208.

Grabowski R, Hofmeister AE, Buckel W (1993); Trends Biochem Sci 18: 297–300.

Graf-Hausner U, Wilson KJ, Christen P (1983): J Biol Chem 258: 8813–8826.

Gramatikova S, Christen P (1996): J Biol Chem 271: 30583–30586.

Gramatikova S, Christen P (1997): J Biol Chem 272: 9779–9784.

Gribskov M (1992): Gene 119: 107–111.

Gribskov M, Lüthy R, Eisenberg D (1990): Meth Enzymol 183: 146–159.

Grishin NV, Phillips MA, Goldsmith EJ (1995): Prot Sci 4: 1291–1304.

Hartl FU, Neupert W (1990): Science 247: 930–938.

Hartmann C, Christen P, Jaussi R (1991a): Nature 352: 762–763.

Hartmann C, Lindenmann J-M, Christen P, Jaussi R (1991b): Biochem Biophys Res Commun 174: 1232–1238.

Hayashi H, Wada H, Yoshimura T, Esaki N, Soda K (1990): Annu Rev Biochem 59: 87–110.

Healy MJ, Christen P (1972): J Am Chem Soc 94: 7911–7916.

Hedstrom L (1994): Curr Opin Struct Biol 4: 608–611.

Hennig M, Grimm B, Contestabile R, John RA, Jansonius JN (1997): Proc Natl Acad Sci U S A 94: 4866–4871.

Hester G, Stark W, Moser M, Kallen J, Markovic-Housley Z, Jansonius JN (1999): J Mol Biol 286: 829–850.

Hill RE, Himmeldirk K, Kennedy IA, Pauloski RM, Sayer BG, Wolf E, Spenser ID (1996): J Biol Chem 271: 30426–30435.

Hofmeister AE, Berger S, Buckel W (1992): Eur J Biochem 205: 743–749.

Horecker BL, Pontremoli S, Ricci C, Cheng T (1961): Proc Natl Acad Sci U S A 47: 1949–1955.

Huber B (1995): Diploma Thesis, University of Zurich, Zurich.

Hyde CC, Ahmed SA, Padlan EA, Miles EW, Davies DR (1988): J Biol Chem 263: 17857–17871.

Isupov MN, Antson AA, Dodson EJ, Dodson GG, Dementieva IS, Zakomirdina LN, Wilson KS, Dauter Z, Lebedev AA, Harutyunyan EH (1998): J Mol Biol 276: 603–623.

Jäger J, Moser M, Sauder U, Jansonius JN (1994): J Mol Biol 239: 285–305.

Jansonius JN (1998): Curr Opinion Struct Biol 8: 759–769.

Jaussi R (1995): Eur J Biochem 228: 551–561.

Jencks WP (1969): "Catalysis in Chemistry and Enzymology." New York: McGraw-Hill, pp. 133–146.

Jhee K-H, Yang L, Ahmed SA, McPhie P, Rowlett R, Miles EW (1998): J Biol Chem 273: 11417–11422.

John RA (1995): Biochim Biophys Acta 1248: 81–96.

Juretić J, Jaussi R, Mattes U, Christen P (1987): Nucl Acids Res 15: 10083–10087.

Juretić N, Mattes U, Ziak M, Christen P, Jaussi R (1990): Eur J Biochem 192: 119–126.

Kern AD, Oliveira MA, Coffino P, Hackert ML (1999): Structure 7: 567–581.

Kirsch JF, Finlayson WL, Toney MD, Cronin CN (1987): In "Biochemistry of Vitamin B_6," Korpela T, Christen P (eds). Basel: Birkhäuser, pp. 59–67.

Kirsch JF, Eichele G, Ford GC, Vincent MG, Jansonius JN, Gehring H, Christen P (1984): J Mol Biol 174: 497–525.

Kochhar S, Hunziker PE, Leong-Morgenthaler P, Hottinger H (1992): J Biol Chem 267: 8499–8513.

Lam HM, Winkler ME (1990): J Bacteriol 172: 6518–6528.

Lamont HC, Staudenbauer WL, Strominger JL (1972): J Biol Chem 247: 5103–5106.

Liao D-I, Breddam K, Sweet RM, Bullock T, Remington SJ (1992): Biochemistry 31: 9796–9812.

Liu J-Q, Dairi T, Itoh N, Kataoka M, Shimizu S, Yamada H (1998): J Biol Chem 273: 16678–16685.

Mahon MM, Gramatikova SI, Christen P, Fitzpatrick TB, Malthouse JPG (1998): FEBS Lett 427: 74–78.

Malashkevich VN, Onuffer JJ, Kirsch JF, Jansonius JN (1995): Nat Struct Biol 2: 548–553.

Malcolm BA, Kirsch JF (1985): Biochem Biophys Res Commun 132: 915–921.

McPhalen CA, Vincent MG, Jansonius JN (1992): J Mol Biol 225: 495–517.

Mehta PK, Christen P (1994): Biochem Biophys Res Commun 198: 138–143.

Mehta PK, Hale TI, Christen P (1989): Eur J Biochem 186: 249–253.

Mehta PK, Hale TI, Christen P (1993): Eur J Biochem 214: 549–561.

Mehta PK, Argos P, Barbour AD, Christen P (1999): Proteins 35: 387–400.

Metzler DE (1977): "Biochemistry." New York: Academic Press, pp. 444–461.

Momany C, Ernst S, Ghosh R, Chang N, Hackert ML (1995a): J Mol Biol 252: 643–655.

Momany C, Ghosh R, Hackert ML (1995b): Prot Sci 4: 849–854.

Morino Y, Osman AM, Okamoto M (1974): J Biol Chem 249: 6684–6692.

Mouratou B, Kasper P, Gehring H, Christen P (1999): J Biol Chem 274: 1320–1325.

Nakai T, Okada K, Akutsu S, Miyahara I, Kawaguchi S-I, Kato R, Kuramitsu S, Hirotsu K (1999): Biochemistry 38: 2413–2424.

Neupert W (1997): Annu Rev Biochem 66: 863–917.

Nomiyama H, Fukuda M, Wakasugi S, Tsuzuki T, Shimada K (1985): Nucleic Acids Res 13: 1649–1658.

Obaru K, Tsuzuki T, Setoyama C, Shimada K (1988): J Mol Biol 200: 13–22.

Okada K, Hirotsu K, Sato M, Hayashi H, Kagamiyama H (1997): J Biochem 121: 637–641.

Okamoto A, Higuchi T, Hirotsu K, Kuramitsu S, Kagamiyama H (1994): J Biochem 116: 95–107.

Okamoto A, Kato R, Masui R, Yamagishi A, Oshima T, Kuramitsu S (1996): J Biochem 119: 135–144.

Okamoto A, Nakai Y, Hayashi H, Kirotsu K, Kagamiyama H (1998): J Mol Biol 280: 443–461.

Onuffer JJ, Kirsch JF (1994): Prot Engin 7: 413–424.

Onuffer JJ, Kirsch JF (1995): Prot Sci 4: 1750–1757.

Oue S, Okamoto A, Yano T, Kagamiyama H (1999): J Biol Chem 274: 2344–2349.

Ouzounis C, Kyrpides N (1996): FEBS Lett 390: 119–123.

Palm D, Klein HW, Schinzel R, Buehner M, Helmreich EJM (1990): Biochemistry 29: 1099–1107.

Pascarella S, Bossa F (1994): Prot Sci 3: 701–705.

Pascarella S, Schirch V, Bossa F (1993): FEBS Lett 331: 145–149.

Peisach D, Chipman DM, van Ophem PW, Manning JM, Ringe D (1998): Biochemistry 37: 4958–4967.

Perona JJ, Craik CS (1997): J Biol Chem 272: 29987–29990.

Pfister K, Sandmeier E, Berchtold W, Christen P (1985): J Biol Chem 260: 11414–11421.

Reardon D, Farber G (1995): FASEB J 9: 497–503.

Recsei PA, Snell EE (1970): Biochemistry 9: 1492–1497.

Renwick SB, Snell K, Baumann U (1998): Structures 6: 1105–1116.

Rétey J (1990): Angew Chem Int Ed Engl 29: 355–361.

Rhee S, Parris KD, Ahmed SA, Miles EW, Davies DR (1996): Biochemistry 35: 4211–4221.

Riley WO, Snell EE (1968): Biochemistry 7: 3520–3528.

Rosenthaler J, Guirard BM, Chang GW, Snell EE (1965): Proc Natl Acad Sci U S A 54: 152–158.

Rubinstein PA, Strominger JL (1974): J Biol Chem 249: 3776–3781.

Rutter WJ (1964): Federation Proc 23: 1248–1257.

Salzmann D, Christen P, Mehta PK, Sandmeier E (to be published).

Sandmeier E, Hale TI, Christen P (1994): Eur J Biochem 221: 997–1002.

Scarsdale JN, Kazanina G, Radaev S, Schirch V, Wright HT (1999): Biochemistry 38: 8347–8358.

Schatz G, Dobberstein B (1996): Science 271: 1519–1526.

Schonbeck ND, Skalski M, Shafer JA (1975): J Biol Chem 250: 5343–5351.

Shaw JP, Petsko GA, Ringe D (1997): Biochemistry 36: 1329–1342.

Shen BW, Hennig M, Hohenester E, Jansonius JN, Schirmer T (1998): J Mol Biol 277: 81–102.

Sonderegger P, Christen P (1978): Nature 275: 157–159.

Sonderegger P, Gehring H, Christen P (1977): J Biol Chem 252: 609–612.

Sonderegger P, Jaussi R, Christen P (1985): "Transaminases." Christen P, Metzler DE (eds) New York: Wiley, pp. 502–510.

Storici P, Capitani G, De Biase D, Moser M, John RA, Jansonius JN, Schirmer T (1999): Biochemistry 38: 8628–8634.

Sugio S, Patsko GA, Manning M, Soda K, Ringe D. (1995): Biochemistry 34: 9661–9669.

Sundararaju B, Antson AA, Phillips RS, Demidkina TV, Barbolina MV, Gollnick P, Dodson GG, Wilson KS (1997): Biochemistry 36: 6502–6510.

Taguchi H, Ohta T (1991): J Biol Chem 266: 12588–12594.

Taoka S, West M, Banerjee R (1999): Biochemistry 38: 2738–2744.

Tobler HP, Gehring H, Christen P (1987): J Biol Chem 262: 8985–8989.

Toney MD, Kirsch JF (1991): J Biol Chem 266: 23900–23903.

Toney MD, Hohenester E, Cowan SW, Jansonius JN (1993): Science 261: 756–759.

Toney MD, Hohenester E, Keller JW, Jansonius JN (1995): J Mol Biol 245: 151–179.

Tormay P, Wilting R, Lottspeich F, Mehta PK, Christen P, Böck A (1998) Eur J Biochem 254: 655–661.

Tsai MD, Byrn SR, Chang C, Floss HG, Weintraub JR (1978): Biochemistry 17: 3177–3182.

Tsuzuki T, Obaru K, Setoyama C, Shimada K (1987): J Mol Biol 198: 21–31.

Tumanyan VG, Mamaeva DK, Bocharov AC, Ivanov VI, Karpeisky M, Yakovlev GI (1974): Eur J Biochem 50: 119–127.

Vacca RA, Giannattasio S, Graber R, Sandmeier E, Marra E, Christen P (1997): J Biol Chem 272: 21932–21937.

Vallee BL, Falchuk KH (1993): Physiol Rev 73: 79–118.

Vederas JC, Floss HG (1980): Acc Chem Res 13: 455–463.

Voet JG, Hindenlang DM, Blanck TJ, Ulevitch RJ, Kallen RG, Dunathan HC (1973): J Biol Chem 248: 841–842.

Watanabe N, Sakabe K, Sakabe N, Higashi T, Sasaki K, Aibara S, Morito Y, Yonaha K, Toyama S, Fukutani H (1989): J Biochem Tokyo 105: 1–3.

Webb EC (1992): "Enzyme Nomenclature 1992: Recommendations of the Nomenclature Committee of the International Union of Biochemistry and Molecular Biology." Orlando: Academic Press.

Williamson JM, Brown GM (1979): J Biol Chem 254: 8074–8082.

Wilmanns M, Hyde CC, Davies RD, Kirschner K, Jansonius JN (1991): Biochemistry 30: 9161–9169.

Wilson AC, Carlson SS, White TJ (1977): Ann Rev Biochem 46: 573–639.

Yamauchi T, Choi SY, Okada H, Yohda M, Kumagai H, Esaki H, Soda K (1992): J Biol Chem 267: 18361–18364.

Yano T, Kuramitsu S, Tanase S, Morino Y, Kagamiyama H (1992): Biochemistry 32: 1810–1815.

Yano T, Oue S, Kagamiyama H (1998): Proc Natl Acad Sci U S A 95: 5511–5515.

Yoshimura T, Nishimura K, Ito J, Esaki N, Kagamiyama H, Manning JM, Soda K (1993): J Am Chem Soc 115: 3897–3900.

Ziak M, Jaussi R, Gehring H, Christen P (1990): Eur J Biochem 187: 329–333.

Ziak M, Jäger J, Malashkevich VN, Gehring H, Jaussi R, Jansonius JN, Christen P (1993): Eur J Biochem 211: 475–484.

O-ACETYLSERINE SULFHYDRYLASE

By CHIA-HUI TAI and PAUL F. COOK, *Department of Chemistry and Biochemistry, University of Oklahoma, 620 Parrington Oval, Norman, Oklahoma 73019*

CONTENTS

Advances in Enzymology and Related Areas of Molecular Biology, Volume 74: Mechanism of Enzyme Action, Part B, Edited by Daniel L. Purich
ISBN 0-471-34921-6

I. Cysteine Synthesis

A. PATHWAYS FOR CYSTEINE SYNTHESIS

Sulfur exists in its elemental state (S^0) and in a wide range of oxidation states. Oxidized and reduced forms of sulfur can be interconverted in various organisms, and the assimilation of inorganic sulfur into organic acids is achieved in different ways in living organisms. The two common sulfur-containing amino acids, cysteine and methionine, are related with respect to routes for their synthesis. Cysteine can be synthesized by transsulfuration of homocysteine, a product of methionine degradation, or by incorporation of the sulfur from inorganic sulfate into cysteine (Scheme 1) (Giovanelli, 1987). In the following brief overview, the latter pathway will be discussed exclusively.

De novo synthesis of L-cysteine utilizes inorganic sulfate as the sulfur source. Sulfate is taken up and reduced intracellularly to sulfide via the sulfate reduction pathway (Scheme 2) (Kredich, 1996). The metabolism of sulfate, the most highly oxidized form of sulfur known, begins with the entrance of sulfate into the cell. Both *Escherichia coli* and *Salmonella typhimurium* have an efficient sulfate permease that also transports thiosulfate (Hryniewicz and Kredich, 1991). The uptake of sulfate into the periplasmic space takes place via a membrane channel, activated by a membrane-associated nucleotide binding protein (Sirko et al., 1990). In the periplasm there are two separate binding proteins for sulfate (*sbp*; Helinga and Evans, 1985) and thiosulofate (*cysP* gene product; Hryniewicz et al., 1990). Each of these proteins has a leader sequence to aid transport across the inner membrane into the cytoplasm. Once in the cell, sulfate is first activated by forming a mixed anhydride with AMP by the enzyme ATP-sulfurylase (Robbins and Lipmann, 1958), a heterooctamer of four catalytic and four regulatory subunits (Liu et al., 1994). The enzyme catalyzes the conversion of MgATP and sulfate to adenosine 5′-phosphosulfate (APS) and PP_i, which has a solution equilibrium constant of approximately 10^{-8} (Robbins and Lipmann,

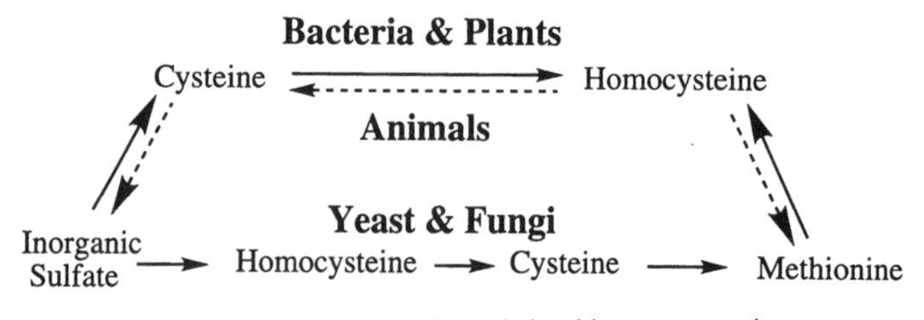

Scheme 1. Sulfur cycle—relationship among species.

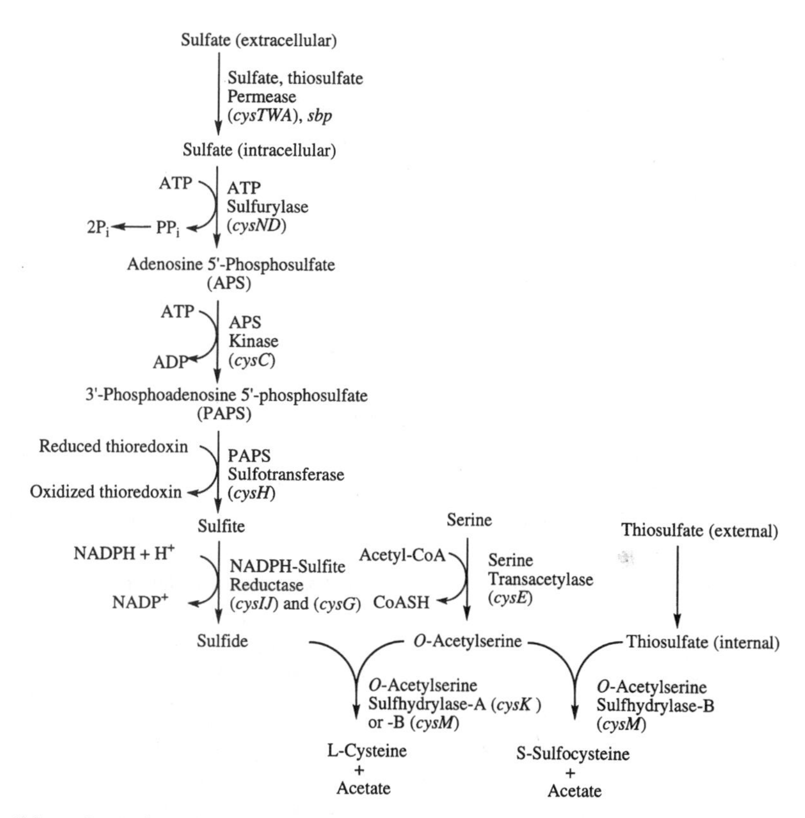

Scheme 2. Pathway for L-cysteine biosynthesis in Enterobacteriaceae. Genes responsible for their gene products are provided below the enzyme name.

1958). To compensate for the low K_{eq}, inorganic pyrophosphate is hydrolyzed to help drive the reaction, and the hydrolysis of GTP by the Ras-like regulatory subunit is coupled to the formation of a AMP-(ATP sulfurylase) complex, from which APS is formed upon attack by sulfate (Leyh et al., 1988, 1992; Leyh and Suo, 1992; Leyh, 1993). The APS thus formed is then converted to 3′-phosphoadenosine 5′-phosphosulfate (PAPS) via a second MgATP-dependent phosphorylation, catalyzed by the dimeric APS kinase. This reaction is thought to proceed via the intermediacy of a phosphoenzyme intermediate (Satishchandran and Markham, 1989; Satishchandran et al., 1992). Sulfite is then generated by the homodimeric PAPS sulfotransferase

(Krone et al., 1991), which catalyzes the transfer of the sulfonyl moiety of PAPS to one of the thiols of a redox active disulfide of thioredoxin to form acceptor–S–SO_3^-, which generates oxidized thioredoxin and sulfite (Tsang and Schiff, 1978; Tsang, 1983). The final step is then the generation of sulfide via a six-electron reduction of sulfite catalyzed by sulfite reductase. The sulfite reductase is a complex enzyme, composed of two different polypeptides, α, a flavoprotein with either FAD or FMN bound, and β, a hemoprotein with Fe_4S_4 cluster and siroheme components (Siegel and Davis, 1974). The overall stoichiometry of the complex is $\alpha_8\beta_4$, with four FAD and four FMN bound to the eight α-subunits. The pathway for electron transfer is from NADPH (the physiologic electron donor) → FAD → FMN → Fe_4S_4 → siroheme → sulfite (Siegel et al., ,1973, 1974). Another protein needed for sulfate reduction (at the level of sulfite reduction) is the *cysG* gene product, an S-adenosyl methionine-dependent uroporphyrinogen III methylase, responsible for synthesis of siroheme (Warren et al., 1990). The sulfide thus produced is the precursor for the synthesis of cysteine in enteric bacteria.

The biosynthesis of L-cysteine in enteric bacteria, such as *S. typhimurium* and *E. coli*, and plants proceeds via a two-step pathway (Scheme 3). The amino acid precursor of L-cysteine is L-serine, which is first acetylated at its β-hydroxyl by acetyl-CoA to give *O*-acetyl L-serine (OAS), catalyzed by the enzyme serine acetyltransferase (SAT) (EC 2.3.1.30) (Kredich and Tomkins, 1966). The final step in cysteine synthesis, replacement of the acetate side chain by a thiol, is catalyzed by *O*-acetylserine sulfhydrylase (OASS) (EC 4.2.99.8) with inorganic sulfide as the thiol donor (Becker et al., 1969). In addition to the utilization of sulfate to form L-cysteine, more reduced forms of sulfur, such as thiosulfate can be used in place of sulfide to give S-sulfocysteine (Nakamura et al., 1984), which is subsequently reduced to L-cysteine.

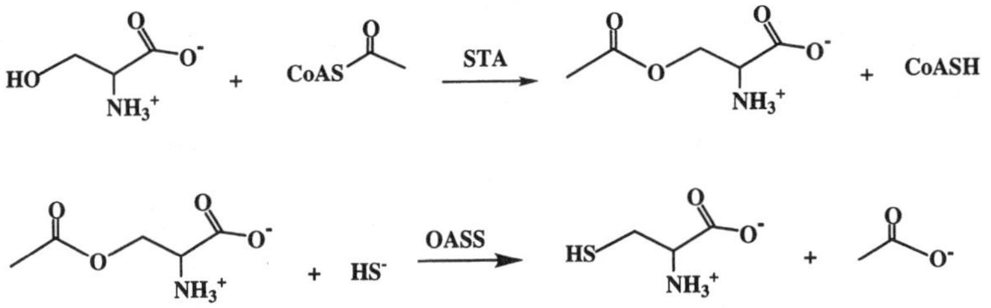

Scheme 3. Reactions responsible for L-cysteine biosynthesis in *Escherchia coli* and *Salmonella typhimurium*, formally the replacement of the β-hydroxyl of L-serine with a thiol.

B. REGULATION OF CYSTEINE BIOSYNTHESIS IN ENTERIC BACTERIA

Regulation of cysteine biosynthesis in enteric bacteria is achieved by several means including gene regulation, feedback inhibition by cysteine, and enzyme degradation (Kredich, 1996). The genes encoding the enzymes involved in the cysteine biosynthetic pathway of *S. typhimurium* form a regulon (Table 1). The cysteine regulon of *S. typhimurium* includes the following genes: *cysTWA* (sulfate permease), *sbp* (periplasmic sulfate-binding protein), *cysND* (ATP sulfurylase), *cysC* (APS kinase), *cysH* (PAPS sulfotransferase), *cysIJ* (sulfite reductase), *cysG* (siroheme synthesis), *cysK* (*O*-acetylserine sulfhydrylase-A), *cysM* (*O*-acetylserine sulfhydrylase-B), *cysE* (serine acetyltransferase), and *cysB* (regulatory protein). Overall, the genes are arranged in clusters of positively regulated operons, the negatively autoregulated *cysB* gene, the *cysE*, and *cysG* genes.

Transcription initiation is facilitated by binding of a tetramer of the CysB protein upstream from the positively regulated promoters, while CysB also regulates its own synthesis by binding to the promoter of the *cysB* gene and inhibiting transcription (Kredich, 1996). The *cysG* and *cysE* genes are not considered part of the regulon as they are not regulated by the CysB protein, although *cysE* is required for OAS synthesis. High levels of expression of the cysteine regulon require three factors: sulfur limitation, the presence of OAS, and the CysB protein (Kredich, 1996). Regulation of gene expression occurs via both derepression and induction. High levels of cysteine inhibit serine acetyltransferase, and this is relieved by a decrease in the level of cys-

TABLE 1
Cysteine Regulon and Related Genes in *Salmonella typhimurium*

Genes		Minutes[a]
cysB	Regulatory gene	33
cysK		54.5
cysPTWAM	Operon	54.7
cysC		61.8
cysND	Operon	61.9
cysHIJ	Operon	62.2
sbp		88.4
cysG	Related[b]	75.3
cysE	Related	81.4

[a]Minutes on *S. typhimurium* chromosome.

[b]Not under CysB control.

Source: Adapted from Kredich (1996).

teine. Activities of the enzymes in the pathway decrease progressively with growth on sulfate, sulfide, cysteine, or cystine, while djenkolic acid and reduced glutathione have shown maximum derepression. *O*-Acetyl-L-serine, the precursor of cysteine, is also the nonenzymatic precursor of *N*-acetyl-L-serine. *N*-Acetyl-L-serine enhances transcription by binding to the CysB protein, and this action is opposed by sulfide.

Finally, feedback inhibition of SAT by the end product L-cysteine represents the major physiologically significant form of kinetic regulation in the pathway. Cysteine is a potent inhibitor, with a K_i of 1 μM at 0.1 mM acetyl CoA for SAT either free or in complex with OASS (Kredich et al., 1969). Thus, cysteine regulates its own biosynthesis. It was observed that SAT and OASS are physically associated in vivo to form a multienzyme complex called cysteine synthase. The complex accounts for all of the SAT activity, but only 5% of the total cellular OASS activity, while the remaining 95% of the OASS activity is uncomplexed (Kredich et al., 1969). Concentrations of OAS of $\geq 10^{-4}$ M causes cysteine synthase to dissociate to free SAT and OASS-A. Cysteine synthase complex can be reconstituted by mixing resolved SAT and OASS-A in the absence of OAS (Kredich et al., 1969). The complex has a molecular weight of over 200,000, but it aggregates to multimers of twice and four times that size (Cook and Wedding, 1978). The stoichiometry of complex dissociation indicates that it is composed of a single ~90,000 Da trimer (S. Roderick, personal communication) of SAT and two 69,000 Da dimers of OASS-A. The multienzyme complex, unlike tryptophan synthase, does not channel its intermediate product OAS between the active sites of the two component enzymes. Rather, OAS is released into solution and must reassociate with OASS to be converted to L-cysteine (Cook and Wedding, 1977a). Given, that the predominant form of OASS is uncomplexed, the function of the multienzyme complex is unclear.

However, based on kinetic and analytical ultracentrifugal studies of the multienzyme complex in the absence and presence of effectors, a mechanism for the regulation of cysteine synthesis has been suggested (Cook and Wedding, 1977a, 1978). In the absence of small molecules, multiple aggregates ranging in molecular weight (MW) from a monomer of the complex to a tetramer are present in equilibrium. Any of the reactants that bind to the active sites of either of the two enzymes, SAT or OASS, shifts the equilibrium toward the monomer, with OAS most effective. Intermediary plateaus are observed in the saturation curves for L-serine and acetyl CoA when OAS synthesis is monitored in the multienzyme complex. Data have been interpreted in terms of the sum of several species with different saturation isotherms. The presence of L-cysteine as a negative effector, binding to an allosteric site on SAT, shifts the equilibrium completely toward the tetramer that is highly positively cooperative and with a $S_{0.5}$ of 1.5 mM for acetyl CoA. Under these conditions, the multienzyme complex is inactive given the cellular levels of acetyl CoA.

II. PLP-Dependent β-Replacement Reactions

Pyridoxal 5′-phosphate (PLP) dependent enzymes, with the exception of phosphorylase, which uses the 5′-phosphate as a general base (Klein, et al., 1982), catalyze their reaction via a number of covalent Schiff base intermediates of amino acids with PLP (Miles, 1986). In Figures 1, 2, 5 and 6, a number of the intermediates that are known to occur in all PLP-dependent reactions are pictured in terms of their structure and spectral properties (Miles, 1986). The pyridoxal 5′-phosphate-dependent enzymes as isolated have the PLP bound in Schiff base linkage to the ε-amino group of an enzymic lysine, that is, an internal aldimine (Fig. 1) (Jenkins and Sizer, 1957). The internal aldimine will absorb in the visible region at 400–430 nm if the imine nitrogen is protonated allowing the formation of an intramolecular hydrogen bond and bringing the π system of the imine into conjugation with the pyridine ring (Davis and Metzler, 1972). A resonance form of the protonated imine results from delocalization of electrons from O3′ to the imine nitrogen giving the visible conjugate termed the ketoeneamine. If the proton resides on O3′, however, the above resonance form is disallowed, and a species absorbing in the near ultraviolet at 310–340 nm called the enolimine is generated. Predominant resonance and tautomeric forms of the protonated internal aldimine are shown. Rotation about the C4–C4′ bond will also prevent the π system of the imine to be in conjugation with the pyridine ring giving absorbance at the lower wavelength of 310–340 nm (Davis and Metzler, 1972). All possibilities in terms of the tautomeric equilibrium of the internal aldimine have been observed, dependent on the enzyme. The prototypical example remains that of aspartate aminotransferase, which exhibits a pH-dependent equilibrium between the protonated (λ_{max}, 430 nm) and unprotonated (λ_{max}, 362 nm) tautomers of the internal aldimine (Jenkins and Sizer, 1957).

Addition of the amino acid substrate results in transimination, that is formation of a new Schiff base between the amino acid and PLP, and displacement of the enzyme's ε-amino group. The first step in the formation of a geminal diamine intermediate occurs via nucleophilic attack of the α-amine of the substrate at C4′ of the PLP imine of the internal aldimine. Intramole-

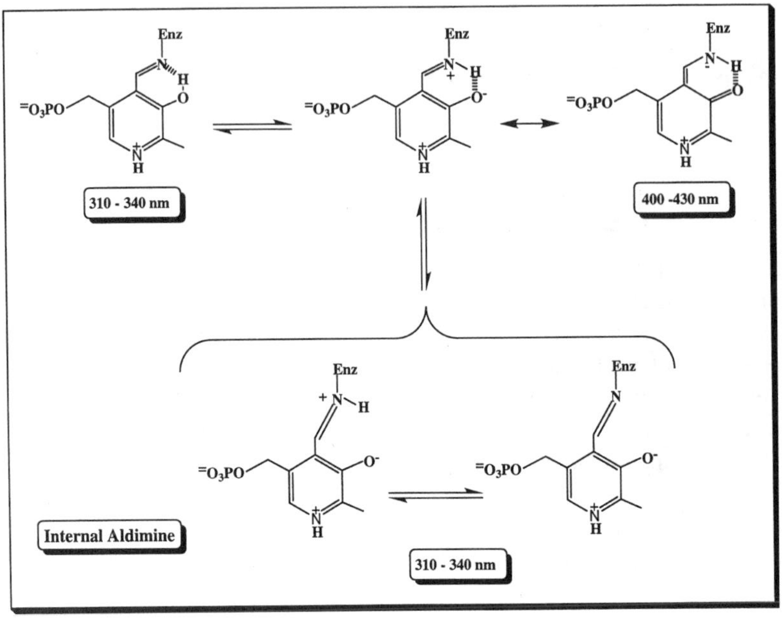

Figure 1. Tautomeric and resonance forms of the internal aldimine. The structure at *top left* represents the enolimine, while those resonance forms on the *right* represent the ketoeneamine. Structures at the *bottom* of the figure are generated by rotation about the C4–C4′ bond of the imine.

cular proton transfer from the incoming α-amino group to the departing ε-amino group and rotation about C4–C4′ to place the leaving ε-amino into a position orthogonal to the pyridine ring must precede collapse of the *gem*-diamine to the external aldimine between PLP and the amino acid substrate (Miles, 1986). Putative geminal diamine intermediates have been reported for only a few PLP-dependent enzymes, for example, serine hydroxymethyltransferase (Scarsdale et al., 1999) and the β-subunit of tryptophan synthase (Roy et al., 1988). Geminal diamine intermediates are shown in Figure 2 along with a form of the external aldimine intermediate. It should be noted that all tautomeric and resonance forms that exist for the internal aldimine may also exist for the external aldimine, and have been documented for other PLP enzymes (Dolphin et al., 1986).

The configuration of atoms around C_α in the bound external aldimine may differ, leading to a difference in the type of reaction catalyzed by PLP-

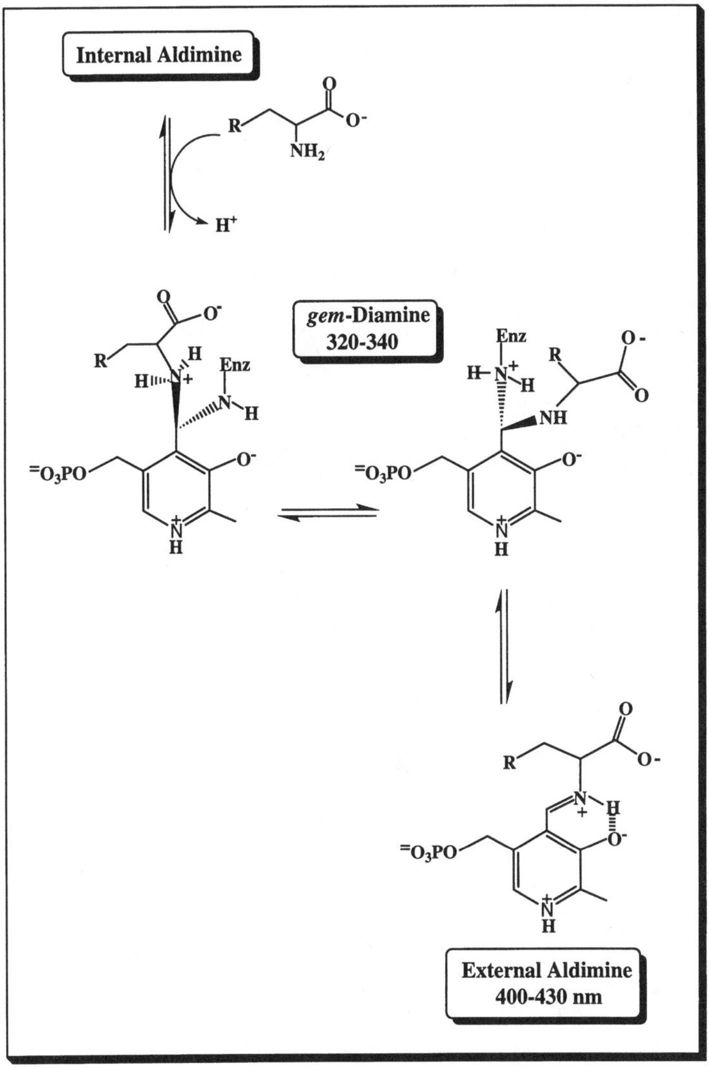

Internal Aldimine

gem-Diamine 320-340

External Aldimine 400-430 nm

Figure 2. Two structures of the *gem*-diamine are pictured in the *middle* of the figure. The structure on the *left* shows attack by the incoming α-amine of the amino acid, while that to the *right* has been rearranged in order to expel the Schiff base lysine of the internal aldimine. At the *bottom* of the figure is a ketoeneamine tautomer of the external Schiff, which is obtained upon collapse of the *gem*-diamine. It should be noted that all tautomeric and resonance forms shown in Figure 1 are also possible for the external Schiff base.

dependent enzymes (Fig. 3) (Dunathan, 1971). In order for a given bond to be labile it must be orthogonal to the plane of the PLP ring. Thus, as shown in Figure 3, the bond to X is potentially labile. If the α-carboxylate is positioned in the X location in Figure 3, the enzyme will catalyze a decarboxylation, while positioning the α-proton in this position results in a variety of different reaction types including aminotransferase, racemase, and elimination, dependent on the positioning and orientation of the remaining two functional groups. In Figure 3, the structure at the bottom right is optimized for a β-elimination reaction with the side chain positioned such that the leaving group, R, is orthogonal to the new π-bond to be formed as the β-substituent is eliminated. A number of PLP-dependent enzymes catalyze the overall elimination of a substituent at the β-position of an amino acid, and in some cases this is followed by addition of a new β-substituent. A general mechanism for the PLP-dependent β-elimination reaction beginning with the external aldimine is given in Figure 4 (Miles, 1986).

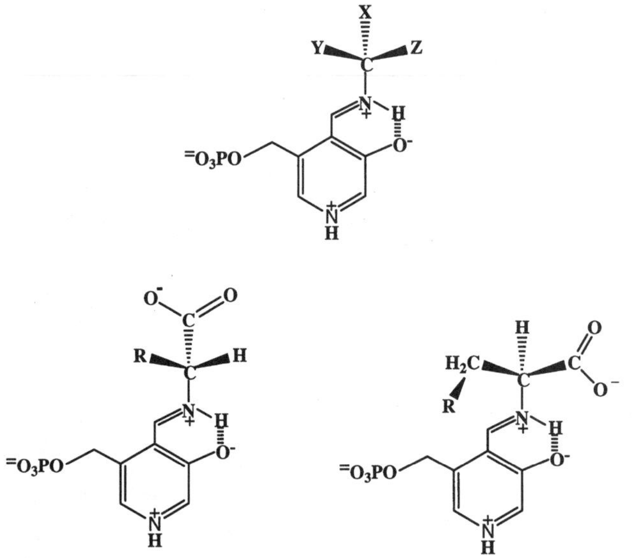

Figure 3. Structures illustrating that lability of a bond about the α-carbon depends on its ability to be orthogonal to the plane of the pyridine ring. In the top structure, X is labile, while in the bottom left an α-decarboxylation reaction is indicated. The structure on the bottom right is set for an α,β-elimination reaction, with the new π-bond to be formed parallel to the plane of the pyridine ring.

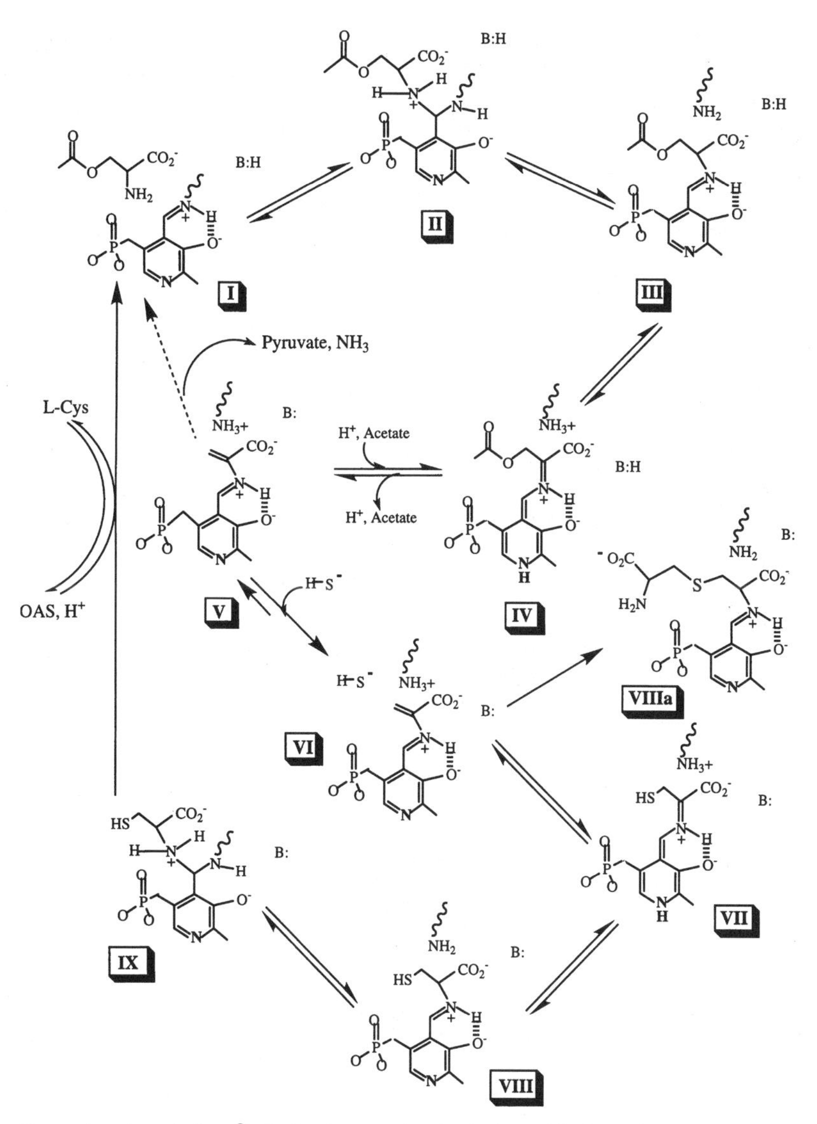

Figure 4. A typical α,β-elimination reaction beginning with the external Schiff base. α-Proton abstraction by the Schiff base lysine and delocalization of electrons into the pyridine ring occurs prior to elimination of the β-substituent. In other words, these reactions are thought to be a two-step process. ESB, Q, and AA represent the external Schiff base, quinonoid, and α-aminoacrylate intermediates, respectively.

The abstraction of the α-proton of the external aldimine and delocalization of electrons into the pyridine ring gives a quinonoid intermediate (Fig. 5), which absorbs at a wavelength higher than the external aldimine (Morozov, 1986). In order to form a quinonoid intermediate, N1 of the pyridinium ring must either be protonated, or become protonated by or ion-paired to an enzyme side chain as the intermediate forms. Whether a quinonoid intermediate is obligatory for all PLP-dependent β-elimination reactions is open to question.

Elimination of the β-substituent occurs next with the formation of α-aminoacrylate in external aldimine linkage with PLP (Fig. 6). The α-aminoacrylate external aldimine absorbs maximally between the wavelengths of the substrate external aldimine and the quinonoid intermediates if the intramolecular hydrogen bond between the imine nitrogen and O3′ of the cofactor exists. Otherwise, the intermediate absorbs at lower wavelength, 330–350 nm. Once elimination of the β-substituent has taken place, the active site lysine can displace the aminoacrylate intermediate, which will produce pyruvate and ammonia in solution, or, in the case of β-replacement reactions, another nucleophile attacks at the β-position of α-aminoacrylate to give the new amino acid by reversal of the steps discussed thus far. The α-aminoacrylate intermediate is very susceptible to nucleophilic attack in solution, but is quite stable in many PLP enzymes. The reason for its increased stability must reside in its environment within the active site of the enzyme, with the site perhaps partially closed to sequester the intermediate away from solvent.

O-Acetylserine sulfhydrylase is a member of the β-family of PLP-dependent enzymes that specifically catalyze β-replacement reactions. En-

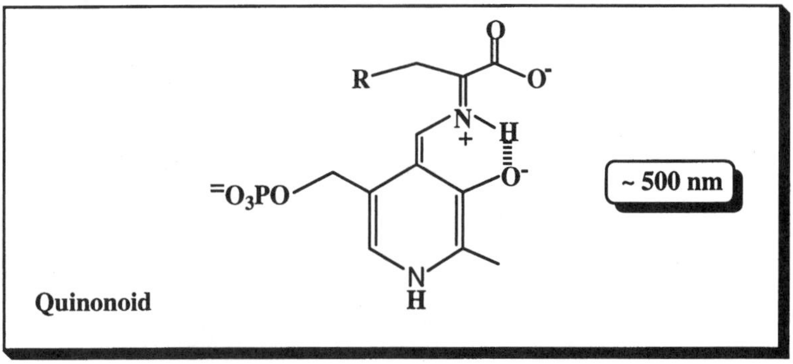

Figure 5. Structure of the quinonoid intermediate. Its extensive conjugation allows its absorbance at lower enegry.

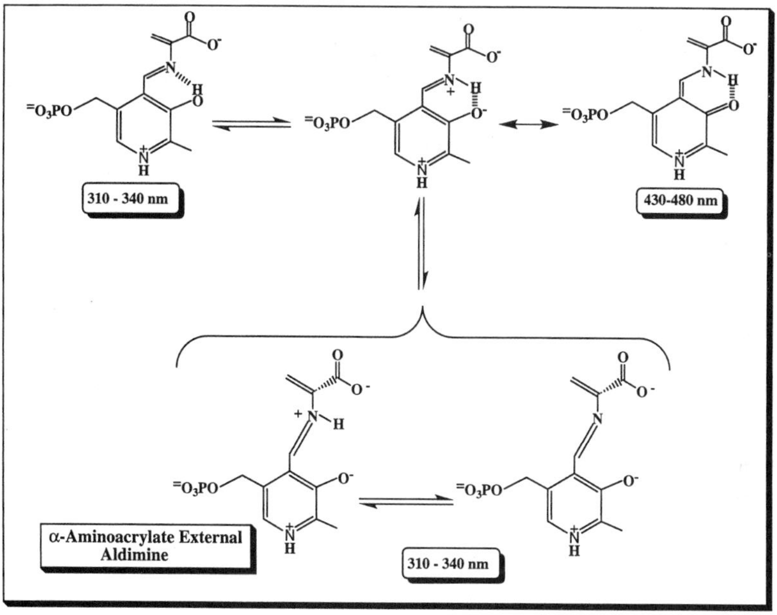

Figure 6. Tautomeric and resonance forms of the α-aminoacrylate intermediate. Its possible structures mimic those of the internal and external Schiff bases. It can be considered the external Schiff base of α-aminoacrylate.

zymes in this class include the β-subunit of tryptophan synthase (β-TRPS), cystathionine β-synthase, β-cyanoalanine synthase, and cysteine lyase (Alexander et al., 1994). Other than OASS, only β-TRPS has been extensively studied, and thus mechanistic comparisons will be limited to it.

III. *O*-Acetylserine Sulfhydrylase

There are two isozymes of OASS, A and B, in enteric bacteria (Becker et al., 1969), and those from *S. typhimurium* are best studied. Both are dimeric with the A isozyme, 68,900 Da (Byrne et al., 1988) and the B isozyme, 64,000 Da (Tai et al., 1993). *O*-Acetylserine sulfhydrylases are pyridoxal 5′-phosphate dependent with 1 mole of PLP/subunit (Kredich and Tomkins, 1966). The A and B isozymes are thought to be expressed under aerobic and anaerobic conditions, respectively. Both isozymes catalyze the formation of cysteine from OAS and sulfide, but the cell can more efficiently utilize more reduced forms of sulfur (e.g. thiosulfate) when the B-isozyme is expressed. Thiosulfate uptake provides an alternative means for cysteine biosynthesis, which eliminates the need for sulfate reduction (Nakamura et al., 1983). The exact mechanism by which S-sulfocysteine is converted to cysteine has not been determined, but could involve hydrolysis to cysteine and sulfate or reduction by glutathione to cysteine and sulfite (Woodin and Segel, 1968).

A sequence alignment of the A- and B-isozymes from *S. typhimurium* is shown in Figure 7. Note that the overall sequence identity is about 40%, with another 20% of the residues highly similar, and another 11% similar. All of the residues involved in PLP and substrate binding are identical in both enzymes. The mechanisms of the two isozymes of OASS are similar but not identical (Tai et al., 1993, 1995), and it is of interest to determine how such similar enzymes differ and why. However, the following thesis will concentrate on OASS-A, the most completely studied isozyme from

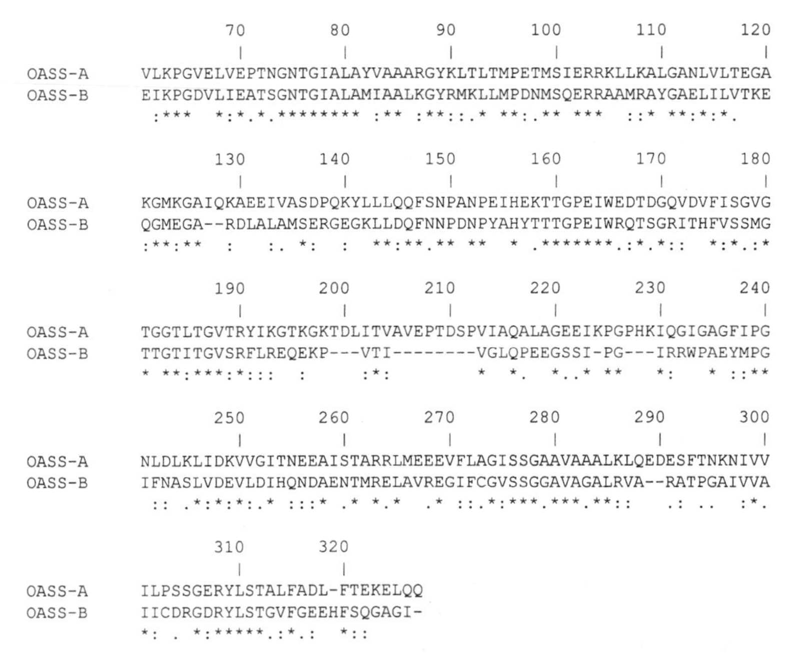

Figure 7. Sequence alignment between the OASS-A and OASS-B isozymes from *S. typhimurium*. An overall 70% homology is obtained between the two enzymes (*, identical; :, highly similar; ., similar.

S. typhimurium. Purification procedures have been developed for both isozymes and continuous assays have been developed that are applicable to both isozymes (Hara et al., 1990; Tai et al., 1993). In one of the assays the disappearance of sulfide is monitored using the sulfide ion selective electrode, and in the second the disappearance of 5-thio-2-nitrobenzoate, an analog of sulfide is monitored at 412 nm.

A. OVEREXPRESSION, PURIFICATION, AND PREPARATION OF APOENZYME

Wild-type recombinant OASS-A is obtained by overexpression from the vector pRSM40 (originally and kindly provided by Professor N.M. Kredich at Duke University Medical School). The plasmid, which contains the *cysK* gene behind its natural promoter, is transduced into *S. typhimurium* strain DW378 (*trpC109*, *cysK1772*, *cysM1770*) via P22 lysates. Overnight growth in the fermentor is in the presence of Vogel-Bonner medium (Vogel and Bonner, 1956), with 0.5% glucose, 1% LB, 40 μM L-tryptophan, 500 μM glutathione, and 100 μg ampicillin. Glutathione is the sole sulfur source and acts by derepressing the cysteine biosynthetic pathway (Kredich, 1971).

The OASS-A is purified by conventional means of sonication, streptomycin sulfate precipitation, ammonium sulfate fractionation, and chromotographies on DEAE 5pw and phenyl 5pw (Tai et al., 1993). The enzyme is >95% pure based on SDS-PAGE, but has only half the specific activity as does the native enzyme. Addition of PLP results in a restoration of the wild-type specific activity, suggesting the bacteria has made so much of the enzyme, it has become cofactor-limited. Overall, an overnight fermentor growth gives about 50 g of wet cell paste, which yields about 1 g of pure OASS-A. The above data suggested that the apoenzyme should be quite stable, and that the PLP cofactor is added as a posttranslational event. To determine whether these hypotheses might be correct, a scheme to prepare the apoenzyme of OASS-A was developed.

Apo-OASS-A can be prepared by dialysis of the α-aminoacrylate intermediate against 5 M GnHCl, at pH 6.5, 100 mM Mes (Schnackerz and Cook, 1995). Once prepared, the apoenzyme is stable for months. Reconstitution with PLP and PLP analogs such as pyridoxal 5′-phosphate, 2′-methyl, and pyridoxal 5′-deoxymethylenephosphonate can easily be accomplished (Cook et al., 1996). Reconstitution was verified by activity in the cases where analog-substituted enzymes are active, and additionally by ^{31}P NMR chemical shift. The apo-OASS-A has been utilized for a number of experiments, most notable those dealing with protein dynamics discussed below.

B. STRUCTURE

O-Acetylserine sulfhydrylase-A was crystallized from PEG 4000 in the presence of Li_2SO_4 and Tris, pH 8 in two crystal forms, one is in a hexagonal space group P6, and the other in the orthorhombic space group $P2_12_12_1$ (Rao et al., 1993). The structure of crystalline orthorhombic OASS-A was solved to 2.2 Å resolution using the technique of multiple isomorphous replacement (Burkhard et al., 1998). A monomer of the OASS dimer, Figure 8, is composed of an N-terminal (residues 1–145) and a C-terminal (residues 146–315) domain. Both domains consist of a central β-sheet structure surrounded by α-helices. One stretch of the N-terminal domain (residues 13–34) "crosses over" into the C-terminal domain, forming the first two strands of its central β-sheet. The overall structure of OASS-A is similar to that of β-TRPS (Hyde et al., 1988), with the exception that the first two helices in β-TRPS are missing in OASS-A, replaced by the first seven residues of the N-terminal domain (the first seven residues make up part of the dimer interface), and that the coiled region of β-TRPS (residues 260–310) that

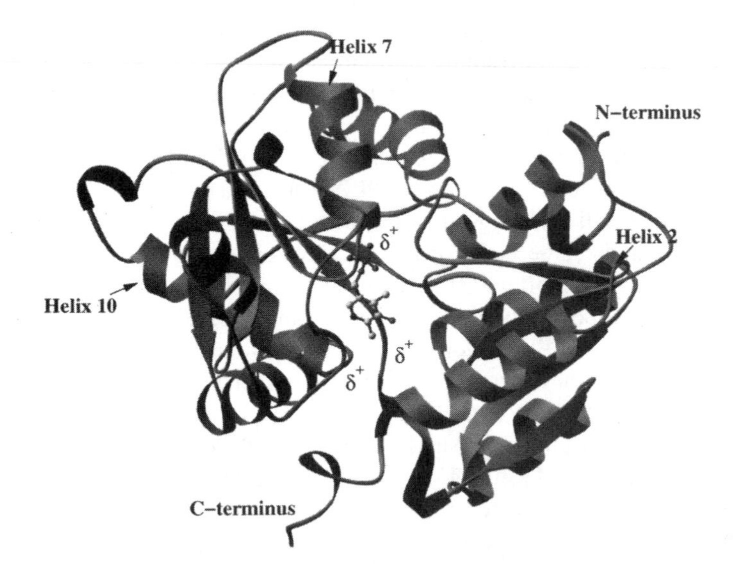

Figure 8. Structure of OASS-A monomer. Three helix dipoles are directed toward the PLP cofactor: helix 2 directed toward the 3′-oxygen, helix 7 directed toward the 5′-phosphate, and helix 10 directed toward N1. α-Helices are in *light blue*, 3_{10} helices are in *blue*, β-sheet is *green*, and coil is *brown*. The cofactor is portrayed in stick and ball with N1 *blue*; O, *red*; and P, *magenta*. (See Color Plates.)

forms extensive interactions with α-TRPS and contributes to the hydrophobic indole channel (Hyde et al., 1988), is missing in OASS-A. In addition, α-helix 10 of β-TRPS, immediately following the extended coiled region 260–310, is also missing in OASS-A, replaced by a 3_{10} helix. It is clear, however, that both enzymes share the same general fold.

The PLP in OASS-A is located at the interface between the two structural domains, deeply buried within the protein. A close-up of PLP in the active site is shown in Figure 9. The 5'-phosphate of PLP is tightly hydrogen-bonded to a glycine/threonine rich loop, with each of the phosphate oxygens accepting two hydrogen bonds from the protein. In good agreement are ^{31}P NMR studies of OASS-A (Cook et al., 1992; Schnackerz and Cook, 1995; Schnackerz et al., 1995). The observed ^{31}P chemical shift is 5.2 ppm with a linewidth of 20.5 Hz, consistent with a tightly bound phosphate with motion restricted to that of the protein. The value of the chemical shift also indicates it is dianionic as bound to the protein. The PLP C4′ is in Schiff base linkage with the ε-

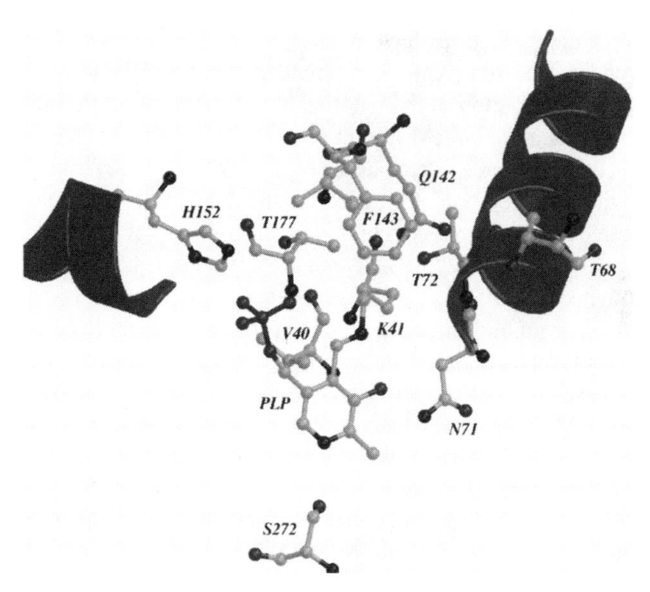

Figure 9. The close-up view of PLP in the active site of OASS-A. The phosphate group of the cofactor is bound at the N-terminus of helix 7, which interacts with the positive end of the dipole with the negative charge of the phosphate portion of the protein (Gly176 to Thr180). Four of the H-bound donors are peptide NH groups. The cofactor, and enzyme side chains are portrayed in stick and ball with N *blue*; O, *red*; C, *yellow*; and P, *magenta*. (See Color Plates.)

amino of K41 in OASS-A and is within hydrogen-bonding distance to O3′ of PLP, which is also hydrogen-bonded to N71. The pyridinium nitrogen of PLP is within hydrogen-bonding distance of S272. In this regard, N1 of PLP in OASS-A is located close to the positive dipole of a α-helix. The *re* face of the cofactor internal Schiff base is directed toward the entrance to the active site.

The exact OAS binding mode has not yet been determined, but information has been obtained from a structure of the K41A mutant of OASS-A. The K41A mutant is isolated in a closed conformation as an external Schiff base with free methionine, and thus represents an analog of the OAS external Schiff base (Burkhard et al., 1999). The conformational change to close the site occurs in the region of residues 85–136 of the N-terminal domain, which rotates by 15° along an axis from residues 136 to 87 to 106. Residues 66–70, a loop in the N-terminal, also undergo a change in conformation, altering the active site cavity to a slight degree. The mutant protein provides some indication of the location of the side chain of OAS, but Met is not a perfect mimic of OAS, and some ambiguity remains. Nonetheless, the side chain of Met extends away from the PLP cofactor toward the entrance to the active site (Fig. 10). Very similar structures have been published for the K87T mutant of β-TRPS with serine and tryptophan bound. A similar rigid body rotation is observed upon formation of the external Schiff base, but positioning of the substrate side chain is quite different in β-TRPS, leading to changes in the orientation of the bound cofactor (Rhee et al., 1997). The above structures aid greatly in the interpretation of the mechanistic data that have been obtained.

The active site of OASS-A is relatively unforgiving of even small structural changes. The A-isozyme has a single cysteine, C42, adjacent in primary sequence to the Schiff base lysine, K41. The cysteine side chain is directed away from the active site into the protein structure from the *si* side of the cofactor internal Schiff base. A conservative change of C42, to serine, has significant consequences to the overall function of the enzyme (Tai et al., 1998). Ultraviolet-visible, fluorescence, circular dichroism, and ^{31}P NMR spectral properties are identical to those of the wild-type enzyme, as are most of the kinetic properties using either 5-thio-2-nitrobenzoate (TNB) (see below) or sulfide as the nucleophilic substrate. However, the α-aminoacrylate intermediate is 50–fold lower in stability compared to that of wild type as reflected by an increase in the OAS:acetate lyase activity. In addition, there is a 17-fold decrease in V. Thus, the subtle rearrangement caused by the smaller serine hydroxyl as it hydrogen bonds with hydrophilic groups in its local environment are transmitted to PLP, affecting its orientation and resulting in the changes in the overall reaction.

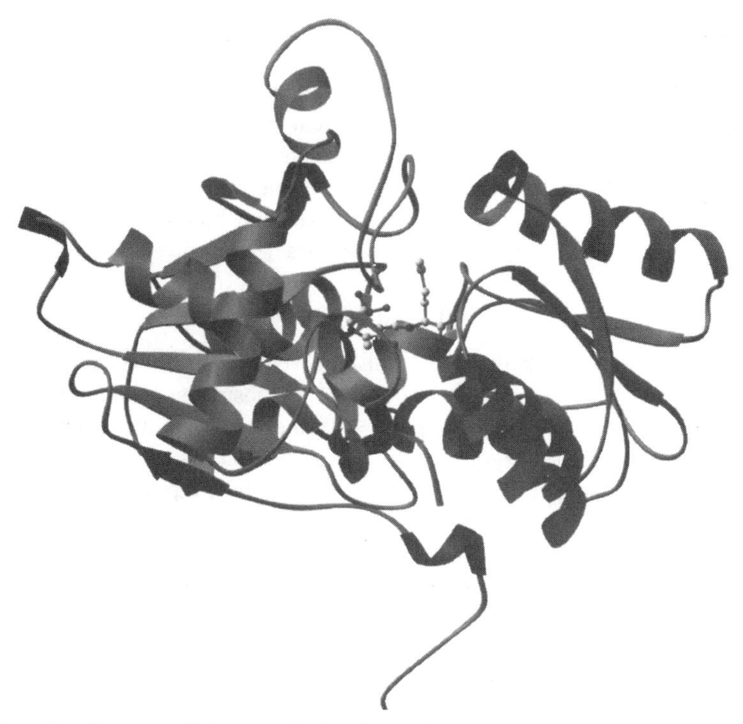

Figure 10. Structure of K41 mutant OASS. The methionine (S in *green*) in external Schiff base linkage with the active site PLP is clearly seen extending back toward the entrance to the site. (See Color Plates.)

Several different orthorhombic crystals of OASS-A have now been obtained, and studied using single crystal microspectrophotometry (Mozzarelli et al., 1998). One of these crystal forms, when soaked with OAS gives the α-aminoacrylate intermediate, but with a K_d for OAS that is 500 times greater that obtained in solution (Cook et al., 1992). Data suggest interference as a result of either the inability to undergo the conformational change required to form the external Schiff base with OAS, the binding of OAS, or both. The α-aminoacrylate intermediate formed in the crystal decays slowly at pH 7, consistent with the OAS:acetate lyase activity observed in solution (see below). Addition of azide, a sulfide analog, to the crystalline α-aminoacrylate intermediate results in a regeneration of the free enzyme spectrum, and thus the crystals are capable of catalyzing the overall reaction.

Addition of L-serine or L-cysteine to the enzyme crystals results in the formation of the external Schiff base as observed in solution, and K_d values much closer to those obtained in solution. The difference between these resides in the side chain acetyl, which is absent in both cases, suggesting the acetyl contributes to the conformational change that closes the site upon formation of the external Schiff base. Some of the other crystal forms also react with OAS to give the α-aminoacrylate intermediate, but with an even higher K_d for OAS, and in the process the crystals are destroyed.

C. SUBSTRATE SPECIFICITY

O-Acetylserine sulfhydrylase-A utilizes several β-substituted amino acids (analog of OAS) with good leaving groups as substrates. Among these are *O*-propionyl- and *O*-butyryl-L-serine, L-serine-O-sulfate, and β-chloro-L-alanine. However, the enzyme is much more promiscuous with respect to the nucleophilic (analog of sulfide) substrate. Nucleophilic substrate analogs that have been identified include the following: NH_3 (1 mM), NH_2OH, 2-aminopyrimidine, 3-aminoquinoline, phenol, *p*-methylaminophenol, *trans*-cinnamic acid, formate (100 mM), 5-thio-2-nitrobenzoate (TNB), 2-mercaptoethanesulfonic acid, thiolactate, 2-mercaptoethanol, thiophenol (1 mM), 2-mercaptopyrimidine (5 mM), mercaptosuccinate, 2-mercaptobenzoxazole, 2-mercaptoimidazole, mercaptoacetate, thiosalicylate, and 2-mercaptopyridine. Note that nitrogen, oxygen, and thiol moities serve as nucleophilic substrates. Substrate specificity appears to depend more on the pK_a of the nucleophile than its structure. The best OAS analog is *O*-propionyl-L-serine, which has a V/E_t of 60% and a K_m similar to that of OAS (Cook and Wedding, 1977b), while the best sulfide analog (unpublished) characterized is mercaptoacetate with a V/E_t of 3% and a K_m 1000-fold greater than that of sulfide.

D. KINETIC MECHANISM

1. *Steady State Kinetics*

One of the first and most important pieces of information to be obtained for a bireactant enzyme reaction is the kind of kinetic mechanism (i.e., sequential or ping pong) and a quantitative description of the reaction pathway. The information obtained is vital to an interpretation of subsequent kinetic data, since it allows one to be certain of the enzyme form that predominates, and thus ensures a structure-based data analysis. The kinetic mechanism of OASS-A was first determined qualitatively using a sulfide ion selective electrode assay as BiBi ping-pong with competitive substrate

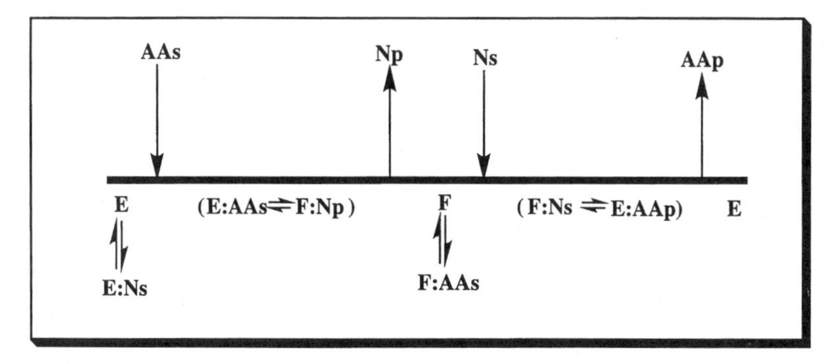

Figure 11. Predominant ping-pong BiBi pathway for *O*-acetylserine sulfhydrylase. The E and F forms are two stable enzyme forms, while AA_s, AA_p, N_s, and N_p represent the amino acid substrate and product and the nucleophilic substrate and product, respectively. Note that both substrates give competitive inhibition by binding to the incorrect enzyme form, typical for a classical ping-pong kinetic mechanism.

inhibition by both OAS and sulfide (Cook and Wedding, 1976). Subsequent assessment of the assay indicated that it overestimated the rate of the OASS reaction. Thus, the kinetic mechanism was redetermined and found to be qualitatively identical (Fig. 11), but quantitatively different than that determined previously (Tai et al., 1993). The turnover number for OASS at pH 7 is 280 s^{-1}, while the second order rate constants that reflect the first and second half reaction are 2×10^5 $M^{-1}s^{-1}$ ($V/K_{OAS}E_t$) and 2×10^7 $M^{-1}s^{-1}$ ($V/K_{sulfide}E_t$) (Tai et al., 1993, 1995). Data thus suggest a rate-limiting first half reaction, given the near diffusion-limited second-order rate constant for sulfide and the F form of the enzyme to produce L-cysteine. All of the product and dead-end inhibition patterns are consistent with the ping-pong mechanism. Sulfide gives only partial competitive substrate inhibition (Cook and Wedding, 1976; Tai et al., 1993), suggesting that a pathway in which sulfide binds prior to OAS is allowed. In agreement with the alternate sequential kinetic mechanism, the first product of the reaction, acetate, exhibits S-parabolic inhibition versus OAS, indicative of acetate binding to both E and F forms of the enzyme which are reversibly connected (Tai et al., 1993). In addition, acetate induces partial substrate inhibition by OAS, suggesting that the sulfide inhibitory site can be occupied by acetate, and the reaction will still proceed at a slower rate (Fig 12). Qualitatively identical results are obtained with β-chloro-L-alanine, an analog of OAS, with TNB, an analog of sulfide, in any combination. However, the *V*/*K* value for the amino acid substrate is

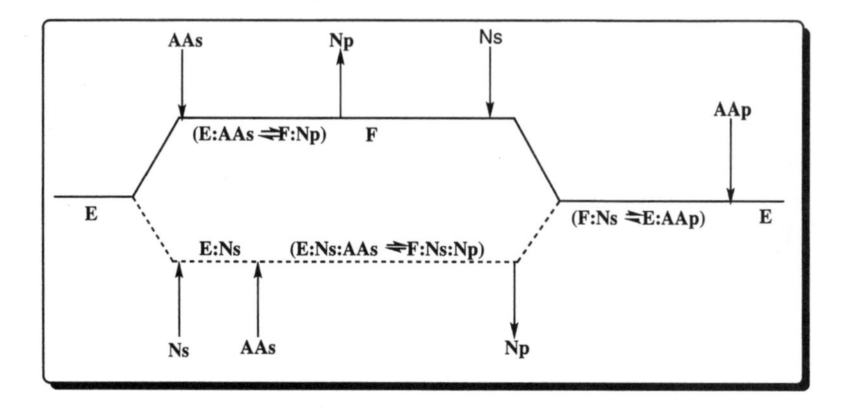

AAs, AAp - Amino acid substrate, product
Ns, Np - Nucleophilic substrate, product

Figure 12. Complete kinetic mechanism for OASS-A. The top, solid pathway is that seen in Figure 11. The bottom pathway allows for some randomness in the mechanism. At high concentrations of the nucleophilic substrate, such that it is bound, the amino acid substrate can still be bound and processed at a finite but low rate (see text).

not constant (as predicted for a ping-pong kinetic mechanism) as the nucleophilic substrate is changed from sulfide to TNB. These data are also consistent with the random component of the kinetic mechanism and a change in the stability of the α-aminoacrylate intermediate. Results are summarized in Table 2. The reaction is practically irreversible in the direction of conversion of cysteine to OAS. In contrast to OASS, β-TRPS and all of the other members of the β-replacement family proceed by a sequential, or ternary complex mechanism (Miles, 1986), although β-TRPS is capable of producing the α-aminoacrylate intermediate in the absence of cosubstrate. The ca-

TABLE 2
Kinetic Parameters for Amino Acid/Nucleophilic Substrate Pairs

	OAS/sulfide	OAS/TNB	BCA/sulfide	BCA/TNB
$V/E_t (s^{-1})$	130 ± 17	0.56 ± 0.08	2.0 ± 0.1	0.016 ± 0.002
$V/K_{OAS}E_t$ $(M^{-1}s^{-1})$	$(1.4 \pm 0.7) \times 10^5$	37 ± 5		
$V/K_{BCA}E_t$ $(M^{-1}s^{-1})$			$(3.7 \pm 0.6) \times 10^3$	1.2 ± 0.2
$VK_{sulfide}E_t$ $(M^{-1}s^{-1})$	$(2.4 \pm 0.8) \times 10^7$		$(3.9 \pm 0.2) \times 10^6$	
$V/K_{TNB}E_t$ $(M^{-1}s^{-1})$		950 ± 55		590 ± 130

Source: Adapted from Tai et al. (1993).

pability of isolating the two halves of the OASS-A reaction make it one of the most useful and important of PLP-dependent enzymes that catalyze a β-replacement reaction.

Under certain circumstances, patterns that appear intersecting can be obtained, see below, with cofactor-analog substituted OASS, or with some of the mutant enzymes (e.g., C42S). The reason for the apparent change in mechanism is an increase in the rate of the OAS:acetate lyase activity, which results in a dependence of the two half reactions on each other. Theory has been developed for this scenario in Cook et al. (1996).

Evidence of randomness with sulfide (or analog) binding to E followed by OAS to give a productive although much less efficient Michaelis complex (Tai et al., 1993) represents an apparent paradox, since sulfide (or analog) presumably must occupy the site for the leaving acetate group of OAS. It is possible that OAS could be bound and processed without the leaving group in its optimum location, generating F, which would be positioned to accept the already bound nucleophile, or that the sulfide inhibitory site is not the same as its substrate site, if there is such a site. The answer to this paradox bears on the rigidity/flexibility of the active site of PLP enzymes, and the chemistry dictated by them. The structure of OASS-A indicates the presence of a hole behind the cofactor in the vicinity of H152. It is possible that an alternative binding mode for the nucleophilic substrate is behind the cofactor such that OAS could still bind, the α-aminoacrylate intermediate could be formed by elimination of acetate and its diffusion away from the site, and then the nucleophilic substrate could add from the opposite face of the α-aminoacrylate intermediate. The reaction under these conditions would, however, give the opposite stereochemistry at C3 (see below).

2. *Ultraviolet-Visible Spectral Studies*

The absorbance spectrum of OASS-A, the E form of the enzyme according to Figures 11 and 12, in the absence of reactants is pH independent over the range 5.5–11, and is shown in Figure 13. The E form of the enzyme represents PLP in internal aldimine linkage with an active site lysine (K41; Rege et al., 1996). The absorbance maximum at 412 nm indicates a predominance of the protonated imine in the ketoeneamine tautomer (Becker et al., 1969; Cook and Wedding, 1976: Cook et al., 1992), but finite absorbance at 330 nm suggests a small contribution from the enolimine tautomer (Fig. 1). Similar spectra are obtained for β-TRPS (Goldberg and Baldwin, 1967).

Addition of OAS to OASS-A results in a decrease in the absorbance at 412 nm with concomitant increases in absorbance at 330 and 470 nm as a re-

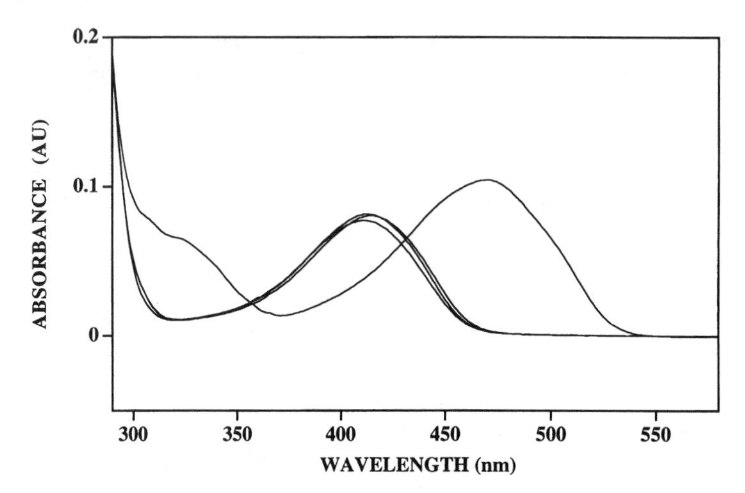

Figure 13. Absorbance spectra of 10 μM OASS-A alone at pH 6.5, 8, and 9.5 and in the presence of 1 mM OAS (λ_{max}, 470 nm) at pH 6.5.

sult of formation of two tautomers of an α-aminoacrylate Schiff base (Figs. 6 and 13) after elimination of the elements of acetic acid from the OAS external aldimine (Cook and Wedding, 1976; Schnackerz et al., 1979; Cook et al., 1992). Addition of L-serine to TRPS results in a mixture of external aldimine and α-aminoacrylate intermediates (Drewe and Dunn, 1985, 1986; Miles, 1986). In this case, the external aldimine absorbs maximally at 422 nm, reflecting the ketoeneamine tautomer, while the α-aminoacrylate intermediate absorbs maximally at 350 nm with a shoulder at 460 nm, reflecting predominately the enolimine tautomer. The difference between the two enzymes likely reflects differences in the hydrophilicity of their active sites, with the two likely stabilizing different charged forms of the α-aminoacrylate intermediate. In addition, the concentration of the α-aminoacrylate intermediate is lower in the β-TRPS reaction compared to that of OASS-A.

In order to determine whether L-cysteine interacts with OASS-A, the spectrum of OASS-A was obtained in the presence of different concentrations of the amino acid product (Schnackerz et al., 1995). A shift in the λ_{max} from 412 to 418 nm is obtained indicative of external aldimine formation, predominately as the ketoeneamine tautomer. No evidence of the α-aminoacrylate intermediate is observed. Thus, it appears that formation of the L-cysteine external aldimine is the irreversible step in the direction of cysteine formation. This result was unexpected since β-TRPS can form the aminoacrylate intermediate from cysteine. However, the natural substrate for β-TRPS is L-serine, and one would expect general acid catalysis in the elimination of hydroxide. Elimination of acetate by OASS requires no such catalysis. It was thus of interest to determine whether L-serine is a substrate.

No substrate activity was detected with L-serine using our conditions at very high enzyme concentrations. However, addition of L-serine to OASS-A results in a decrease in the absorbance at 412 nm concomitant with an increase in the absorbance at 320 nm and a shift in the λ_{max} of the remaining visible absorbance to 418 nm (Schnackerz et al., 1995). Data suggest that the external aldimine with L-serine can be formed as an equilibrium mixture between ketoeneamine and enolimine tautomers. Data from Flint et al. (1996) have measured activity with L-serine and OASS-A using a mass spectrometric assay, but with a turnover number $<5 \times 10^{-6}$ s^{-1}. It is reasonable to assume then that the inability to eliminate the thiol of L-cysteine is a result of the lack of a suitable general acid in the OASS-A reaction. This aspect will be considered further below.

3. *Stopped-Flow Kinetic Studies*

Given the spectral signature of a number of intermediates along the reaction pathway, the use of pre-steady-state kinetic studies can be utilized to ask a number of questions concerning the overall reaction. Rapid-scanning and single wavelength stopped-flow kinetic studies of the OASS-A reaction (Cook et al., 1996; Woehl et al., 1996) have been used (1) to obtain estimates for rate constants along the reaction pathway and (2) to locate slow steps in the overall reaction, and each of the two half reactions. Reaction of OASS-A with OAS in the absence of sulfide is pseudo first order and dependent on the concentration of OAS. At early times, a rapid formation of the OAS external aldimine is observed, and behaves in a pre-equilibrium manner, giving a K_{ESB} of 5 mM, similar to the steady-state K_{OAS} of 2 mM (Tai et al., 1993). Data indicate the first half reaction (E plus OAS to F plus acetate) is rate-determining overall, and abstraction of the α-proton from the OAS external Schiff base is the slowest step in the first half reaction. The rate constant for formation of the OAS external Schiff bases is ≥1000 s^{-1}, and for the subsequent formation of the α-aminoacrylate intermediate is 400 s^{-1}, similar to the value of 280 s^{-1} measured for V/E_t (see above). No evidence was obtained for the presence of either *gem*-diamine or quinonoid intermediates in the OASS-A reaction. For TRPS, α-proton abstraction also limits the overall reaction (Miles and McPhie, 1974; Lane and Kirschner, 1983; Drewe

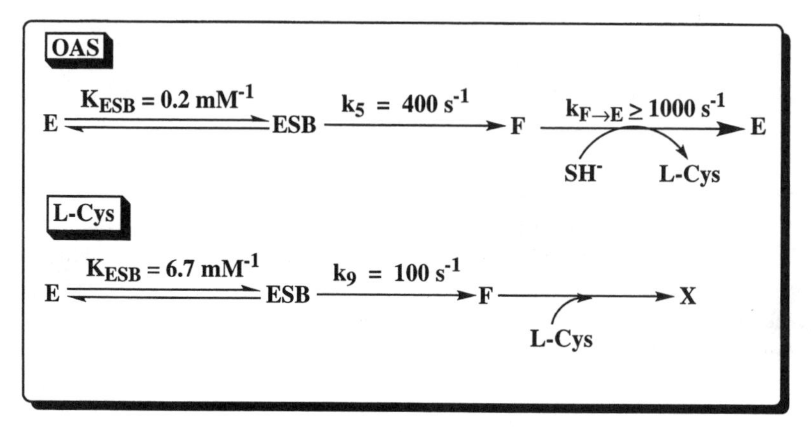

Figure 14. Summary of the stopped-flow kinetic data. Note that the generation of the α-aminoacrylate intermediate from cysteine occurs at one-fourth the rate observed with OAS, and is transient in the absence of a nucleophile other than cysteine. Note also the rapidity of the second half reaction, and rate-limitation of the α-proton abstraction. The definition of X is given in Figure 15, and in Scheme 4.

and Dunn, 1985, 1986). In the TRPS reaction, however, one does observe a quinonoid intermediate, while the OASS reaction appears concerted (see below). A minimal mechanism for the first half reaction and a summary of the estimated rate constants is given in Figure 14.

Reaction of preformed α-aminoacrylate intermediate (addition of OAS to the OASS syringe) with sulfide gives a rapid bleaching of α-aminoacrylate intermediate to give the free enzyme spectrum, which remains until sulfide is depleted. One then observes a first-order regeneration of the α-aminoacrylate intermediate spectrum. Thus, all steps in the second half reaction are very rapid, >1000 s^{-1}, even at the lowest concentration of sulfide used (5 μM; K_{sulfide} is 6 μM). The second half reaction for β_2-TRPS is very fast as is true for OASS, but for $\alpha_2\beta_2$-TRPS one observes rapid quinonoid formation followed by appearance of the external Schiff base.

Based on the above, the second half reaction was thought to lie far toward the formation of L-cysteine as a result of a practically irreversible addition of sulfide to the α-aminoacrylate intermediate. Addition of L-cysteine to OASS-A results in the apparent formation of the external aldimine (see above). To determine whether the rate of external Schiff base formation could be measured, the reaction of L-cysteine with OASS-A was measured by rapid-scanning stopped-flow spectrophotometry. Interesting is the observation that the external Schiff base with L-cysteine forms rapidly, giving a K_{ESB} of 0.15 mM, similar to the K_{cys} measured from spectral titrations (Schnackerz et al., 1995). Following formation of the external Schiff base, transient formation of the α-aminoacrylate intermediate occurs with a rate constant 100 s^{-1}. Overall data suggest attack of the thiol of cysteine on C3 of the α-aminoacrylate intermediate, resulting in formation of the L-lanthionine external Schiff base, which the enzyme binds tightly (Fig. 15). Thus in the absence of another nucleophile, the product thiol is reasonable and attacks the aminoacrylate intermediate to give the new product. However, in the presence of a better competing nucleophile, the second half reaction is reversible. In agreement with this suggestion, Flint et al. (1996) isolated several enzymes from bacteria that are responsible for generating the sulfide required for FeS cluster formation. Two of the enzymes isolated were the two isozymes, A and B, of *O*-acetylserine sulfhydrylase, indicating the in vivo reversibility of the second half reaction. The competing nucleophile in this case was not identified.

4. *Isotope Effect Studies*

Isotope effects are a versatile probe of enzyme mechanism. Given the suggested rate-determining α-proton abstraction, primary deuterium kinetic isotope effects were measured in the steady state and pre-steady state by direct comparison of the initial rates obtained with OAS and OAS-2-D (Hwang et al., 1996). The value of $^{D}(V/K_{OAS})$ is pH dependent with a value of 2.8 at low pH and a lower value of 1.7 at high pH. Measurement of the isotope effect on the isolated rate constant for α-proton abstraction in the presteady state gives a value about equal to the low pH value of 2.8 suggesting this value is the intrinsic isotope effect that can be used to interpret transition state structure (see below). Analysis of the data indicate that the OAS external Schiff base partitions toward the α-aminoacrylate intermediate 1.5-times faster than it does back toward the internal Schiff base. As a result, estimates of most of the rate constants along the reaction pathway of the first half of the reaction have been obtained.

5. *Summary*

The kinetic mechanism for OASS-A is predominantly ping-pong (Cook and Wedding, 1976; Tai et al., 1993) with a slow first half reaction (Woehl et al., 1996) and overall limitation by C_{α}–H bond cleavage (Hwang et al., 1996); the overall turnover number is 280 s^{-1}. All product and dead-end inhibition patterns are consistent with the ping-pong mechanism.

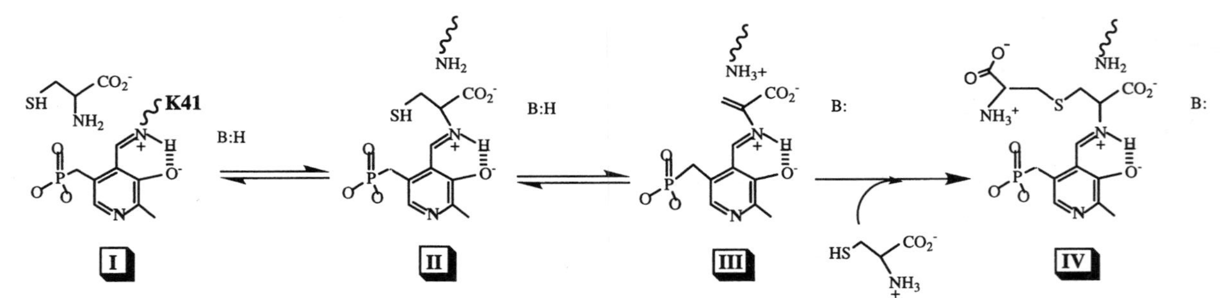

Figure 15. The bottom pathway shown in Figure 14 is shown mechanistically in this figure. Formation of the L-cysteine external Schiff base precedes elimination of SH^- to give the α-aminoacrylate intermediate transiently. In the absence of another nucleophile, the thiol of another L-cysteine serves as the nucleophile to generate the L-lanthionine external Schiff base, for which the sulfhydrylase has a reasonably high affinity.

The resting enzyme contains PLP in internal aldimine linkage with K41 (Rege et al., 1996) predominantly as the ketoeneamine tautomer (Cook et al., 1992). Addition of OAS results in the formation of acetate and the α-aminoacrylate intermediate as a mixture of ketoeneamine and enolimine tautomers at the end of the first half reaction (Cook and Wedding, 1976; Cook et al., 1992).

There is evidence for a random pathway to the ping-pong mechanism, in that sulfide gives only partial competitive inhibition, and acetate induces partial substrate inhibition by OAS (Tai et al., 1993). Initial velocity patterns can sometimes appear intersecting to the left of the ordinate dependent on the relative rate of the second half reaction $((V/K_{nuc})[Nuc])$ and the rate of the OAS:acetate lyase activity, that is, the stability of the F form of the enzyme (Cook et al., 1992). The best example of this behavior is observed with the C42S mutant enzyme (Tai et al., 1998).

Stopped-flow experiments corroborate and expand on the steady-state studies. Rapid-scanning stopped-flow exhibits the rapid formation of the OAS external Schiff base prior to a slow, rate-limiting formation of the α-aminoacrylate intermediate (Woehl et al., 1996). The second half reaction is very rapid, and lies farther to the right than the first. However, it is reversible as shown by stopped-flow studies (see below). Isotope effect data corroborate a rate-limiting C_{α}–H proton abstraction. A quantitative analysis of the data suggests the OAS external Schiff base partitions 1.5–times faster toward the α-aminoacrylate intermediate than it does toward the internal Schiff base (Hwang et al., 1996).

E. CHEMICAL MECHANISM

1. Spectral Studies

As indicated above, the predominant form of the external Schiff base is thus the protonated ketoeneamine tautomer, which has an intramolecular hydrogen bond between the Schiff base nitrogen and O3' of the cofactor PLP. The α-aminoacrylate intermediate, on the other hand, exists as a mixture of two species, the ketoeneamine and enolimine tautomers. The ultraviolet-visible absorbance spectrum of OASS-A, in the absence and presence of OAS, is independent of pH over the ranges 5.5–11 and 5.5–9.2, respectively (Cook et al., 1992). The respective tautomeric forms of the two species predominate in resting (E) and α-aminoacrylate intermediate (F) forms of the enzyme over the entire pH range. Any ioinzations observed in pH-rate profiles obtained under conditions where these enzyme forms predominate cannot therefore be attributed to contributions from the cofactor. In agreement with the UV-visible spectral studies, ^{31}P-NMR spectra of the free enzyme and α-aminoacrylate intermediate are also pH independent. Both species give a δ of 5.2 ppm and line width of 20.5 Hz, indicative of a tightly bound dianioinc phosphate, with its motion restricted to that of the protein (Cook et al., 1992).

As discussed above, the dissociation constants for the L-cysteine and L-serine external Schiff bases were determined based on the spectral changes that attained upon their interaction with OASS-A. Monitoring the change at several wavelengths gave the same dissociation constant, attesting to the fact that the equilibrium could be treated as one between two enzyme species, the internal and external Schiff bases (Schnackerz et al., 1995). Repeating the dissociation constants as a function of pH over the range 6.5–10.5 provided different results dependent upon whether L-serine or L-cysteine is the amino acid analog. The p$K_{\mathrm{I\ serine}}$ pH-rate profile gave a limiting slope of +2, giving pK values of 7.6 and 8.4, with a pH independent value of 4.2 mM obtained for $K_{\mathrm{I\ serine}}$. With L-cysteine, the pH-rate profile decreases at both high and low pH with limiting slopes of +2 and –1, giving an average pK of 7.4 at low pH and a pK of 9.6 on the basic side, with a $K_{\mathrm{I\ cysteine}}$ of 120 μM. Data are interpreted in terms of optimum protonation states for binding these amino acids to form the external Schiff base. In both cases there is a requirement for the neutral form of the α-amino group of the amino acid, the pK at 8.4 for serine and at 9.6 for cysteine. The extra group observed with cysteine reflects the thiol group, pK of 7.4, which must be protonated for binding. In addition, there is a requirement for a second enzyme group, with a pK of 7.4–7.6, that must be unprotonated for binding. It was originally thought that this group would have an entropic contribution, maintaining the leaving group in the proper orientation for the elimination. However, repeat of the experiment with L-methionine (unpublished), which has a $K_{\mathrm{I\ methionine}}$ of 78 μM, similar to that of cysteine, gave data identical to those of serine, so it is doubtful the enzyme group with a pK of 7.4–7.6 interacts with the leaving group.

The first half reaction is reversible (see above). The equilibrium position of the first half reaction is provided by monitoring the α-aminoacrylate intermediate absorbance at 470 nm. Recording spectra as a function as a function of OAS/acetate concentrations (Fig. 16), allows a determination of the equilibrium constant. The equilibrium constant is pH dependent, with a plot of log K_{eq} versus pH giving a slope of 1, indicative of a single proton produced in the first half reaction, required to neutralize the negatively charged acetate product (Fig. 16). The overall K_{eq} is 1.6×10^{-3} M, giving a value of 16,000 in favor of the α-aminoacrylate intermediate at pH 7 (Tai et al., 1995).

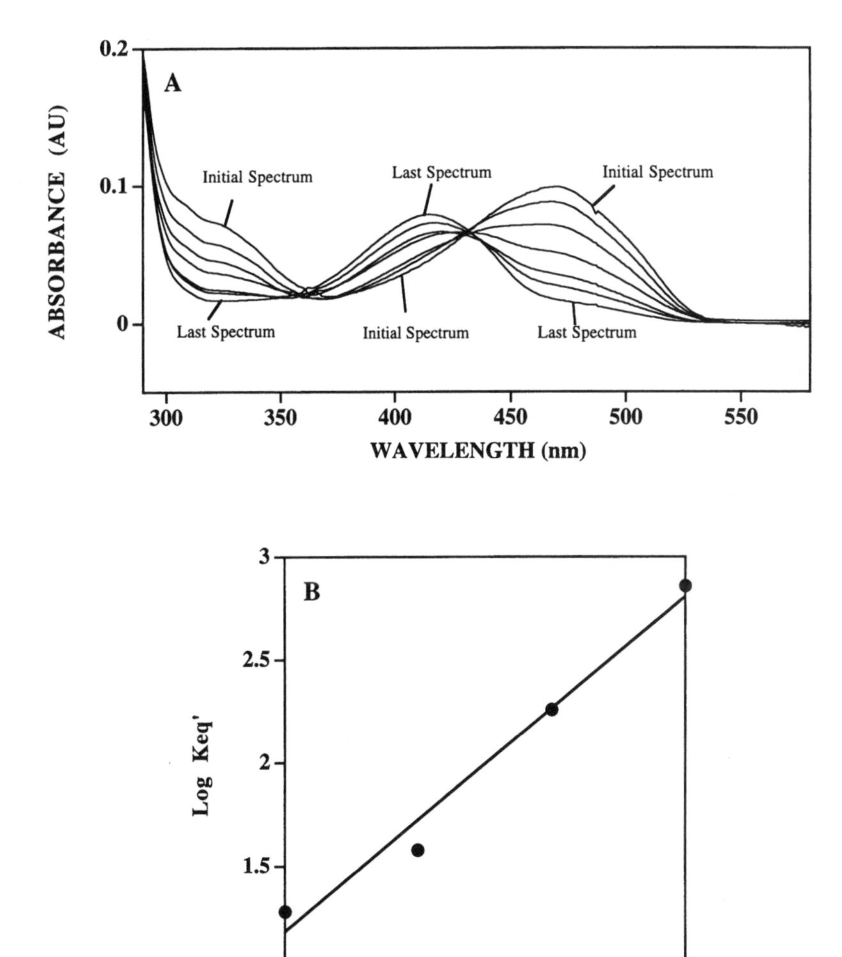

Figure 16. **(A)** Absorbance spectra of 10 μM OASS-A in the presence of 20 μM OAS with stepwise addition of 25, 50, 100, 150, 200, and 300 mM of acetate in 100 mM Mes, pH 6.5. The initial spectrum reflects the α-aminoacrylate Schiff base intermediate. **(B)** Plot of log *K*eq′ versus pH for the first half-reaction catalyzed by OASS-A.

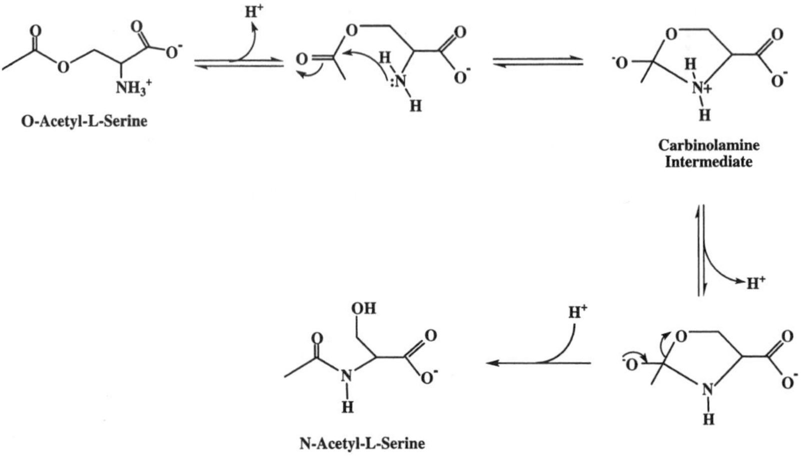

Figure 17. Proposed mechanism for the rearrangement of OAS to NAS.

2. pK *for OAS*

The nucleophile in the first half reaction for formation of the external Schiff base is the α-amine of OAS. Since its p*K* was not reported, and in order to interpret data obtained from pH-rate profiles, the p*K* for the α-amine of OAS was measured using the pH dependence of the first order rate constant for its nonenzymatic conversion to N-acetyl-L-serine (NAS) (Tai et al., 1995). The first order rate constant was measured from the amount of OAS remaining (via formation of the α-aminoacrylate intermediate at 470 nm) with time at each pH value. The pH rate profile gives a limiting slope of +2 with estimated p*K* values of 8.7 ± 0.2 and 7.7 ± 0.2. The p*K*s were interpreted based on the mechanism shown in Figure 17, with the value of 8.7 assigned to the carbinolamine and the value of 7.7 to the α-amine of OAS, based on its similarity to the α-amine p*K* of β-chloro L-alanine (7.9).

3. *OAS:Acetate Lyase Reaction*

If the α-aminoacrylate intermediate of OASS is preformed and allowed to incubate in the absence of a nucleophile, the intermediate exhibits a pH-dependent decay to free enzyme with the overall reaction as given in Figure 18 (Cook et al., 1992). The pH-rate profile exhibits a limiting slope of +1 with a pH-independent first order rate constant of about 0.07 s^{-1} and a pK of 8.1 ± 0.2. A proposed mechanism for the reaction is given in Figure 18, with the pK of 8.1 reflecting the Schiff base lysine (K41; Rege et al., 1996) in the α-aminoacrylate intermediate.

As suggested above, the rate of the lyase activity can be modulated in at least two ways. Anytime the lyase activity begins to equal or exceed the value of V/K_b the initial velocity pattern will no longer appear parallel. This was found in the case of the pyridoxal 5′-deoxymethylenephosphonate-substituted OASS (Cook et al., 1996). In this case, the lyase activity is only 10% that observed for the wild-type enzyme, but the activity of the PDMP-OASS is quite low with TNB as a reactant, and it cannot effectively compete with the lyase activity at the highest concentrations of TNB used. Finally, mutation of C42 to serine, gives an enzyme that appears in almost every way wild-type (Tai et al., 1998). However, subtle changes have occurred at the active site that affect the geometry of the bound PLP as suggested by a num-

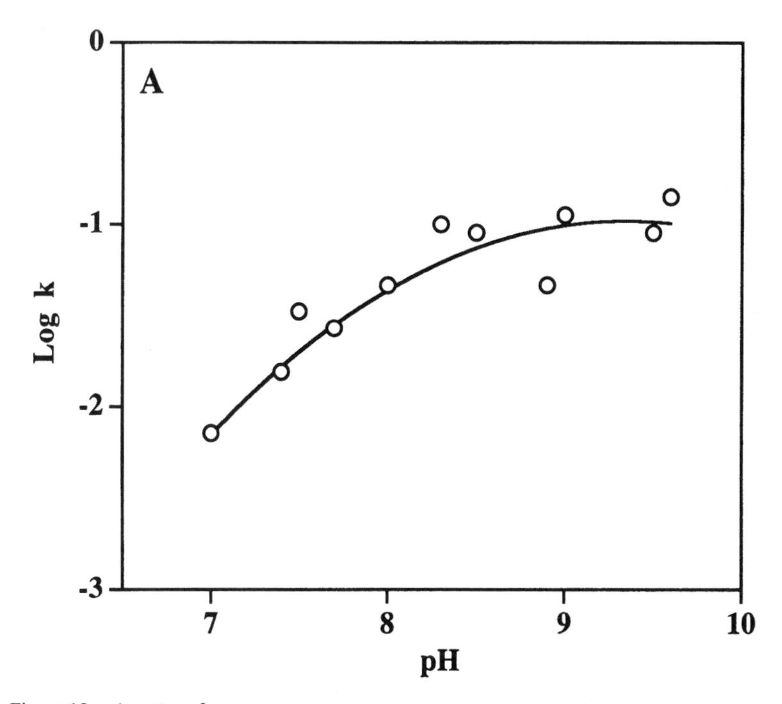

Figure 18. (*continued*)

B. Proposed Mechanism for the Lyase Activity of OASS

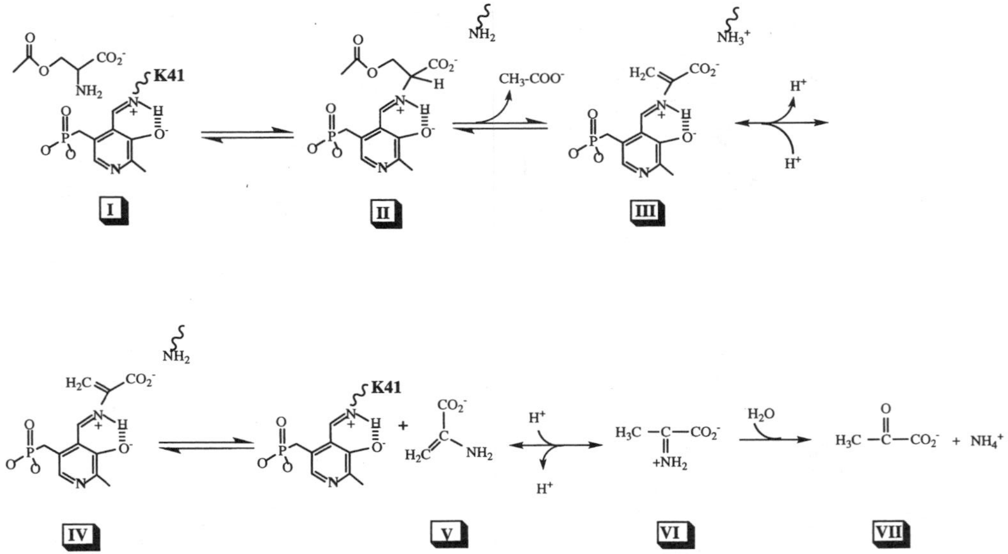

Figure 18. **(A)** pH Dependence of the disappearance of absorbance of the α-aminoacrylate intermediate. The solid line is theoretical from the fit, and the points are experimental values. **(B)** Proposed mechanism for the lyase activity of OASS. OAS forms an external Schiff base (II) with PLP, with concomitant release of the active site lysine from the internal Schiff base (I). The α-proton is abstracted from external Schiff base via a general base, initiating the elimination of the acetate and the formation of the α-aminoacrylate intermediate (III). As the pH increases, the ε-amino group of the active-site lysine becomes deprotonated (IV) and in a transaldimination reaction generates free enzyme and free α-aminoacrylate (V). The free α-aminoacrylate tautomerizes to iminopyruvate (VI) and is hydrolyzed nonenzymically to pyruvate and ammonia (VII).

ber of spectral probes. The largest change is a 50-fold increase in the lyase activity of OASS to a value of 1 s^{-1}. As a result, TNB, although quite efficient in the mutant enzyme reaction, still cannot effectively and completely compete with the lyase activity at the highest concentrations of TNB used.

4. *Acid-Base Chemistry*

The pH dependence of kinetic parameters is an invaluable tool in determining the optimum protonation state of functional groups on reactants and enzyme important for catalysis and/or binding. pH-rate profiles have been measured for three reactant pairs of different overall rates to assess the affects on leaving group and nature of the nucleophilic substrate (Tai et al., 1995). Thus data were obtained with OAS/sulfide, OAS/TNB, and BCA/TNB. The pH independent values of the kinetic parameters give an indication of the rates of the overall reaction (V/E_t), and the two half reactions (V/K_{aa} and V/K_{nuc}); data are listed in Table 3. The near diffusion-limited value of $V/K_{sulfide}$ suggests a rapid second half reaction with OAS/sulfide as the substrate pair. With TNB as the nucleophilic substrate, it is again the second half reaction that has a rate constant 10 times (OAS/TNB) and 200 times (BCA/TNB) greater than that of the first reaction. The disagreement between the values of OAS with sulfide and TNB as nucleophilic substrates has been discussed above under kinetic mechanism. The estimated pK values provide information on the number of groups involved in catalysis and/or binding and their optimum protonation state. Data are summarized in Table 4. The V/K_{OAS} and V/K_{BCA}, reflecting the first half reaction, have bell-shaped pH-rate profiles, but the base-side pK is well-defined in only one of them, V/K_{BCA}, because of its increased stability compared to OAS (see above). The base-side pK reflects the pK of the α-amine of OAS or BCA, which must be unprotonated for nucleophilic attack on C4′ of the Schiff

TABLE 3
pH-Independent Values of V/K Values for Amino Acid/Nucleophile Pairs

Parameter	Amino Acid	Nucleophile
$V/K_{OAS}E_t$ ($M^{-1}s^{-1}$) (TNB)	125	
$V/K_{BCA}E_t$ ($M^{-1}s^{-1}$) (TNB)	5	
$V/K_{sulfide}E_t$ ($M^{-1}s^{-1}$) (OAS)		8×10^7
$V/K_{TNB}E_t$ ($M^{-1}s^{-1}$) (OAS)		1000
$V/K_{TNB}E_t$ ($M^{-1}s^{-1}$) (BCA)		1000

Source: Adapted from Tai et al. (1995).

TABLE 4
Summary of pK Values

Condition or Measurement	pK_a[a]	pK_b
OAS/TNB (V/K_{OAS}	7.5 ± 0.7	–
OAS/TNB (V/K_{TNB})	7.1 ± 0.1	8.2 ± 0.1
OAS/sulfide (V/K_{OAS})	~7	–
OAS/sulfide ($V/K_{sulfide}$)	7.3 ± 0.3	–
BCA/TNB (V/K_{BCA})	6.7 ± 0.1	7.4 ± 0.1
BCA/TNB (V/K_{TNB})	6.9 ± 0.2	8.3 ± 0.2
OAS:acetate lyase	8.1 ± 0.1	
Fluor. Enhance.—acetate	7.2 ± 0.1	
Fluor. Enhance.—cysteine	8.0 ± 0.1	
$K_{1\ cysteine}$—Spectral Titration	7.4 ± 0.2 (2)	9.6 ± 0.4
$K_{1\ serine}$—Spectral Titration	7.6 ± 0.2	
	8.4 ± 0.2	
t_{av}—Phosphorescence	~8	
${}^{D}(V/K_{OAS})$ vs. pH	6.6 ± 0.2[b]	

[a]If more than one pK is indicated from the titration on either acid or base side of the profile, the second is listed below the first. If an average value is obtained, a (2) is given.

[b]This is the best-defined value of the kinetic pK with OAS available. Corrected for kinetic perturbation, a value of 7 is obtained. Note the agreement with the pK_a for V/K_{BCA}.

base linkage of the cofactor. The group observed on the acid side, however, with a pK of 6.7 –7.4 may have a structural role (see above under spectral studies). The latter group is also observed in the second half reaction with a value of 6.9–7.3, but must be unprotonated, while a second group is observed with a pK of 8.2–8.3, which reflects the ε-amino of K41, which formed the internal Schiff base with the cofactor (see OAS:acetate lyase activity above). K41 must be protonated to begin the second half reaction in order to protonate the external aldimine upon attack by the nucleophilic substrate. Finally, the binding constants for acetate and SCN^-, competitive inhibitors of the nucleophilic substrate (e.g., sulfide) are pH independent, suggesting either no discrete binding site for the nucleophilic substrate, or a site made up of nonionizable protein residues.

5. *Isotope Effects and Transition State Structure*

As discussed above, the intrinsic primary kinetic deuterium isotope effect on C_α–H cleavage is 2.8 (Hwang et al., 1996). However, the value of the primary deuterium KIE alone tells one little concerning the nature of the transition state. As discussed by Westheimer (1961), the primary hydrogen isotope effect for C–H bond cleavage behaves as a bell-shaped curve. An early transition state (one where there is very little bond cleavage from the donor) gives a value of 1, as does a transition state with a large amount of bond formation to the acceptor. The Westheimer curve goes through a maximum where the H is symmetrically placed between the donor and acceptor, with a value of 6–8 calculated based on semiclassical considerations. As a result the value of 2.8 measured for OASS could reflect either an early or late transition state for C_α–H bond cleavage.

In an effort to pin down further the transition state structure for C_α–H bond cleavage, the β-secondary deuterium kinetic isotope effect was measured using OAS-3,3-(h_2,d_2) assuming a concerted transition state for C_α–H and C_β–O bond cleavages, that is a concerted α,β-elimination. A concerted α,β-elimination seemed a perfectly logical assumption given the lack of an observed quinonoid intermediate and a good leaving group in acetate for C–O bond cleavage. The β-secondary deuterium kinetic isotope effect is small (1.1 compared to an experimentally determined equilibrium secondary effect of 1.8 reflecting protons at C3 of OAS and at C3 of the α-aminoacrylate intermediate) (Hwang et al., 1996). Taken together with the small primary deuterium effect, data suggest a concerted α,β-elimination that is substrate-like with little C_α–H or C–O bond cleavage (shown schematically in Figure 19).

6. *Summary*

Based on the above information, a mechanism has been proposed for the sulfhydrylase, Scheme 4. The internal Schiff base, I, is protonated to begin the reaction (Cook et al., 1992) and OAS binds as the monoanion (α-amine $pK = 7.7$; Tai et al., 1995). In addition, an enzyme residue with a p*K* of about 7 must be protonated for optimum catalysis and binding. (The function of this residue is unclear but the most reasonable possibility will be discussed below under Dynamics.) It is speculated that Ser272 donates a hydrogen bond to N1 and N1 is thus unprotonated. The 5′-phosphate of PLP is dianionic during the course of the reaction (Cook et al., 1992) and is rigidly bound and tumbling with the protein.

The α-amine of the amino acid substrate approaches C4′ of the imine for nucleophilic attack at the *re* face (see Structure of the K41 mutant above). The OAS external Schiff base, III, is formed presumably via the intermediacy of *gem*-diamine intermediates, one of which is shown, II. The *gem*-diamine intermediates are generally not stable, but have been observed

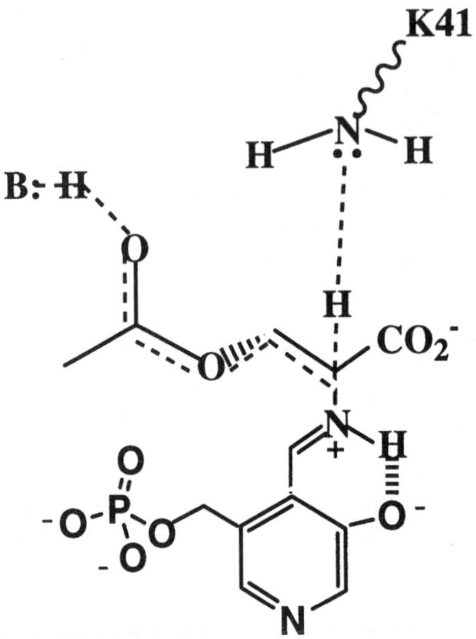

Figure 19. Schematic description of the transition state for a concerted α,β-elimination of the elements of acetic acid from OAS. Note that the transition state is early with respect to C_α–H and C_β–O bond cleavage.

transiently in a few cases (e.g., Brzovic et al., 1992). The problem with spectral visualization of the *gem*-diamine intermediate is that its spectral properties overlap with those of some external aldimine tautomers. It would certainly be of interest if one could stabilize the *gem*-diamine intermediates long enough that they could be adequately characterized. The ε-amine of K41 (Rege et al., 1996), the internal aldimine lysine, is unprotonated (pK = 8.2; Cook et al., 1992; Tai et al., 1995) and acts as a general base to accept the α-proton of OAS to generate the α-aminoacrylate intermediate, V. In agreement with the suggestion that K41 is the general base, mutagenesis studies (Rege et al., 1996) of OASS-A in which the internal aldimine lysine K41 was changed to alanine, eliminate its function as a catalytic base, halting the reaction at the OAS external Schiff base, III. The Schiff base lysine, as expected, also facilitates the transimination of amino acid reactants between internal and external Schiff's bases. Similar studies and results have been obtained for the K87T mutant of TRPS-β_2 (Lu et al., 1993). Based on the pH de-

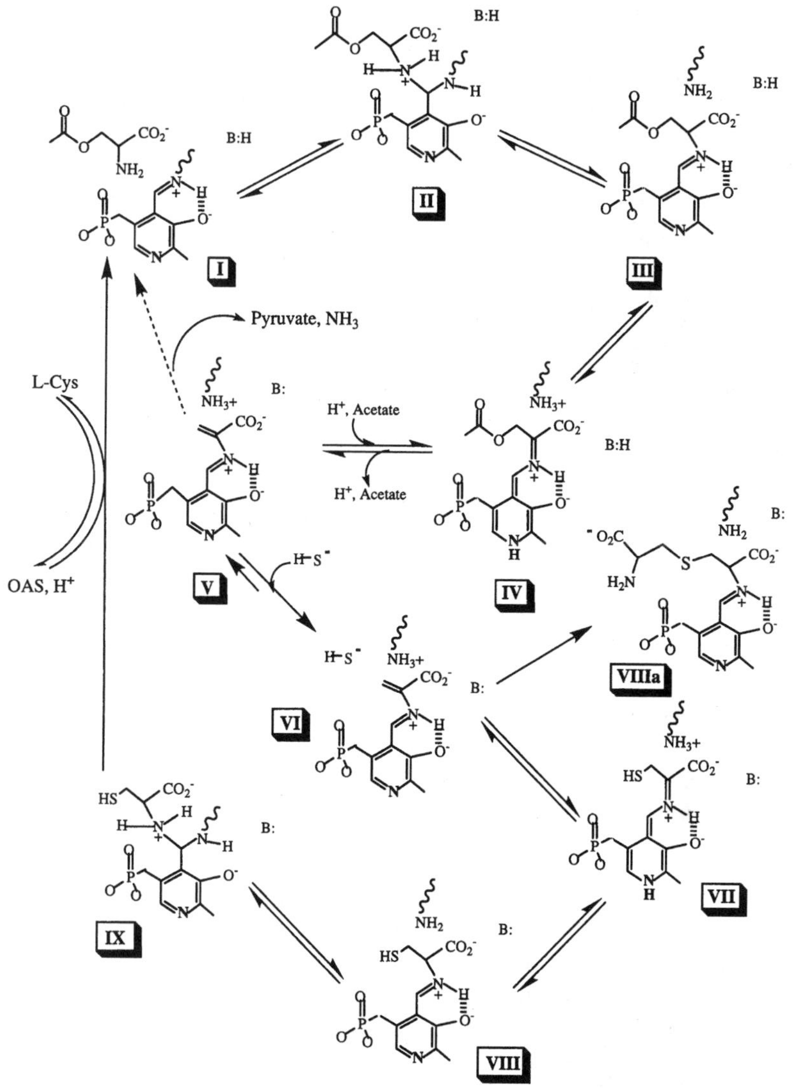

Scheme 4. Proposed mechanism for OASS-A as discussed in the Summary. Note that the overall reaction from OAS to L-cysteine goes from I $\rightarrow$ IX $\rightarrow$ I. Also included are the lyase reaction (*dotted arrow*), and formation of the L-lanthionine external Schiff base (VIIIa). Note also that the path from I with L-cysteine to V is possible but requires a good nucleophile to compete with the thiol of L-cysteine.

pendence of kinetic parameters, a proton is likely released with acetate. It is not possible to determine whether a quinonoid intermediate, IV, forms upon deprotonation of the external aldimine. However, no evidence has ever been obtained for such an intermediate, under a variety of experimental conditions, even though one has been observed in the closely related TRPS-β_2 reaction (Goldberg and Baldwin, 1967; Lane and Kirschner, 1983). The two reactions differ in that the leaving hydroxide in the TRPS-β_2 reaction must be protonated, while acetate requireds no acid-base chemistry. Thus, the protonated imine is likely to produce enough electron withdrawing ability in the elimination of acetate, but not in the elimination of hydroxide.

As the pH increases in the absence of a nucleophile, the α-aminoacrylate intermediate, V, is transiminated to regenerate I and free α-aminoacrylate, which decomposes to pyruvate and ammonia. The abortive β-elimination reaction of OASS-A is very slow, 10^{-4}-fold (Cook et al., 1992), compared to the β-substitution reaction, 240 s^{-1} (Woehl et al., 1996). Although OASS-A and TRPS catalyze a β-elimination reaction, it is very slow compared to the overall β-replacement reaction as suggested above, even though the α-aminoacrylate intermediate "waits" for the nucleophile once formed in both reactions. The enzymes that catalyze strict β-elimination reactions such as tyrosine phenol lyase, proceed via the same aminoacrylate intermediate, but are structurally very different (Sundararaju et al., 1997), and catalyze a very rapid β-elimination reaction. The difference between the two classes of reaction, then, must reside in the ability of OASS-A and TRPS to stabilize the reactive α-aminoacrylate Schiff base. It will be of interest to determine the molecular mechanism for the increased stability of the intermediate.

Once the aminoacrylate intermediate has been formed, and with the exception of the very slow β-elimination reaction, the first half, or ping portion, of the reaction is complete. Sulfide binds as SH^- (Tai et al., 1995), and generates the L-cysteine external Schiff base, VIII, with K41 acting as a general acid to protonate the α-carbon. It is again unlikely a quinonoid intermediate, VII, is involved in this reaction. Transimination via K41 then occurs to release L-cysteine, and regenerate free enzyme, I.

There is a twist on the mechanism, however, if one initiates the reaction with the product cysteine. A transient formation of the α-aminoacrylate intermediate, V, is observed, after which the lanthionine external Schiff base is formed (Woehl et al., 1996). The lanthionine external Schiff base results from attack of the β-thiol of L-cysteine on C-3 of the α-aminoacrylate intermediate, V. The latter occurs unless there is a better nucleophile present, such as sulfide, or some other thiol, or azide (Flint et al., 1996). In the presence of the better nucleophile, the reaction behaves more like the first half reaction, albeit slower.

F. STEREOCHEMISTRY

The stereochemistry of a reaction is extremely useful in determining the overall pathway taken by the chemical reaction. The stereochemistry of the overall OASS-A reaction was determined by Floss et al. in 1976. These authors showed that the reaction from OAS to L-cysteine proceeds with retention of configuration at C-3 indicating that sulfide adds to the same face of the aminoacrylate intermediate that acetate departed. Shorthly after this timely work, the report of the ping pong kinetic mechanism appeared (Cook and Wedding, 1976). Both studies were of import since the replacement of the serine hydroxyl with indole also proceeds with retention of configuration at C3 (Schleicher et al., 1978), but the enzyme TRPS gives intersecting initial velocity patterns. The OASS results aided the interpretation of the stereochemical results obtained with TRPS.

A number of questions still remain concerning stereochemistry. For example, although the overall stereochemistry is retention, one can obtain this result via a *syn* or *anti* α,β-elimination reaction. The K41A mutant structure with L-methionine bound in external Schiff base linkage allows us to speculate on the elimination reaction (Fig. 10) (Burkhard et al., 1999). The mutant protein provides some indication of the location of the side chain of OAS relative to that of K41, the general base. The lysine is directed toward the *si* face of the cofactor and the C_α–H it is to abstract, while the side chain of *met* is directed away from the *re* face of the cofactor toward the entrance to the active site. Data thus suggest an *anti* elimination of the elements of acetic acid to generate the α-aminoacrylate intermediate. The latter is not a great surprise since the anti elimination is slightly favored (Maskill, 1985) under conditions where proton transfer is not required, as is true, for example, in the OASS reaction.

G. DYNAMICS

Movements made by the protein as the reaction pathway is traversed represent an indication of the flexibility of the protein itself and provide an indication as to how the protein participates in and facilitates catalysis. Overall, these protein movements can be considered part of the dynamics of the system. We have used several spectral probes to monitor the dynamics of

OASS-A, including ^{31}P NMR, phosphorescence, and static and time-resolved fluorescence.

1. ^{31}P- NMR Spectral Studies

The value of the ^{31}P NMR chemical shift (δ) is an excellent indicator of the tightness of binding around the 5'-phosphate in PLP-dependent enzymes (Schnackerz, 1986). The internal Schiff base of OASS-A has a δ of 5.2 ppm relative to 85% H_3PO_4, and a linewidth of 20.5 Hz, indicative of a dianionic phosphate tightly bound and tumbling with the protein in solution (Cook et al., 1992). Addition of L-cysteine generates the L-lanthionine external Schiff base (see above), and an increase in the ^{31}P δ to 5.3 ppm with the same 20.5 Hz linewidth (Schnackerz et al., 1995). Data suggest an even tighter interaction between enzyme and phosphate, and thus a change in the conformation of the external Schiff base compared to the internal Schiff base. The latter is certainly borne out by the difference in structures between the native enzyme and the K41A mutant protein (see Structure above). Finally, the serine external Schiff base is a mixture of ketoeneamine and enolimine tautomers in equilibrium. In this case a broad ^{31}P δ is observed at 4.4 ppm with a linewidth of 50 Hz suggesting two species in intermediate exchange. The value of 4.4 is an average with a value of 5.3 ppm reflecting the ketoeneamine tautomer, and a value of 3.5 ppm for the enolimine tautomer. Obviously, the structure of the cofactor in these two species is quite different with much looser binding of the phosphate in the enolimine tautomer. Finally, addition of OAS to the enzyme gives a δ of 5.2 ppm and a linewidth of 20.5 Hz for the α-aminoacrylate intermediate, identical to that observed for the internal Schiff base (Cook et al., 1992).

2. Phosphorescence

Using apo-OASS-A, holo-OASS-A, and a mutant, W50Y, that eliminates one of the two tryptophans of OASS-A, assignment of the 0,0 vibronics was accomplished (Strambini et al., 1996). At 170 K, apo-OASS-A has 0,0 vibronics at 405 (polar), and 410 (hydrophobic) nm. Addition of PLP to apo-OASS-A decreases intrinsic tryptophan fluorescence by 40–45% and results in the appearance of ketoeneamine fluorescence. Phosphorescence, on the other hand, is quenched 70% in intensity and lifetime. The holoenzyme phosphorescence spectrum has a single 0,0 vibronic at 405 nm, plus a wide band at 450–550 resulting from delayed fluorescence of the cofactor enhanced by

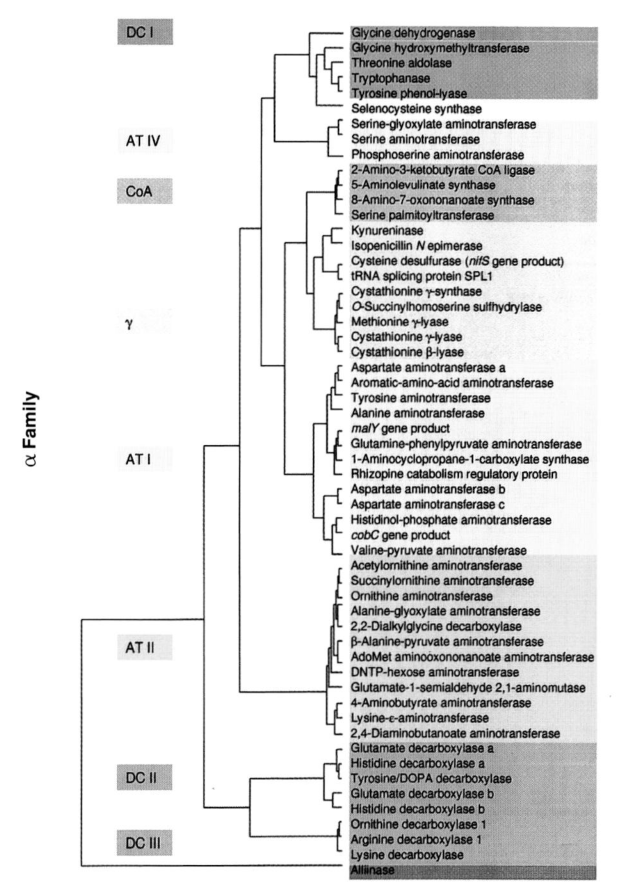

Chapter 4 Figure 5. Evolutionary pedigree of the α family. The evolutionary tree, in which, of course, only enzymes with known amino acid sequences are included, was constructed with the GrowTree program on the basis of FPA data (for details, see Databases and Methodology). The branch lengths represent the relative evolutionary distances as measured in reciprocal values of the mean F-Zscores between the branching points. The color code in the left column defines the subfamilies. The subfamilies are numbered according to previously published work (Mehta et al., 1993; Sandmeier et al., 1994; Alexander et al., 1994). The aminotransferase subfamilies (AT I, II, and IV) are shown in different shades of *yellow* and the amino acid decarboxylase subfamilies (DC I–III) in variations of *magenta.* Within subfamilies, some enzymes have been subdivided into different groups, that is, aspartate aminotransferase groups *a* to *c* in subfamily AT I, glutamate decarboxylase groups *a* and *b,* as well as histidine decarboxylase groups *a* and *b* in decarboxylase subfamily DC II, because the sequences of the different groups could not be aligned with each other due to distant relationship within the same subfamily. A different situation is found with ornithine and arginine decarboxylases. The members of ornithine decarboxylase group 1 have an α family fold and are all of procaryotic origin. The single known amino acid sequence of arginine decarboxylase group 1 is from *Escherichia coli* (biodegradative isoenzyme). However, other forms of these enzymes belong to the alanine racemase family with a $(\alpha/\beta)_8$-barrel fold (ornithine decarboxylase 2 and arginine decarboxylase 2; Fig. 6)

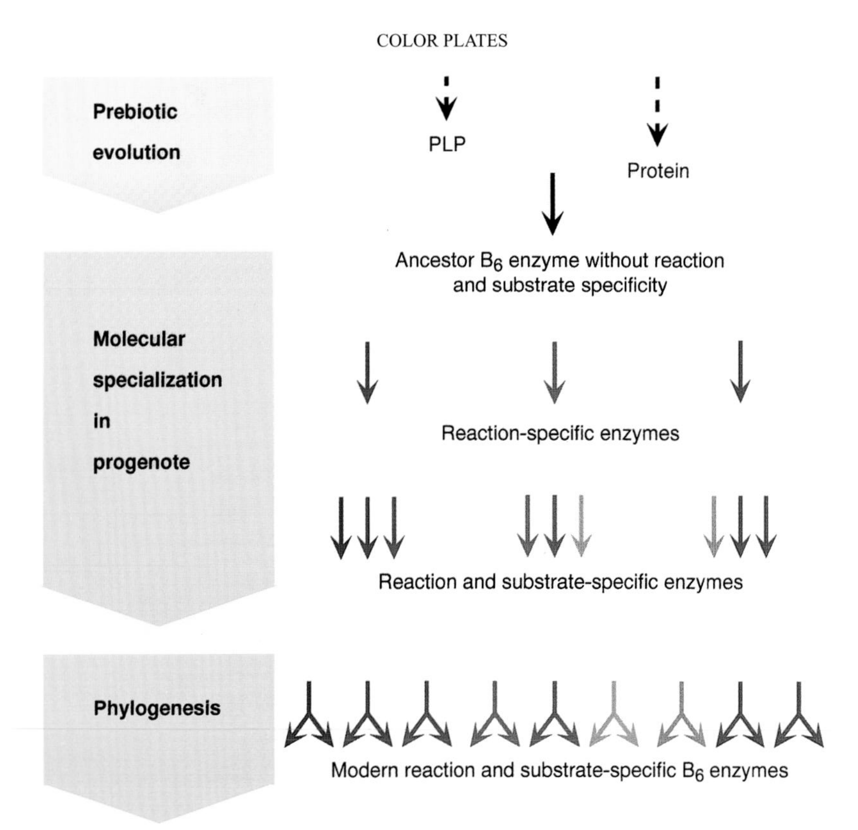

Chapter 4 Figure 13. Synopsis of the molecular evolution of B_6 enzymes. The scheme depicts the temporal sequence of the processes that led from the ancestor of a given B_6 enzyme family, which very likely was an allrounder catalyst, to the reaction and substrate-specific enzymes of amino acid metabolism in recent species. The B_6 enzymes are of multiple evolutionary origin, the course of events outlined in the scheme has occurred in parallel a couple of times, each time starting from a different ancestor protein with another fold. For brevity's sake, the development of only one B_6 enzyme family is shown. The main features of molecular evolution are most clearly evident from the data on the large α family (Fig. 5). However, the data on the other much smaller B_6 enzyme families (Fig. 6) (i.e., the β family, as well as the D-alanine aminotransferase and the alanine racemase families) comply with the same scheme. The first cells and the last common ancestor of prokaryotes and eukaryotes are estimated to have been in existence about 3500 and 1500–2000 million years ago, respectively (Doolittle et al., 1996).

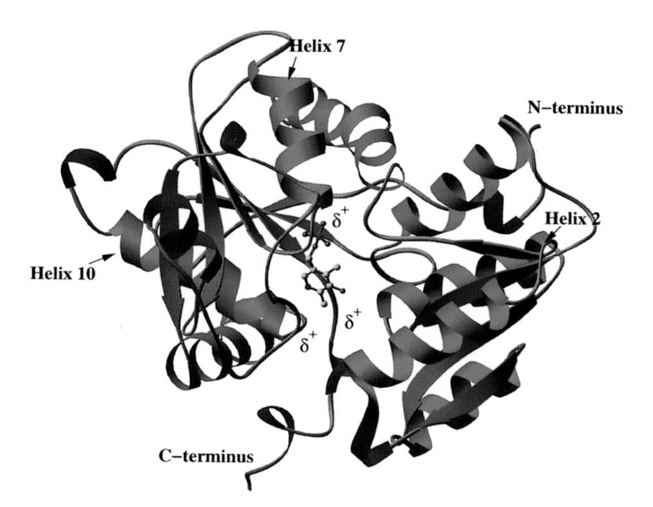

Chapter 5 Figure 8. Structure of OASS-A monomer. Three helix dipoles are directed toward the PLP cofactor: helix 2 directed toward the 3′-oxygen, helix 7 directed toward the 5′-phosphate, and helix 10 directed toward N1. α-Helices are in *light blue*, 3_{10} helices are in *blue*, β-sheet is *green*, and coil is *brown*. The cofactor is portrayed in stick and ball with N1 *blue*; O, *red*; and P, *magenta*.

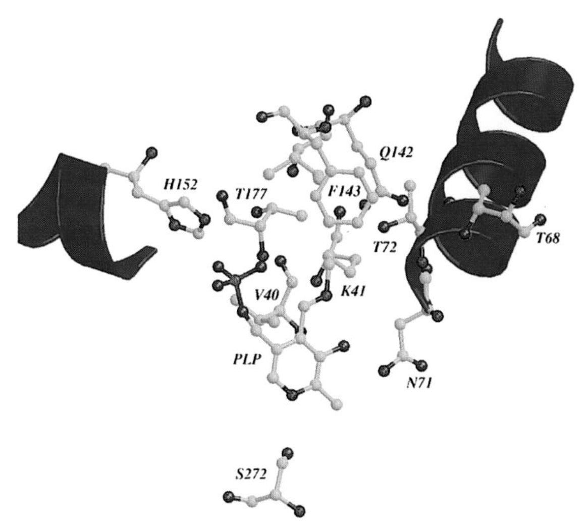

Chapter 5 Figure 9. The close-up view of PLP in the active site of OASS-A. The phosphate group of the cofactor is bound at the N-terminus of helix 7, which interacts with the positive end of the dipole with the negative charge of the phosphate portion of the protein (Gly176 to Thr180). Four of the H-bound donors are peptide NH groups. The cofactor, Gly176, Thr177, Gly178 and Thr180 are portrayed in stick and ball with N1 *blue*; O, *red*; and P, *magenta*.

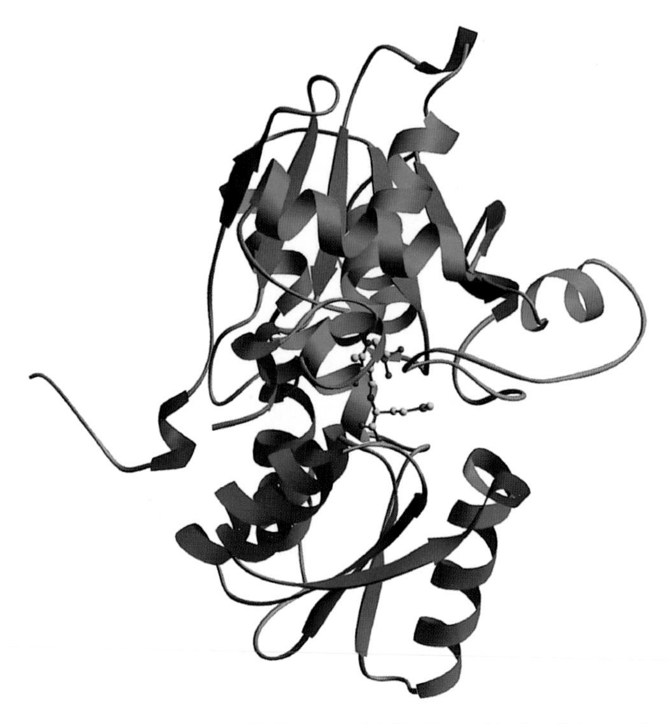

Chapter 5 Figure 10. Structure of K41 mutant OASS. The methionine (S in *green*) in external Schiff base linkage with the active site PLP is clearly seen extending back toward the entrance to the site.

triplet to singlet energy transfer. The W50Y mutant enzyme has a phosphorescence spectrum identical to that of the holoenzyme, and thus W50 fluorescence and phosphorescence are completely quenched by triplet to singlet energy transfer to PLP or by the local polar environment or both. (In support of the quality of the data, a separation between W161 and PLP of 25Å was calculated in excellent agreement with the structural data, see above.)

Addition of reactants such as cysteine or serine exhibits only a change in the intensity of the cofactor delayed fluorescence. The addition of OAS splits the 0,0 vibronic of W161 to 406 and 408.5 nm, which results from a change in the change in environment of W161, that is the α-aminoacrylate intermediate must exist in two conformers. Whether these conformers represent the two tautomers of the α-aminoacrylate intermediate or subtle differences in the two subunits of the dimer will have to await further study.

A brief study at 273 K exhibits an increase in the wavelength of fluorescence and phosphorescence. PLP binding to apo-OASS increases the rigidity near W161, while OAS and L-serine give the opposite effect, suggesting structural changes as one goes from internal Schiff base to external Shiff base or the α-aminoacrylate intermediate.

Finally, an increase in the pH from 6.5 to 9 gives a change in the amplitudes of the two conformers with an estimated pK of 8. This pH change may reflect the same enzyme residue observed in the V/K_{OAS} pH-rate profile with a pK in the range 6.9–7.5 (Tai et al., 1995). That is, the pH-dependent change monitored may well be a conformational change with deprotonation of the enzyme group as the trigger. More will be discussed concerning this aspect below.

3. *Static and Time-resolved Fluorescence*

When the fluorescence emission spectrum of OASS-A is measured, with excitation at 298 nm two maxima are observed, one at 337 nm, resulting from intrinsic tryptophan fluorescence, and a second at 498 nm, resulting from delayed fluorescence of cofactor and triplet to singlet energy transfer (McClure and Cook, 1994). Enhancement of the long wavelength (498 nm) fluorescence is observed upon binding of L-cysteine to form an external Schiff base or acetate, which binds abortively to the α-carboxyl subsite. The enhancement requires an unprotonated enzyme group with pK of 7, consistent with triplet to singlet energy transfer resulting from a pH-dependent conformational change, as discussed above.

We have characterized the fluorescence properties of the two tryptophan residues of the enzyme, Trp 50 and 161, in the native state and after the binding of substrates and products, as probes of the conformational changes that take place in the apo- to holoenzyme transformation and during the catalytic process (Benci et al. 1997, 1999a, 1999b). Data in all cases mimic those obtained via phosphorescence. Upon excitation at 298 nm, the emission of the apo- and holoenzymes are centered at 343 and 338 nm, respectively, and are characterized by biexponential decays. The emission of the holoenzyme is reduced by about 60% with respect to that of the apoenzyme. The deconvolution of peaks and time-resolved fluorescence data indicate that (1) the emission of Trp 50 is centered at 342 and 333 nm in the apo- and holoenzyme, respectively; (2) the emission of Trp 161 is centered at 357 nm in both the apo- and holoenzyme; (3) an energy transfer process quenches the fluorescence of the two tryptophan residues, being more efficient for Trp50, which is less exposed to the solvent; (4) the quenching of Trp161 emission is mainly due to conformational changes accompanying the apo- to holoenzyme transition.

Steady-state and time-resolved fluorescence, collected in the presence of products, acetate or L-cysteine, the product analog L-serine, and the substrate OAS or its analog β-chloro-L-alanine, indicate that tryptophan emissions are significantly affected by variations of the equilibrium distributions between the enolimine and ketoeneamine tautomers, which are differently populated during catalysis.

4. *Summary*

The ^{31}P NMR data suggest slight differences in the structures around the 5′-P for the internal Schiff base and the lanthionine external Schiff base (both largely ketoeneamine) and a large difference for enolimine portion of the serine external Schiff base. Addition of cysteine or serine increase delayed fluorescence and triplet to singlet energy transfer. Addition of OAS exhibits a splitting of the 0,0 vibronic, the result of two distinct conformations, likely enolimine and ketoeneamine tautomers. Nonetheless, the α-aminoacrylate Schiff base conformation differs from either the internal or external Schiff base conformations. All of the time-resolved fluorescence data are consistent with conformation changes reflecting redistribution of ketoeneamine and enolimine tautomers as catalysis occurs.

It is important to remember that the structural changes are substantial. The native structure (internal Schiff base) is active site open, while the K41A mutant enzyme (ketoeneamine external Schiff base) is active site closed. The trigger for the conformational change from open to closed as one goes from the internal to external Schiff base is the occupancy of the α-carboxyl subsite of the active site (Burkhard et al., 1999). Associated with this, as observed in pH-rate profiles, pH-dependent changes in phosphorescence, and pH-dependent changes in fluorescence enhancement upon binding acetate or cysteine is an enzyme group with a pK in the range 7–8. Dependent on the protonation state of the enzyme group, structural changes likely occur that also reflect a redistribution of the tautomeric equilibrium.

Finally, the minimal catalytic cycle can likely be pictured as shown in Fig. 20. The changes may be pH dependent, and the open conformations for the internal Schiff base and the α-aminoacrylate Schiff base are not identical structurally, as expected because of the increased stability of the latter.

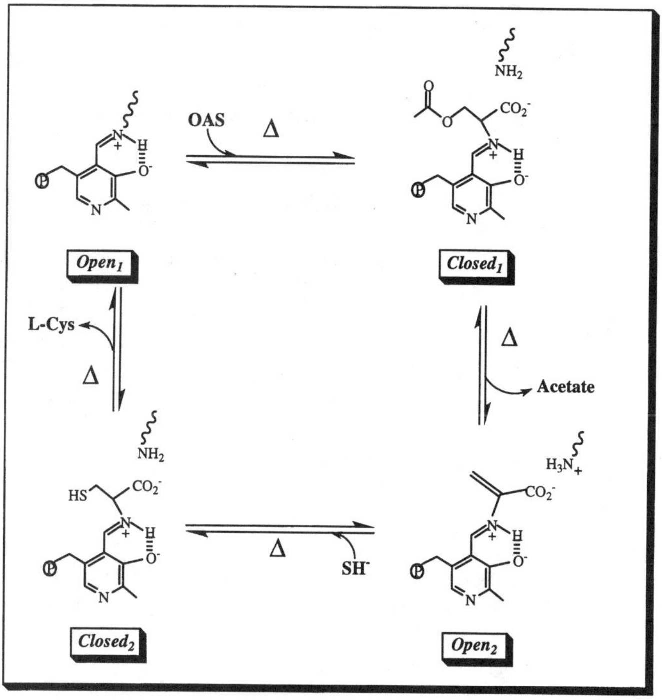

Figure 20. Schematic representing the likely structural changes that take place along the reaction pathway for OASS-A. The Δ represents a conformational change. The two open and two closed states are not identical, but more work will be required to sort out this aspect.

References

Alexander FW, Sandmeier E, Mehta PK, Christen P (1994): Evolutionary relationship among pyridoxal 5′-phosphate-dependent enzymes. regiospecific alpha, beta, and gamma families. Eur J Biochem 219:953–960

Becker MA, Kredich NM, Tomkins GM (1969): The purification and characterization of *O*-acetylserine sulfhydrylase A from *Salmonella typhimurium*. J Biol Chem 244:2418–2427.

Benci S, Vaccari S, Mozzarelli A, Cook PF (1997): Time-resolved fluorescence of *O*-acetylserine sulfhydrylase catalytic intermediates. Biochemistry 36:15419–15427.

Benci S, Vaccari S, Mozzarelli A, Cook PF (1999a): Time-resolved fluorescence of *O*-acetylserine sulfhydrylase. Biochim Biophysica Acta 1429:317–330.

Benci S, Bettati S, Vaccari S, Schianchi G, Mozzarelli A, Cook PF (1999b): Conformational probes of *O*-acetylserine sulfhydrylase: fluorescence of tryptophans 50 and 161. J Photochem Photobiol 48:17–26.

Brzovic PS, Kayastha AM, Miles EW, Dunn MF (1992): Substitution of glutamic acid 109 by aspatic acid alters the substrate specificity and catalytic activity of the β-subunit in the tryptophan synthase bienzyme complex from *Salmonella typhimurium*. Biochemistry 31:1180–1190.

Burkhard P, Rao GSJ, Hohenester E, Cook PF, Jansonius JN (1998): Three dimensional structure of *O*-acetylserine sulfhydrylase from *Salmonella typhimurium* at 2.2 Å. J Mol Biol 283:111–120.

Burkhard P, Tai C-H, Ristroph CM, Cook PF, Jansonius JN (1999): Ligand binding induces a large conformational change in *O*-acetylserine sulfhydrylase from *Salmonella typhimurium*. J Mol Biol 291:941–953.

Byrne CR, Monroe RS, Ward KA, Kredich NM (1988): DNA sequences of the cysK regions of *Salmonella typhimurium* and *Escherichia coli* and linkage of the cysK regions to ptsH. J Bacteriol 190:3150–3157.

Cook PF, Wedding RT (1976): A reaction mechanism from steady state kinetic studies for *O*-acetylserine sulfhydrylase from *Salmonella typhimurium*. J Biol Chem 251:2023–2029.

Cook PF, Wedding RT (1977a): Initial characterization of the multifunctional complex, cysteine synthase. Arch Biochem Biophys 179:293–302.

Cook PF, Wedding RT (1977b): Overall mechanism and rate equation for *O*-acetylserine sulfhydrylase. J Biol Chem 252:3549.

Cook PF, Wedding RT (1978): Cysteine synthase from *Salmonella typhimurium*: aggregation, kinetic behavior and effect of modifiers. J Biol Chem 253:7874–7879.

Cook PF, Hara S, Nalabolu S, Schnackerz KD (1992): pH dependence of the absorbance and ^{31}P NMR spectra of *O*-acetylserine sulfhydrylase in the absence and presence of *O*-acetyl-L-serine. Biochemistry 31:2298–2303.

Cook PF, Tai C-H, Hwang C-C, Woehl EU, Dunn MF, Schnackerz KD (1996): Substitution of pyridoxal 5′-phosphate in the *O*-acetylserine sulfhydrylase from *Salmonella typhimurium* by cofactor analogs provides an effective test of the mechanism of α-aminoacrylate formation. J Biol Chem 271:25842–25849.

Davis L, Metzler DE (1972): Pyridoxal-linked elimination and replacement reactions. In "The Enzymes" 3rd ed. Boyer PD (ed). New York: Academic Press, pp. 33–74.

Dolphin D, Poulson R, Avramovic O (1986): "Vitamine B_6 Pyridoxal Phosphate: Chemical, Biochemical, and Medical Aspects, Part A & B." New York: John Wiley & Sons.

Drewe WF Jr, Dunn MF (1985): Detection and identification of intermediates in the reaction of L-serine with *Escherichia coli* tryptophan synthase via rapid-scanning ultraviolet-visible spectroscopy. Biochemistry 24:3977–3987.

Drewe WF Jr, Dunn MF (1986): Characterization of the reaction of L-serine and indole with *Escherichia coli* tryptophan synthase via rapid-scanning ultraviolet-visible spectroscopy. Biochemistry 25:2494–2501.

Dunathan HC (1971): Stereochemical aspect of pyridoxal phosphate catalysis. Adv Enzymol 35:79–134.

Flint DH, Tuminello JF, Miller TJ (1996): Studies on the synthesis of the Fe-S cluster of dihydroxy-acid dehydratase in *Escherichia coli* crude extract. J Biol Chem 271:16053–16067.

Floss HG, Schleicher E, Potts RG (1976): Stereochemistry of the formation of cysteine by *O*-acetylserine sulfhydrylase. J Biol Chem 251:5478–5483.

Giovanelli J (1987): Sulfur amino acids of plants: an overview. Methods Enzymol 143: 419–426.

Goldberg ME, Baldwin RL (1967): Interaction between the subunits of the tryptophan synthase of *Escherichia coli* : optical properties of an intermediate bound to the $\alpha_2\beta_2$ complex. Biochemistry 6:2113–2119.

Hara S, Payne MA, Schnackerz KD, Cook PF (1990): A rapid purification procedure and computer-assisted sulfide ion electrode assay for *O*-acetylserine sulfhydrylase from *Salmonella typhimurium*. Protein Expr Purif 1:70–76.

Hellinga HW, Evans PR (1985): Nucleotide sequence and high-level expression of the major *Escherichia coli* phosphofructokinase. Eur J Biochem 149:363–373.

Hryniewicz MM, Kredich NM (1991): The *cysP* promoter of *Salmonella typhimurium*: characterization of two binding sites for CysB protein, studies of an in vivo transcription initiation, and demonstration of the anti-inducer effects of thiosulfate. J Bacterol 17:5876–5886.

Hryniewicz M, Sirko A, Palucha A, Böck A, Hulanicka MD (1990): Sulfate and thiosulfate transport in *Escherichia coli* K-12: identification of a gene encoding a novel protein involved in thiosulfate binding. J Bacteriol 172:3358–3366.

Hwang C-C, Woehl EU, Dunn MF, Cook PF (1996): Kinetic isotope effects as a probe of the β-elimination reaction catalyzed by *O*-acetylserine sulfhydrylase. Biochemistry 35:6358–6365.

Hyde CC, Ahmed SA, Padlan EA, Miles EW, Davies DR (1988): Three dimensional structure of the tryptophan synthase $\alpha_2\beta_2$ multienzyme complex from *Salmonella typhimurium*. J Biol Chem 263:17857–17871.

Jenkins WT, Sizer IW (1957): Glutamic aspatic transaminase. J Am Chem Soc 79:2655–2656.

Klein HW, Palm D, Helmreich EJM (1982): General acid-base catalysis of α-glucan phosphorylase stereospecific glucosyl transfer from D-glucal is a pyridoxal 5′-phosphate and orthophosphate (arsenate) dependent reaction. Biochemistry 21:6675–6684.

Kredich NM (1996): Biosynthesis of cysteine. In "*Escherichia coli* and *Salmonella*" 2nd ed. Neidhardt FC (ed). Washington, DC: ASM Press, pp. 514–527.

Kredich NM (1971): Regulation of L-cysteine biosynthesis in *Salmonella typhimurium*. I. Effects of growth on varying sulfur sources and *O*-acetyl-L-serine on gene expression. J Biol Chem 246:3474–3484.

Kredich NM, Tomkins GM (1966): The enzymatic synthesis of L-cysteine in *Escherichia coli* and *Salmonella typhimurium*. J Biol Chem 241:4955–4965.

Kredich NM, Becker MA, Tomkins GM (1969): Purification and characterization of cysteine synthase, a bifunctional protein complex, from *Salmonella typhimurium*. J Biol Chem 244:2428–2439.

Krone FA, Westphal G, Schwenn JD (1991): Characterization of the gene *cysH* and of its product phospho-adenylsulfate reductase from *Escherichia coli*. Mol Gen Genet 225:314–319.

Lane AN, Kirschner K (1983): The mechanisms of binding of L-serine to tryptophan synthase from *Escherichia coli*. Eur J Biochem 129:561–570.

Leyh TS (1993): GTPase mediated activation of ATP sulfurylase. Crit Rev Biochem Mol Biol 28:515–542.

Leyh TS, Suo Y (1992): The physical biochemistry and molecular genetics of sulfate activation. J Biol Chem 267:542–545.

Leyh TS, Taylor J, Markham GD (1988): The sulfate activation locus of *Escherichia coli* K12: cloning, genetic and enzymatic characterization. J Biol Chem 263:2409–2416.

Leyh TS, Vogt TF, Suo Y (1992): The DNA sequence of the sulfate activation locus of *Escherichia coli* K12. J Biol Chem 267:10405–10410.

Liu C, Martin E, Leyh TS (1994): GTPase activation of ATP sulfurylase: the mechanism. Biochemistry 33:2042–2047.

Lu Z, Hagata S, Mcphie P, Miles EW (1993): Lysine 87 in the β-subunit of tryptophan synthase that forms an internal aldimine with pyridoxal phosphate serves critical roles in transimination, catalysis, and product release. J Biol Chem 268:8727–8734.

Maskill H (1985) "The Physical Basis of Organic Reactions." New York: Oxford University Press, pp. 295–300.

McClure GD, Cook PF (1994): Product binding to the α-carboxyl subsite results in a conformational change at the active site of *O*-acetylserine sulfhydrylase-A:evidence from fluorescence spectroscopy. Biochemistry 33:1674–1683.

Miles EW (1986): Pyridoxal phosphate enzymes catalyzing β-elimination and β-replacement reactions. In "Vitamine B_6 Pyridoxal Phosphate:Chemical, Biochemical, and Medical Aspects, Part B." Dolphin D, Poulson R, Avramovic O (eds). New York: John Wiley & Sons, pp. 253–310.

Miles EW, McPhie P (1974): Evidence for a rate-determining proton abstraction in the serine deaminase reaction of the β-subunit of tryptophan synthase. J Biol Chem 249:2852–2857.

Morozov YV (1986): Spectroscopic properties, electronic structure, and photochemical behavior of vitamine B_6 and analogs. In "Vitamine B_6 Pyridoxal Phosphate:Chemical, Biochemical, and Medical Aspects, Part A." Dolphin D, Poulson R, Avramovic O (eds). New York: John Wiley & Sons, pp. 131–222.

Mozzarelli A, Bettati S, Pucci AM, Cook PF (1998): The catalytic competence of *O*-acetylserine sulfhydrylase in the crystal probed by polarized absorption microspectrophotometry. J Mol Biol 283:121–133.

Nakamura T, Kon H, Iwahashi H, Eguchi Y (1983): Evidence that thiosulfate assimilation by *Salmonella typhimurium* is catalyzed by cysteine synthase B. J Bacteriol 156:656–662.

Nakamura T, Iwahashi H, Eguchi Y (1984): Enzymatic proof for the identity of the *S*-sulfocysteine synthase and cysteine synthase B of *Salmonella typhimurium*. J Bacteriol 158:1122–1127.

Rao JGS, Goldsmith EJ, Mottonen J, Cook PF (1993): Crystallization and preliminary x-ray data for the A-isozyme of *O*-acetylserine sulfhydrylase from *Salmonella typhimurium*. J Mol Biol 231:1130–1132.

Rege V, Kredich NM, Tai C-H, Karsten WE, Schnackerz KD, Cook PF (1996): A change in the internal aldimine lysine (K41) in *O*-acetylserine sulfhydrylase to alanine indicates a role for the lysine in transimination and as a general base catalyst. Biochemistry 35:13485–13493.

Rhee S, Parris KD, Hyde CC, Ahmed SA, Miles EW, Davies DR (1997): Crystal structures of a mutant (K87T) tryptophan synthase $\alpha_2\beta_2$ multienzyme complex with ligands bound to the active sites of the α- and β-subunits reveal ligand-induced conformational changes. Biochemistry 36:7664–7680.

Robbins PW, Lipmann F (1958): Enzymatic synthesis of adenosine-5′-phosphosulfate. J Biol Chem 233:686–690.

Roy M, Miles EW, Phillips RS, Dunn MF (1988): Detection and identification of transient intermediates in the reactions of tryptophan. Evidence for a tetrahedral (dem-diamine) intermediate. Biochemistry 27:8661–8669.

Satishchandran C, Markham GD (1989): Adenosine 5′-phosphosulfate kinase from *Escherichia coli* K12. J Biol Chem 264:15012–15021.

Satishchandran C, Hickman YN, Markham GD (1992): Characterization of the phosphorylated enzyme intermediate formed in the adenosine 5′-phosphosulfate kinase reaction. Biochemistry 31:11684–11688.

Scarsdale JN, Kazanina G, Radaev S, Schirch V, Wright HT (1999): Crystal structure of rabbit cytosolic serine hydroxymethyltransferase at 2.8 Å resolution:mechanistic implication. Biochemistry 38:8347–8358.

Schleicher E, Marcuso K, Potts RG, Mann DR, Floss HG (1976) Stereochemistry and mechanism of reactions catalyzed by tryptophanase and tryptophan synthase. J Am Chem Soc 98:1043–1044.

Schnackerz KD (1986): ^{31}P-NMR spectroscopy of vitamine B_6 and derivatives. In "Vitamine B_6 Pyridoxal Phosphate:Chemical, Biochemical, and Medical Aspects, Part A." Dolphin D, Poulson R, Avramovic O (eds). New York: John Wiley & Sons, pp. 245–264.

Schnackerz KD, Cook PF (1995): Resolution of the pyridoxal 5′-phosphate from *O*-acetylserine sulfhydrylase and reconstitution with the native cofactor and analogs. Arch Biochem Biophys 324:71–77.

Schnackerz KD, Erlich JH, Giesemann W, Reed AT (1979): Mechanism of action of D-serine dehydratase–identification of a transient intermediate. Biochemistry 18:3557–3563.

Schnackerz KD, Tai C-H, Simmons JWIII, Jacobson TM, Rao GSJ, Cook PF (1995): Identification and characterization of the external aldimine intermediate of the *O*-acetylserine sulfhydrylase reaction. Biochemistry 34:12152–12160.

Siegel LM, Davis PS (1974): Reduced nicotinamide adenine dinucleotide phosphate-sulfite reductase of Enterobacteria. IV. The *Escherichia coli* hemoflavoprotein:subunit structure and dissociation into hemoprotein and flavoprotein components. J Biol Chem 249:1587–1598.

Siegel LM, Murphy MJ, Kamin H (1973): Reduced nicotinamide adenine dinucleotide phosphate-sulfite reductase of Enterobacteria. I. The *Escherichia coli* hemoflavoprotein:molecular parameters and prosthetic groups. J Biol Chem 248:251–261.

Siegel LM, Davis PS, Kamin H (1974): Reduced nicotinamide adenine dinucleotide phosphate-sulfite reductase of Enterobacteria. III. The *Escherichia coli* hemoflavoprotein:catalytic parameters and the sequence of electron flow. J Biol Chem 249:1572–1586.

Sirko A, Hryniewicz M, Hulanicka MD, Böck A (1990): Sulfate and thiosulfate transport in *Escherichia coli* K-12:nucleotide sequence and expression of the *cysTWAM* gene cluster. J Bacteriol 172:3351–3357.

Strambini G, Cioni P, Cook PF (1996): Tryptophan and coenzyme luminescence as a probe of conformation along the *O*-acetylserine sulfhydrylase reaction pathway. Biochemistry 35:8392–8400.

Sundararaju B, Antson AA, Phillips RS, Demidkina TV, Barbolina MA, Gollnick P, Dodson GG, Wilson KS (1997): The crystal structure of *Citrobacter freundii* tyrosine phenol-lyase complexed with 3–(4′-hydroxyphenyl)propionic acid, together with site-directed mutagensis and kinetic analysis, demonstrates that arginine 381 is required for substrate specificity. Biochemistry 36:6502–6510.

Tai C-H, Nalabolu SR, Jacobson TM, Minter DE, Cook PF (1993): Kinetic mechanisms of *O*-acetylserine sulfhydrylases A and B from *Salmonella typhimurium* with natural and alternate substrates. Biochemistry 32:6433–6442.

Tai C-H, Nalabolu SR, Jacobson TM, Simmons JWIII, Cook PF (1995): pH dependence of kinetic parameters for *O*-acetylserine sulfhydrylases A and B from *Salmonella typhimurium*. Biochemistry 34:12311–12322.

Tai C-H, Yoon M-Y, Rege VD, Kredich NM, Schnackerz KD, Cook PF (1998): A cysteine (C42) immediately N-terminal to the internal aldimine lysine is responsible for stabilizing the α-aminoacrylate intermediate in the reaction catalyzed by *O*-acetylserine sulfhydrylase. Biochemistry 37:10597–10604.

Tsang ML-S (1983): Function of thioredoxin of 3′-phosphoadenosine 5′-phosphosulfate in *E. coli*. In "Thioredoxins—structure and function" Gadal P (ed). Paris:Centre National de Recherche Scientifique, pp. 103–110.

Tsang ML-S, Schiff JA (1978): Assimilatory sulfate reduction in an *Escherichia coli* mutant lacking thioredoxin activity. J Bacteriol 134:131–138.

Vogel HJ, Bonner DM (1956): Acetylornithinase of *Escherichia coli*: partial purification and some properties. J Biol Chem 218:97–106.

Warren MJ, Roessner CA, Santander PJ, Scott AI (1990): The *Escherichia coli* cysG gene encodes S-adenosylmethioine-dependent uroporphyrinogen III methylase. Biochem J 265:725–729.

Westheimer FH (1961): The magnitude of the primary kinetic isotope effect for the compounds of hydrogen and deuterium. Chem Rev 61:265–275.

Woehl EU, Tai C-H, Dunn MF, Cook PF (1996): Formation of the α-aminoacrylate intermediate limits the overall reaction by *O*-acetylserine sulfhydrylase. Biochemistry 35:4776–4783.

Woodin TS, Segel IH (1968): Glutathione reductase-dependent metabolism of cysteine S-sulfate by *Penicillium chrysogenum*. Biochim Biophys Acta 167:78–88.

THE AROMATIC AMINO ACID HYDROXYLASES

By PAUL F. FITZPATRICK, *Department of Biochemistry and Biophysics and Department of Chemistry, Texas A&M University, College Station, Texas 77843-2128*

CONTENTS

I. Introduction

Hydroxylation of the aromatic amino acids phenylalanine, tyrosine, and tryptophan is required for proper function of the central nervous system in multicellular organisms from nematodes to humans. Phenylalanine is converted to tyrosine by the enzyme phenylalanine hydroxylase (PAH) (Fig. 1). Loss of this catalytic activity impairs the catabolism of phenylalanine, leading to the accumulation of phenylpyruvate. The hallmark of such a defect is the genetic disease phenylketonuria, the most common inherited disease of amino acid metabolism in the United States. In the absence of early intervention and diet modification, this disease leads to profound and irreversible

Advances in Enzymology and Related Areas of Molecular Biology, Volume 74: Mechanism of Enzyme Action, Part B, Edited by Daniel L. Purich
ISBN 0-471-34921-6

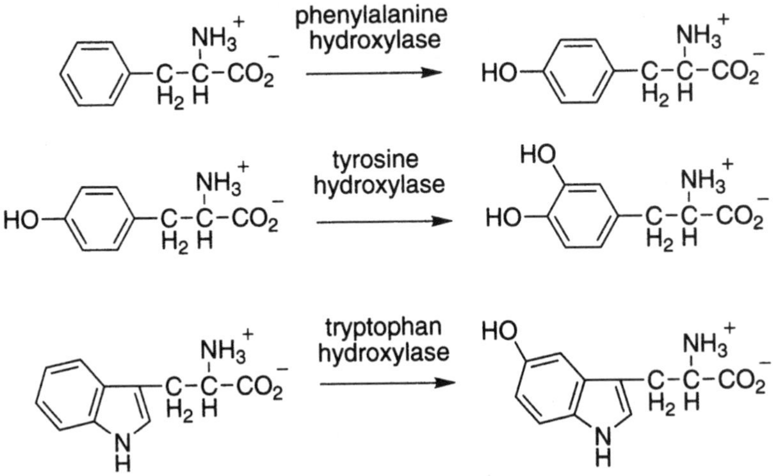

Figure 1. The reactions catalyzed by the aromatic amino acid hydroxylases.

mental retardation. Phenylalanine hydroxylase is a liver enzyme (Udenfriend and Cooper, 1952), but forms of the enzyme are also present in *Chromobacterium violaceum* (Nakata et al., 1979) and *Pseudomonas aeruginosa* (Zhao et al., 1994). Hydroxylation of tyrosine to dihydroxyphenylalanine is catalyzed by tyrosine hydroxylase (TYH). This is the first step in the biosynthesis of the catecholamines dopamine, norepinephrine, and epinephrine. Consequently, it is not surprising that mutations in tyrosine hydroxylase have severely deleterious effects. While earlier suggestions regarding the involvement of TYH in manic depression have not been supported by later studies (Byerley et al., 1992; De bruyn et al., 1994), point mutations in TYH have been found in cases of DOPA-responsive dystonia and parkinsonism (Knappskog et al., 1995; Ludecke et al., 1995, 1996; van den Heuvel et al., 1998). Complete loss of TYH prevents normal embryonic development (Kobayashi et al., 1995; Zhou et al., 1995), consistent with a need for catecholamines for normal neurological function. Tyrosine hydroxylase is found both in brain tissues and in the adrenal gland (Nagatsu et al., 1964). Hydroxylation of tryptophan to 5-hydroxytryptophan is catalyzed by tryptophan hydroxylase (TRH). This is the first step in the biosynthesis of the neurotransmitter serotonin (Boadle-Biber, 1993). Decreased levels of serotonin have been implicated in depression (Roy et al., 1989), impulsive violence (Brunner et al., 1993), and alcoholism (Grahame-Smith, 1992), and a specific allele of TRH is a marker for suicidality (Nielsen et al., 1994).

Tryptophan hydroxylase is a brain enzyme (Lovenberg et al., 1967); it has also been reported to be present in *C. violaceum* (Letendre et al., 1974).

In light of the obvious physiological importance of these three enzymes, it is not surprising that they have been heavily studied since they were first described. The majority of the initial studies of these enzymes was performed with PAH, because of its relative abundance in the liver. There is a large literature on the neurochemical properties of tyrosine hydroxylase, but until recently the characterization of its enzymology lagged that of PAH. Tryptophan hydroxylase has generally been found to be difficult to obtain due to its instability, so that molecular studies of this enzyme have been relatively few. This review will focus on progress during the last decade in understanding the enzymology of these three hydroxylases. Before doing so, pertinent results from earlier literature will be discussed. For more detailed descriptions of the literature on these enzymes before 1988 and for discussion of topics which are not covered here, the reader is directed to one of several comprehensive reviews (Kaufman, 1985, 1995; Shiman, 1985; Kappock and Caradonna, 1996).

The similarities among the three aromatic amino acid hydroxylases are far more extensive than simply the ability to catalyze the reactions in Figure 1.

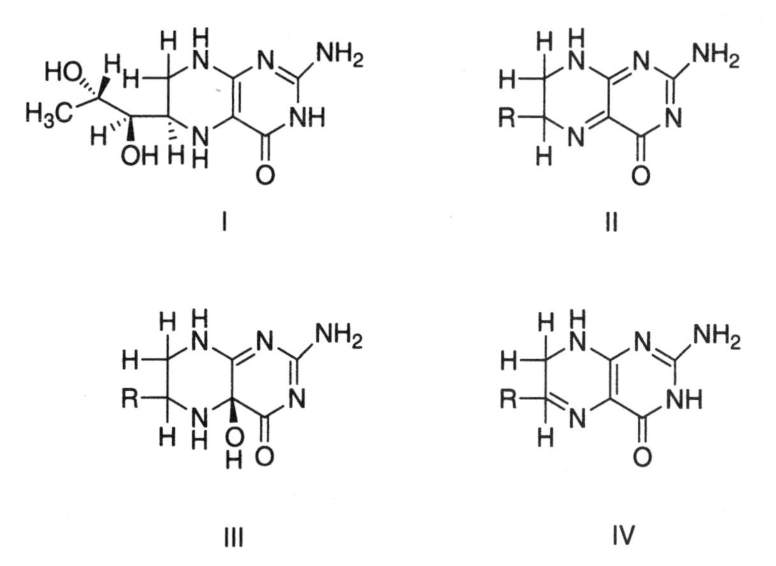

Figure 2. Structures of pterin substrates or products: I, tetrahydrobiopterin; II, quinonoid dihydropterin; III, 4a-hydroxytetrahydropterin; IV, 7,8-dihydropterin.

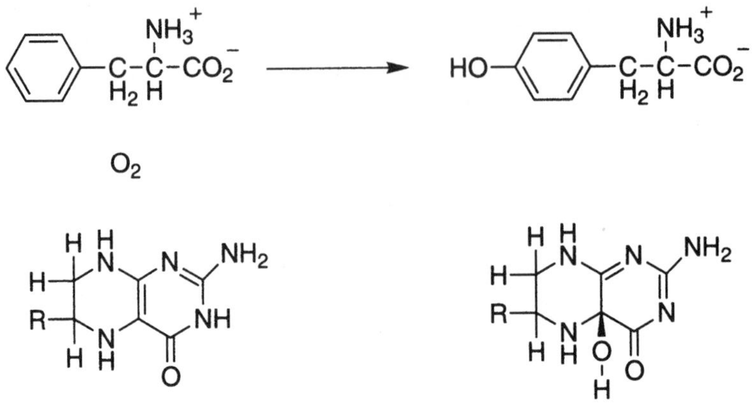

Figure 3. Substrates and products of the phenylalanine hydroxylase reaction.

They are all formally monooxygenases, in that the source of the atom of oxygen incorporated into the amino acid is molecular oxygen (Kaufman et al., 1962; Daly et al., 1968) and the other atom of oxygen is reduced to form water. The immediate source of the electrons necessary for this reduction in all three cases is a tetrahydropterin. The physiological substrate is tetrahydrobiopterin, (6R, 1R′, 2S′)-6-(1′,2′-dihydroxypropyl)-5,6,7,8-tetrahydropterin (I), shown in Figure 2 (Kaufman, 1963; Matsuura et al., 1985). The quinonoid dihydropterin II was initially thought to be the product of the enzyme-catalyzed reaction, since this is the species that accumulates in solution (Kaufman, 1963; Benkovic et al., 1985b). However, the 4a-hydroxytetrahydropterin III was shown to be the initial pterin product of the PAH reaction by Lazarus et al. (1981, 1982) and subsequently established as the initial pterin product in the TYH (Dix et al., 1987; Haavik and Flatmark, 1987) and TRH reactions (Moran et al., 1998). 4a-Hydroxytetrahydropterins rapidly dehydrate in aqueous solution to the quinonoid dihydropterins (Bailey et al., 1995), so that it is the latter form that accumulates over time. Finally, the quinonoid tautomers of dihydropterins are not the thermodynamically most stable form; there is a relatively slow rearrangement to the 7,8-dihydropterin IV in solution (Archer and Scrimgeour, 1970). In the case of the PAH-catalyzed reaction, the stereochemistry of the oxygen in III has been determined to be S as shown (Dix et al., 1985). The source of the oxygen in the hydroxyl group is molecular oxygen (Dix et al., 1985). Thus, the overall reaction catalyzed by the aromatic amino acid hydroxylases is as shown in Figure 3 for PAH.

II. The Active Site Iron

The aromatic amino acid hydroxylases are all nonheme iron enzymes. When isolated, both PAH and TYH contain up to one atom of high-spin ferric iron per monomer (Wallick et al., 1984; Bloom et al., 1986; Haavik et al., 1988). However, studies with all three enzymes have shown that catalysis requires ferrous iron (Marota and Shiman, 1984; Wallick et al., 1984; Fitzpatrick, 1989; Moran et al., 1998). The presence of ferric iron in the enzymes when purified appears to be due to facile oxidation of the iron during purification. The iron in PAH and TYH is readily reduced by a tetrahydropterin (Marota and Shiman, 1984; Wallick et al., 1984; Ramsey et al., 1996), implying that the ferrous state is maintained in the cell by tetrahydrobiopterin. Thus, these enzymes can be classified as mononuclear nonheme ferrous iron enzymes (Que and Ho, 1996).

A variety of spectroscopic methods have been used to probe the structure of the iron site, predominantly focusing on the more accessible ferric forms. Several studies have taken advantage of the fact that both PAH and TYH will form catechol complexes with useful spectroscopic properties. With either enzyme, binding of a catechol to the ferric enzyme generates a complex that exhibits a broad absorbance band in the 400–900 nm range. Indeed, TYH isolated from either bovine adrenal glands or rat pheochromocytoma cells is blue because of the presence of bound catecholamines (Andersson et al., 1988, 1992). The visible absorbance and resonance Raman spectra of catechol complexes of PAH have been used to identify the ligands to the metal (Cox et al., 1988). The spectral properties of the PAH-catechol complexes are similar to those of model complexes in which the iron is chelated by both phenolic oxygens in the catechol, with two pyridine-like nitrogens and a carboxylate providing additional ligands. Studies of TYH using ^{18}O-labeled catechols have confirmed the direct interaction of both catechol oxygens with the iron in complexes formed with that enzyme (Michaud-Soret et al., 1995). The source of the pyridine-like nitrogen ligands would necessarily be imidazole side chains of histidine residues. X-ray absorption spectroscopic studies of ferrous human TYH support a hexacoordinate site containing 3 ± 1 imidazole ligands. In contrast, EXAFS analyses of the copper-containing bacterial PAH have been interpreted in favor of a four-coordinate site, with two of the ligands being imidazole and the other two being either oxygen or nitrogen ligands (Blackburn et al., 1992). Site-directed mutagenesis has been used to more directly determine the number and identity of the histidine ligands. Mutation of either His285 or His 290 of rat PAH to serine results in proteins that contain no iron and lack catalytic activity (Gibbs et al., 1993). Mutation of the homologous His138 or His143 of bacterial PAH similarly abolishes catalytic activity, although the mutant proteins still bind copper. Electron spin echo envelope modulation analyses of the mutant proteins are consistent with only a single imidazole remaining as a ligand in each (Balasubramanian et al., 1994). When each of the five conserved histidines in rat TYH is mutated, only mutations of His331 or His336 produce inactive iron-free enzymes (Ramsey et al., 1995). These residues are homologous to those identified in PAH. This combination of the spectral and mutagenic studies identifies these two histidines as the nitrogenous ligands to the metal.

In light of the catalytic relevance of the ferrous forms of these enzymes, more recent spectroscopic studies have focused on this species. X-ray absorption and Mössbauer spectra of both ferrous PAH and TYH support a hexacoordinate site with distorted octahedral geometry (Meyer-Klaucke et al., 1996; Loev et al., 1997), similar to that found in the ferric enzyme. The spectral properties of human TYH are unchanged upon addition of a tetrahydropterin (Meyer-Klaucke et al., 1996). Similarly, only minor changes in the spectra of rat PAH are seen upon the binding of either phenylalanine or the air stable tetrahydropterin analog 6-methyl-5-deazatetrahydropterin (Fig. 4) (Loev et al., 1997; Kemsley et al., 1999). In contrast, when both phenylalanine and 6-methyl-5-deazatetrahydropterin are added to rat PAH, the spectral properties are more consistent with a five-coordinate, square pyramidal site (Kemsley et al., 1999). This rearrangement from a hexacoordinate to a pentacoordinate site when both substrates are bound would provide an open site to which molecular oxygen could bind in the quaternary complex.

In contrast to the eukaryotic enzymes, PAH from *C. violaceum* contains a stoichiometric amount of copper when isolated (Pember et al., 1986). Consistent with the pattern seen with the other enzymes, the copper is in the cupric form when isolated, and catalysis requires that the copper be reduced. In contrast to the eukaryotic enzymes, dithiothreitol is more effective than

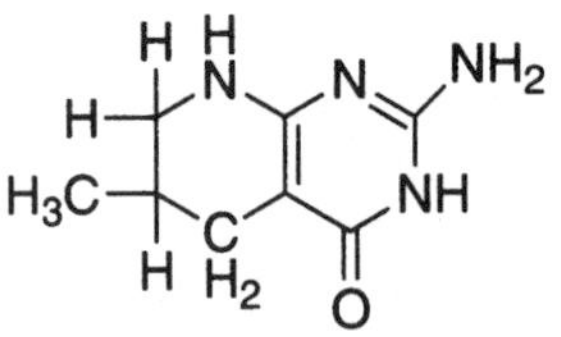

Figure 4. 6-Methyl-5-deazatetrahydropterin.

dihydropterins in activating the cupric enzyme. Reduction of the copper results in dissociation of the metal, forming a metal-free enzyme that is active under standard assay conditions (Carr and Benkovic, 1993). This result led to the suggestion that no metal is required for hydroxylation by the bacterial enzyme (Carr and Benkovic, 1993; Carr et al., 1995), and implies that the metal in the eukaryotic enzyme also is not directly involved in catalysis. To address this problem, Chen and Frey (1998) independently cloned and characterized the PAH from *C. violaceum*. They similarly found that the enzyme contained copper upon isolation, and that the copper could be removed upon reduction with dithiothreitol. In contrast to the earlier reports, these workers found that the metal-free enzyme could not catalyze the formation of tyrosine from phenylalanine, although it could catalyze a phenylalanine-dependent oxidation of tetrahydropterin. The ability to hydroxylate phenylalanine showed a strict requirement for iron, in analogy to the result with the eukaryotic enzymes. Fully active enzyme containing one iron per monomer could be obtained either by incubation of the metal-free enzyme with ferrous sulfate or by growth of the bacteria in iron-supplemented media. These results strongly suggest that the bacterial PAH is a nonheme iron enzyme similar to the eukaryotic enzymes.

III. Structures

Consistent with the common features of the reactions they catalyze, all three hydroxylases show significant structural similarities. All of the eukaryotic enzymes are homotetramers (Nakata and Fujisawa, 1982; Okuno and Fujisawa, 1982; Kappock et al., 1995) with subunit molecular weights of 51,000 to 56,000, while bacterial PAH is a monomer of 30,300 (Nakata et al., 1979; Zhao et al., 1994). In the case of PAH, there is an equilibrium between the homodimer and the homotetramer (Kappock et al., 1995), but the other two enzymes have only been described as tetramers.

The high degree of sequence similarity these enzymes share became apparent when the primary structures of human PAH (Ledley et al., 1985) and rat TYH (Grima et al., 1985) were determined. Alignment of the amino acid sequences showed that the N-termini had little similarity, while the C-terminal two-thirds were clearly homologous. This observation was extended to tryptophan hydroxylase when the sequence of the rabbit enzyme was determined (Grenett et al., 1987). The conclusion drawn at the time was that the C-terminal two-thirds of these proteins constituted a catalytic core with a common ancestor, while the variable N-terminal thirds were responsible for the substrate specificities (Ledley et al., 1985; Grenett et al., 1987). Limited proteolysis and site-directed mutagenesis have been used to more narrowly define the common catalytic cores of the eukaryotic enzymes. In the case of rat PAH, a mutant protein lacking residues 1–121 and 410–453 retains most of the activity of the intact enzyme (Dickson et al., 1994). Similarly, mutants of TYH lacking residues 1–187 and 456–498 are active (Ribeiro et al., 1993; Quinsey et al., 1998), and a mutant rabbit TRH lacking residues 1–100 and 417–444 is fully active (Moran et al., 1998). *Pseudomonas* PAH residues 19–264 are readily aligned with residues 179–429 of rat PAH when several small gaps are introduced into the rat sequence (Zhao et al., 1994). These structural studies and sequence comparisons have led to the present model in which all three eukaryotic enzymes contain homologous catalytic domains of about 290 amino acid residues extending close to the C-termini. In addition, each contains a discrete N-terminal regulatory domain of 100–180 residues.

To date, three-dimensional structures have been described for the catalytic domains of both human PAH and rat TYH. In the case of TYH, the structures are of a mutant protein lacking the N-terminal 155 residues (Goodwill et al., 1997), while structures of human PAH have been described for mutant proteins lacking either the N-terminal 117 residues (Fusetti et al., 1998) or the N-terminal 116 residues and the C-terminal 28 residues (Erlandsen et al., 1997). As noted above, these regions of the proteins contain all of the residues necessary for catalysis. The alpha carbons of residues 116–424 of human PAH can readily be superimposed on residues 162–470 of rat TYH with a root-mean-square deviation of only 0.66 Å (Fig. 5), demonstrating that the catalytic portions of these proteins have identical folds.

The overall structures of the catalytic domains of PAH and TYH are predominantly helical. Residues 473–496 at the C-terminus of TYH form a 40 Å long α-helix; this is the predominant interface for tetramer formation. (The homologous residues were not present in the protein used to obtain the structure of PAH.) The C-terminal helices from four monomers pack in an antiparallel coiled-coil arrangement to form a tetramer. This arrangement had been predicted based upon the observation that removal of the C-terminal 20 residues from a TYH mutant protein lacking the first 155 N-terminal residues converted the protein from a tetramer to a monomer and from the pattern of hydrophobic residues in this region (Lohse and Fitzpatrick, 1993). While this helix is the major interaction between subunits, there are clearly others. Human PAH containing only residues 103–427 lacks the C-terminal helix but still forms dimers (Knappskog et al., 1996), and rabbit TRH con-

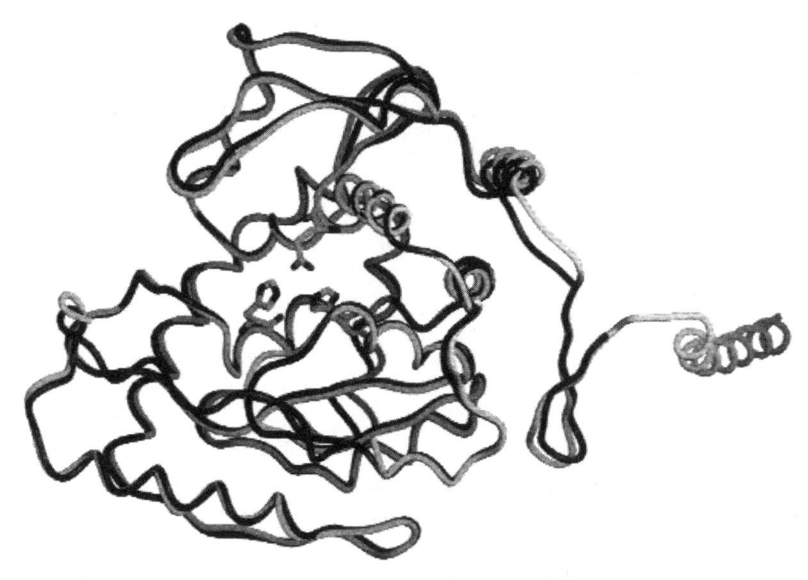

Figure 5. Superposition of the alpha carbon backbones of the catalytic domains of human phenylalanine hydroxylase (*black*) and rat tyrosine hydroxylase (*gray*). The structures are based on the coordinates in the PDB files 1TOH and 1PAH.

taining only the corresponding residues 102–416 is monomeric at low concentrations and tetrameric at high concentrations (Moran et al., 1998). In the TYH and PAH structures, additional interactions between subunits can be seen that involve the two β-strands and the short intervening loop that precede the long helix at the C-terminus.

The overall structure of the monomer is basket-like, with the active site a deep cleft in the center. The walls of the cleft are formed mainly by four helices, with the iron atom located in the cleft 10 Å below the protein surface. The two histidines identified by mutagenesis and a glutamate (residues 330 and 376 in PAH and TYH, respectively) are ligands to the metal. In addition, three water molecules are iron ligands in the PAH structure (Fig. 6), while two or three waters are found as iron ligands in different TYH structures (Erlandsen et al., 1997; Goodwill et al., 1997, 1998). This arrangement is identical to a motif that has been found in several other metalloenzymes, including isopenicillin N synthase and a biphenyl cleaving extradiol dioxygenases (Hegg and Que, 1997). In several of these latter enzymes, a substrate replaces a water molecule as a ligand to the metal during catalysis.

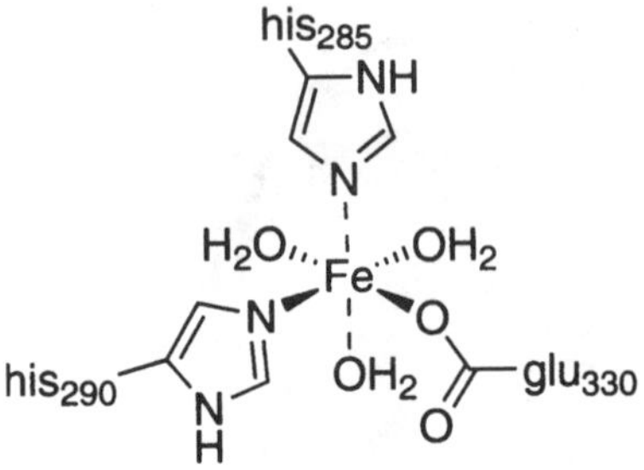

Figure 6. The ligands to the active site iron in human phenylalanine hydroxylase.

In all cases the structures of TYH and PAH have been determined with the iron in the catalytically inactive ferric form. As noted above, spectroscopic data for PAH do not suggest that there are significant differences in the ligand environment of the iron in the ferric and ferrous states. In addition, no differences in structure could be detected between crystals of TYH in the iron-free and iron-bound forms (Goodwill et al., 1997).

Besides the three amino acids that are ligands to the iron, the only other residue in close proximity to the metal is a tyrosine residue (325 in PAH, 371 in TYH), with its hydroxyl oxygen 4.5 Å from the iron and within hydrogen bonding distance (2.8 Å) of one of the metal bound waters. Because of its proximity to the iron, this tyrosine was proposed to have a catalytic role, specifically to stabilize a radical intermediate in catalysis (Erlandsen et al., 1997). Site-directed mutagenesis has been used to test this hypothesis, by replacing Tyr371 in TYH with a phenylalanine (Daubner and Fitzpatrick, 1998). The mutant protein was fully active, ruling out an essential role of this residue in catalysis.

The structure of the catalytic domain of TYH has been described with 7,8-dihydrobiopterin bound as an air-stable analog of tetrahydrobiopterin (Goodwill et al., 1998). The overall structure of the protein is identical to the pterin-free form. Several water molecules are displaced by binding of the pterin, including one found at 3.1 Å from the iron in the unliganded protein. This water is 3.6 Å away when the pterin is bound and may be analogous to the third water molecule seen as a ligand to the iron in PAH. The interactions of the dihydropterin with the protein are depicted in Figure 7. The pterin C4a is 5.6 Å from the iron, an appropriate distance for a peroxo bridge (see below). The carbonyl oxygen of the dihydropterin is 3.1 Å from an oxygen of Glu376 and 3.2 Å from the hydroxyl oxygen of Tyr371, consistent with hydrogen bonds between these residues and the pterin. As noted above, the role of Tyr371 in catalysis has been studied by site-directed mutagenesis. The Y371F protein shows an increase in the K_m value for tetrahydrobiopterin of about twofold at pH 7.1 (Daubner and Fitzpatrick, 1998). This is consistent with a relatively weak hydrogen bond between this residue and the pterin carbonyl oxygen. The side chain hydroxyls of dihydrobiopterin also appear to form hydrogen bonds with the protein. The oxygen at C2' is 3.3 Å from one of the water molecules that are ligands to the iron, and the oxygen at C1' is 3.2 Å from the amide nitrogens of both Leu294 and Leu295. These latter residues are located on a loop that extends over the mouth of the active site cleft (Goodwill et al., 1997).

The aromatic ring of Phe300 is parallel to the pterin at a distance of about 3.5 Å in the TYH structure; this is an appropriate distance for a stacking interaction. Unexpectedly, the structure of the dihydrobiopterin complex of the catalytic domain of TYH showed a peak of electron density 1.4 Å from the meta carbon of Phe300 (Goodwill et al., 1998). Satisfactory modeling of the electron density could be obtained by placing a hydroxyl on this carbon. In retrospect, similar density could also be detected for the pterin-free enzyme, but not for the analogous residue in PAH. This unexpected hydroxylation of Phe300 to form 3–hydroxyphenylalanine was proposed to be due to an autocatalytic reaction involving tetrahydropterin. However, mass spectrometry and amino acid sequencing of the tryptic peptide containing

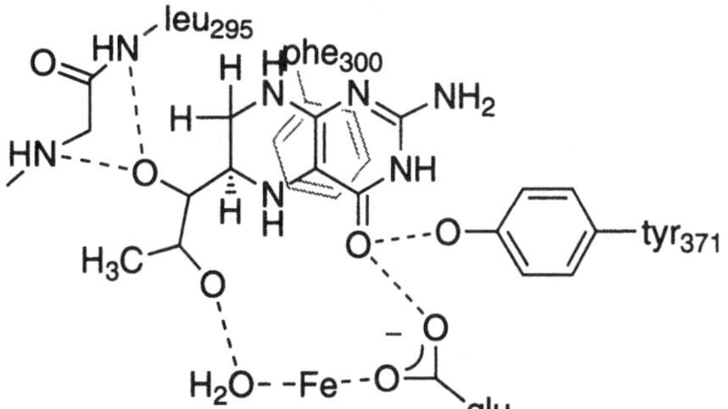

Figure 7. Interactions of dihydrobiopterin in the active site of rat tyrosine hydroxylase.

Phe300 show that Phe300 is unmodified in wild-type recombinant TYH and in the catalytic domain when they are isolated (Ellis et al., 1999b). Phe300 is hydroxylated when the catalytic domain is incubated in the presence of ferrous ammonium sulfate, dithiothreitol, and tetrahydrobiopterin, the conditions used to obtain crystals. This modification does not occur with the wild-type enzyme even in the presence of iron and reductant. The simplest explanation for these results is that the modification of Phe300 occurred during crystal growth due to the presence of high concentrations of ferrous iron and dithiothreitol. Conditions such as these have been shown to result in oxidative modification of protein amino acid residues, with phenylalanine being one of the more readily oxidized (Stadtman, 1993). Replacement of Phe300 of TYH with alanine results in a decrease in the V/K value for 6–methyltetrahydropterin of about fivefold and a decrease in the V_{max} value of about threefold. Both results can be attributed to somewhat looser binding of the pterin as a result of increased size of the binding pocket when the methyl side chain of alanine replaces the phenyl ring of phenylalanine. Even if a stacking interaction does occur in the wild-type protein, most of the energetic contribution to binding is expected to come from hydrophobic rather than π-stacking interactions (Shamoo et al., 1989). The F300A mutant would still have a hydrophobic binding site.

Structures have also been described for the catalytic domain of PAH bound to a series of catechols (Erlandsen et al., 1998). While these are unlikely to be relevant to catalytic intermediates, they are relevant to the role of catechols in the regulation of TYH, as discussed later. All four compounds examined, epinephrine, norepinephrine, dopamine, and DOPA, bind in a bidentate fashion, consistent with conclusions drawn from spectroscopic data, as shown in Figure 8. No significant differences among the binding of any of these compounds were seen, although the electron density for DOPA was not as well defined as those of the other compounds. The catecholate oxygens are within hydrogen bond distance of Glu330 and Tyr325. The aromatic ring of Phe254 lies parallel to the catechol ring at a distance of 4 Å; this is clearly reminiscent of the mode of binding of dihydropterin seen with TYH. While the location of the catechol moieties was well-determined, the position of the side chains was ambiguous, with apparent electron density consistent with either the arrangement shown in Figure 8 or the reverse. The amino groups of the catecholamines were solvent exposed. Since the structures were determined with a protein lacking the N-terminal domain, the lack of interactions seen with the amino group

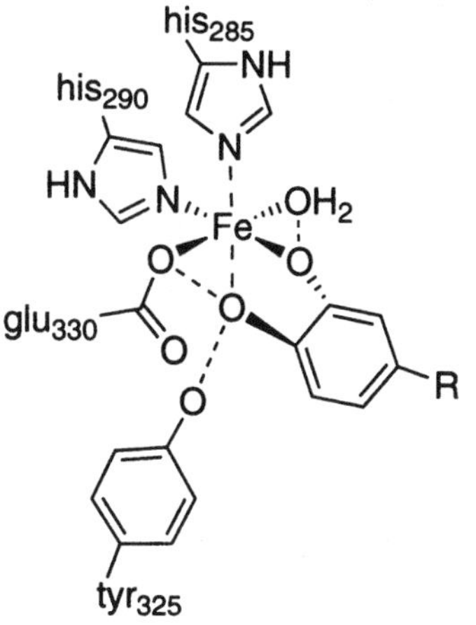

Figure 8. Interactions of catecholamines in the active site of human phenylalanine hydroxylase.

leaves open the possibility that such interactions normally occur with residues in the regulatory domain.

No structure of either PAH or TYH has been described with an amino acid bound, with the exception of DOPA, which binds as a catechol inhibitor rather than a substrate. Since spectroscopic data suggest that there are conformational changes when both tetrahydropterin and amino acid are bound, the lack of such structures means that we lack critical details regarding the relative orientations of the substrates and the iron in a catalytically competent complex. Site-directed mutagenesis of TYH has provided some insight into the identities of amino acid residues involved in binding amino acids (Daubner and Fitzpatrick, 1999). Arg316 and Asp328 of TYH are conserved in all pterin-dependent hydroxylases. The charged moieties of both residues are located about 10 Å from the iron atom. This is an appropriate distance if these residues interact with the carboxyl and amino groups, respectively, of an aromatic amino acid bound in the active site with the aromatic ring extended toward the metal. Replacement of Arg316 with lysine results in a 4000-fold decrease in the V/K value for tyrosine, consistent with an interaction of the arginine guanidino group with the carboxylate of the amino acid substrate. Replacement of Asp328 with serine results in a 200-fold decrease in the V/K value for tyrosine; this would be consistent with an interaction between the aspartyl carboxylate and the substrate amino group.

Thus, the emerging data from structural and mutational studies are beginning to build a picture of the positions of the substrates in the active site during catalysis. The tetrahydropterin binds with the 4a carbon appropriately positioned relative to the iron to accommodate molecular oxygen. The aromatic ring of the amino acid substrate would also be positioned close to the iron, where it could react with a hydroxylating intermediate containing the metal atom.

Only in the case of rat PAH has a structure been described that includes both the regulatory and the catalytic domains (Fig. 9) (Kobe et al., 1999). In this case, structures were determined of enzyme lacking the C-terminal 24 amino acid residues. This protein forms dimers but not tetramers; otherwise it

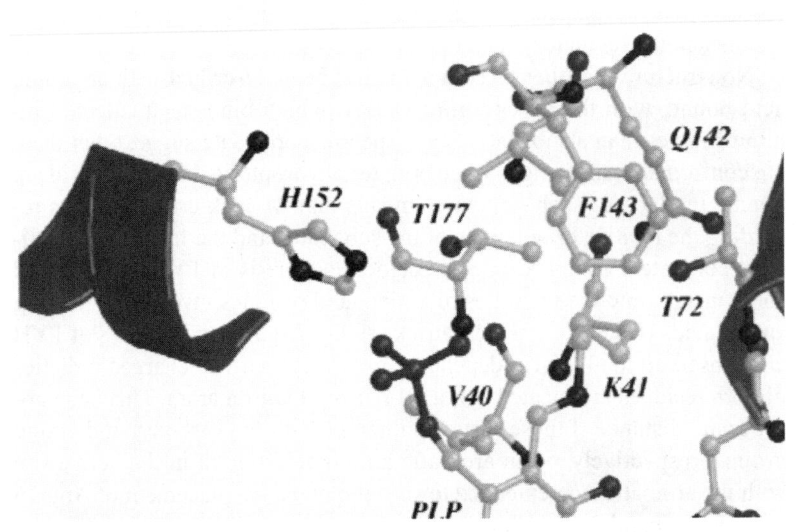

Figure 9. Regulatory and catalytic domains of phosphorylated human phenylalanine hydroxylase. The regulatory domain, residues 19–120, is shown as a dark ribbon. The catalytic domain, residues 121–427, is shown in space-filling format. The structure is based on the coordinates in the PDB file 1PHZ.

has the same catalytic and regulatory properties as the wild-type enzyme (Kobe et al., 1997). The catalytic domain, residues 118–427, has the same structure when the regulatory domain is attached as was seen with the isolated domain. The regulatory domain of PAH, residues 19–117, is an α-β sandwich, with βαββαβ topology. While the structures of the catalytic domains of PAH and TYH do not resemble those of any other proteins, the 60–70 residue core of the PAH regulatory domain has the same topology as the enzyme DCoH, a bifunctional pterin-4a-carbinolamine dehydratase/HNF1 dimerization cofactor (Endrizzi et al., 1995), and the regulatory domain of phosphoglycerate dehydrogenase (Schuller et al., 1995). This similarity provides insights into the structural bases for several regulatory properties of PAH. A binding site for pterin has been identified in DCoH (Köster et al., 1996), while phosphoglycerate dehydrogenase contains a regulatory binding site for serine (Schuller et al., 1995). This suggests that the regulatory domain of PAH contains binding sites for both a pterin and an amino acid. Allosteric binding sites for these ligands had previously been proposed based upon studies of the regulatory properties of PAH (Parniak and Kaufman, 1981; Xia et al., 1994). By analogy to the serine binding site in phosphoglycerate dehydrogenase, a binding site for phenylalanine would be located in the PAH regulatory domain at the interface between the catalytic and regulatory domains (Kobe et al., 1999). In support of such a conclusion, Cys237 is located in the same region; modification of Cys237 by thiol reagents mimics the activation of PAH by phenylalanine (Gibbs and Benkovic, 1991).

Residues 19–33 in the regulatory domain of PAH extend into the active site cleft, physically restricting access to the active site. No density could be detected for the N-terminal 18 residues, which include Ser16, the phosphorylation site. With that caveat, there were no detectable differences between the structures of the phosphorylated and unphosphorylated enzymes. Implications of these structures for the regulatory properties of PAH will be discussed later in this review.

IV. Substrate Specificity

The three pterin-dependent hydroxylases show overlapping substrate specificities, although a complete comparison of the relative specificities of these enzymes has not been done. This lack is due both to the difficulty of obtaining wild-type TRH and the complex regulatory properties of PAH. PAH is activated by binding of phenylalanine at a regulatory site, so that the activity seen with a specific amino acid as substrate depends both on its catalytic efficiency and on its ability to activate the enzyme. The requirement for activation is much less with 6–methyltetrahydropterin than with tetrahydrobiopterin (Zigmond et al., 1989), so that it is possible to analyze the catalytic specificity of PAH using the former pterin instead of the physiological one. The substrate specificity of TRH has been studied with the recombinant catalytic domain rather than the intact enzyme. With TYH and PAH, removal of the regulatory domains decreases the relative specificities of the enzymes for one another's substrates in a quantitative rather than a qualitative fashion (Daubner et al., 1997a). Therefore, conclusions drawn from studies of truncated TRH are likely to be applicable to the intact protein.

PAH and TRH will each hydroxylate both phenylalanine and tryptophan. In the case of activated PAH, the V/K value for phenylalanine is three orders of magnitude greater than that for tryptophan (Fisher and Kaufman, 1973a). In contrast, the catalytic domain of TRH has comparable V/K and V_{max} values for phenylalanine and tryptophan (Moran et al., 1998); this decreased specificity may be due to the lack of the regulatory domain (Daubner et al., 1997a). PAH will not hydroxylate tyrosine (Fisher and Kaufman, 1973a), while the formation of tyrosine by TRH is four orders of magnitude slower than hydroxylation of tryptophan or phenylalanine (Moran et al., 1998). TYH will hydroxylate both phenylalanine and tryptophan. With the intact protein, the V/K value for phenylalanine hydroxylation is one-tenth that for tyrosine; in contrast, the V/K values for the two substrates are nearly identical if the catalytic domains are compared (Daubner et al., 1997a). With intact TYH the V/K value for tryptophan is two orders of magnitude smaller than that for tyrosine (Fitzpatrick, 1991b).

The structural bases for these specificities have not been established. The active sites of the three hydroxylases are nearly identical, with the majority of the active site residues conserved among all three enzymes from eukaryotic and prokaryotic sources. The observation that the relative specificities of PAH and TYH are influenced by the removal of the regulatory domains suggests that subtle changes in the overall shape of the active site rather than the identity of a residue at any single position may be responsible for the specificity. The possible complexity of the structural basis for the specificity is illustrated by the results of mutating tyrosine 371 in TYH to phenylalanine (Daubner and Fitzpatrick, 1998). With tyrosine as substrate, the only effect of the mutation is a twofold increase in the K_m values for tetrahydropterins. In contrast, the K_m for phenylalanine is tenfold lower in the mutant protein than in the wild-type enzyme. As a result, the *V*/*K* value for phenylalanine is greater with this mutant than the value obtained with wild-type PAH.

V. Mechanism

A. IDENTITY OF THE HYDROXYLATING INTERMEDIATE

The nonenzymatic reaction of tetrahydropterins with oxygen has been studied to gain insight into the role of the pterin in the enzyme-catalyzed hydroxylation reactions. At neutral pH, tetrahydropterins rapidly react with oxygen to form hydrogen peroxide and quinonoid dihydropterin. The rate-limiting step in the reaction is proposed to be the one electron transfer from tetrahydropterin to oxygen to form superoxide and the pterin radical cation, as shown in Figure 10 (Eberlein et al., 1984). The radical pair would quickly collapse to form a 4a-hydroperoxypterin, which would rapidly eliminate hydrogen peroxide. Although it has not been possible to detect such peroxypterins directly, 6-methyl-5-deazatetrahydropterin will react with singlet oxygen to form a peroxide (Moad et al., 1979). This peroxide will catalyze the oxidation of thiols to sulfoxides, forming a 4a-hydroxypterin.

The reaction of flavins with oxygen is similar to that of tetrahydropterins, involving a slow single-electron transfer and subsequent radical coupling to form a 4a-peroxyflavin, which eliminates hydrogen peroxide to form the oxidized flavin (Eberlein and Bruice, 1983; Massey, 1994). Peroxyflavins will also oxidize thiols, as well as a variety of other species, forming 4a-hydroxyflavin as product (Bruice et al., 1983). A 4a-peroxyflavin is the most likely candidate for the hydroxylating intermediate in the reactions of flavin phenol hydroxylases (Massey, 1994). Conversion of the peroxyflavin to a 4a-hydroxyflavin occurs concomitantly with substrate hydroxylation in *p*-hydroxybenzoate hydroxylase (Entsch et al., 1976). These results make an analogous 4a-peroxypterin an

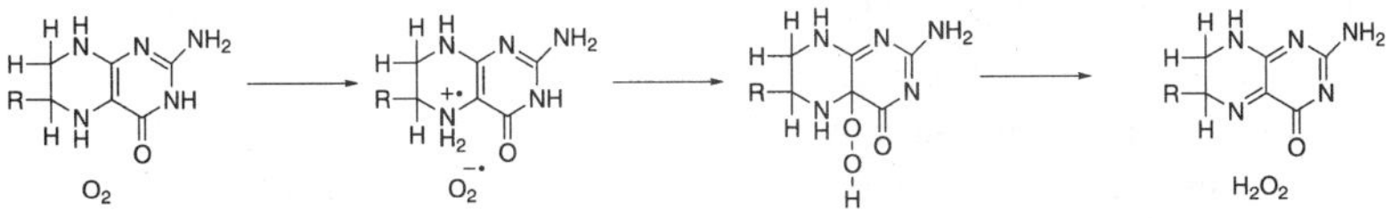

Figure 10. Mechanism of Eberlein et al. (1984) for the reaction of oxygen with tetrahydropterin.

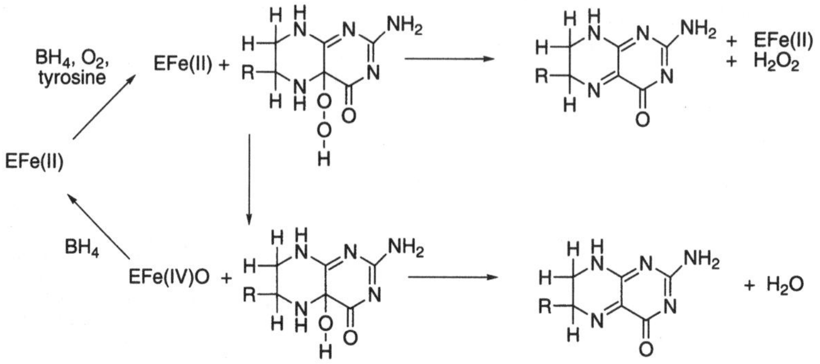

Figure 11. Mechanism for formation of pterin products during unproductive turnover by phenylalanine hydroxylase in the presence of tyrosine, based on Davis and Kaufman (1989).

attractive intermediate in the hydroxylation reactions catalyzed by the pterin-dependent enzymes.

When tyrosine is used as a substrate for PAH, no dihydroxyphenylalanine is produced. Instead, the enzyme catalyzes oxidation of the tetrahydropterin at a greatly reduced rate (Fisher and Kaufman, 1973b; Daubner et al., 1997a). This reaction has been used as a probe for a pterin peroxide as an intermediate in the enzyme-catalyzed reaction. If the enzyme-catalyzed oxidation of tetrahydropterins resembles the autoxidation reaction of tetrahydropterins, a peroxypterin should be formed on the enzyme surface. In the absence of hydroxylation, such a species would be expected to break down to form quinonoid dihydropterin and hydrogen peroxide. Thus, the formation of the latter two products would be evidence for the intermediacy of a peroxypterin in the hydroxylation reaction. Indeed, hydrogen peroxide formation has been reported during oxidation of 6-methyltetrahydropterin by PAH in the presence of tyrosine, *p*-fluorophenylalanine, or *p*-chlorophenylalanine (Davis and Kaufman, 1991). However, it has also been reported that no hydrogen peroxide is produced when *p*-chlorophenylalanine is used as a substrate for PAH (Dix and Benkovic, 1985). The amount of hydrogen peroxide detected in the presence of tyrosine accounted for about 30% of the reducing equivalents consumed. While the less than stoichiometric amount of peroxide formed could be attributed in part to a nonenzymatic reaction between the remaining tetrahydropterin and the peroxide (Blair and Pearson, 1974; Eberlein et al., 1984), an alternative explanation (Fig. 11) was provided by Davis and Kaufman (1989) based on the observation that 4a-hydroxytetrahydrobiopterin could be detected during PAH-catalyzed oxidation of tetrahydrobiopterin in the presence of tyrosine. Reaction of oxygen and the tetrahydropterin in the active site would result in formation of a peroxypterin. The peroxypterin could break down to hydrogen peroxide and the quinonoid dihydropterin. Alternatively, the peroxypterin could transfer an oxygen atom to the active site iron to form the hydroxypterin and a ferryl species, the actual hydroxylating intermediate. If no hydroxylation occurs, as is the case with tyrosine, the hydroxypterin would be released from the enzyme to form the quinonoid dihydropterin in solution. Additional reducing equivalents would be required to reduce the iron back to the ferrous form.

Somewhat different results were obtained with metal-free PAH from *C. violaceum*. In the absence of metal this enzyme is reported to catalyze the oxidation of dimethyltetrahydropterin in a reaction that requires phenylalanine but does not produce tyrosine (Chen and Frey, 1998). Hydrogen peroxide and 7,8-dihydropterin were the only products that could be detected. The amount of hydrogen peroxide produced was less than the amount of tetrahydropterin consumed, as was the case with the tyrosine-dependent oxidation of tetrahydropterin catalyzed by the eukaryotic enzymes. However, no 4a-hydroxypterin could be detected.

Benkovic et al. (1985a) have proposed the peroxypterin iron species shown in Figure 12 as an intermediate in the pterin hydroxylases. This is attractive because of its potential for further reactions. In addition, the 4a carbon of the dihydropterin in the TYH/7,8–dihydrobiopterin structure is an appropriate distance from the iron for such a peroxo bridge (Goodwill et al., 1998). The peroxy pterin-metal complex could be formed as a result of the reaction of the peroxy pterin with the iron. In a sense, the role of the pterin would be to deliver peroxide to the iron, in a manner analogous to the peroxide shunt seen with cytochrome P_{450} (Guengerich and MacDonald, 1984).

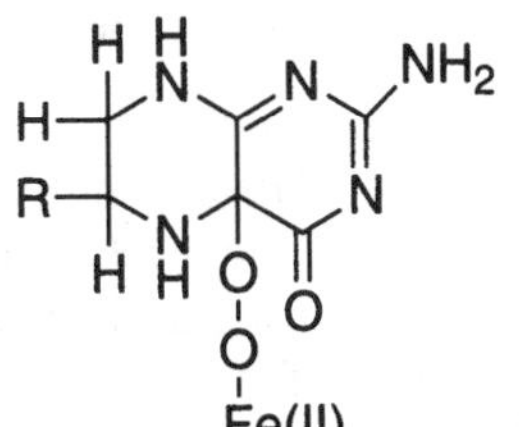

Figure 12. Peroxypterin-iron complex proposed as an intermediate by Benkovic et al. (1985a).

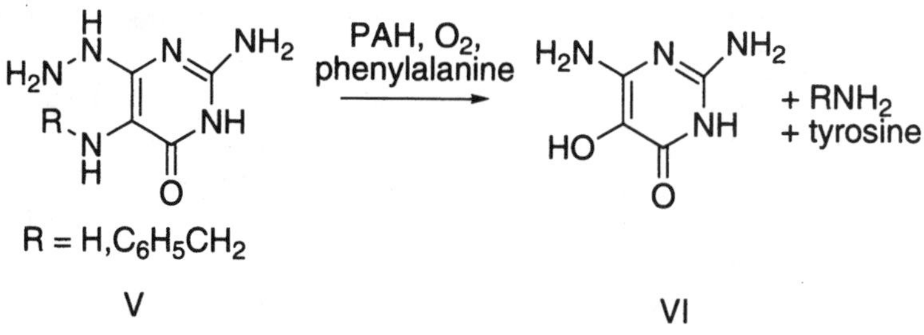

Figure 13. Utilization of 2,5,6-triamino-4-pyrimidinones by phenylalanine hydroxylase.

An alternative approach to understanding the reaction between tetrahydropterins and oxygen during catalysis was initiated by the observation that 2,5,6-triamino-4-pyrimidinones such as V (Fig. 13) can replace tetrahydropterins as substrates for PAH, although hydroxylation occurs at a greatly reduced rate (Bailey and Ayling, 1978; Kaufman, 1979). During the course of the reaction the pyrimidinone is converted to the 2,6-diamino-5-hydroxy-4-pyrimidinone VI (Bailey and Ayling, 1980). The source of the oxygen added to the pyrimidine is O_2 (Bailey et al., 1982). The loss of the 5-amino substituent during this reaction was initially taken as support for a mechanism for PAH involving an oxenoid intermediate formed by opening of the pyrazine ring of a hydroxypterin (Hamilton, 1971). However, the inability of PAH to catalyze ring closure of the product formed upon such a ring opening reaction with tetrahydropterins provided strong evidence against such a mechanism (Lazarus et al., 1981). Still, the isolation of VI and the demonstration that the source of the oxygen was O_2 provided evidence that the initial site of the reaction between tetrahydropterins and molecular oxygen is the 4a position of the tetrahydropterin.

B. MECHANISM OF AMINO ACID HYDROXYLATION

Once the hydroxylating intermediate forms, it must react with the amino acid substrate to form the hydroxylated amino acid product. One of the initial critical observations regarding the mechanism of this was that an NIH shift occurs with all three enzymes. That is, when 4-^{3}H-phenylalanine is used as a substrate for PAH (Guroff et al., 1966) or TYH (Daly et al., 1968) or 5-^{3}H-tryptophan is used as a substrate for TRH (Renson et al., 1966), the amino acid product contains 90–100% of the radioactivity. The label is found on a carbon adjacent to the site of hydroxylation in all cases (Fig. 14), having undergone a 1,2-shift. The observation of an NIH shift in these enzymes and several other monooxygenases was initially interpreted in favor of a stepwise mechanism for hydroxylation (Fig. 15) in which a cationic intermediate such as VII or VIII was formed (Guroff et al., 1967). The cation would undergo the 1,2-shift; subsequent rearomatization would generate the hydroxylated amino acid. Retention of tritium in the product would occur because of the large isotope effect expected for cleavage of the carbon-tritium bond compared to the carbon-hydrogen bond. Similar 1,2-shifts were observed with *p*-methylphenylalanine, *p*-chlorophenylalanine, and *p*-bromophenylalanine (Guroff et al., 1967).

The subsequent detection of arene oxides as products or intermediates in a number of biological oxygenation reactions led to the proposal that similar species were intermediates in the reactions catalyzed by the pterin-dependent hydroxylases (Daly et al., 1972). In the case of phenylalanine hydroxylation, the arene oxide IX in Figure 16 would be formed by the initial reaction with the enzyme oxygenating species. Benzene epoxides spontaneously and rapidly rearrange to the corresponding phenols as shown, via carbocations and NIH shifts (Bruice and Bruice, 1976), so that the intermediacy of an arene oxide would be consistent with the earlier observations.

As a more direct test of the intermediacy of arene oxides in the PAH reaction, [2,5-H_2]-phenylalanine has been characterized as a substrate for the rat liver enzyme (Miller, Benkovic, 1988). Structural analyses of the product are consistent with formation of the epoxide X (Fig. 17). Formation of

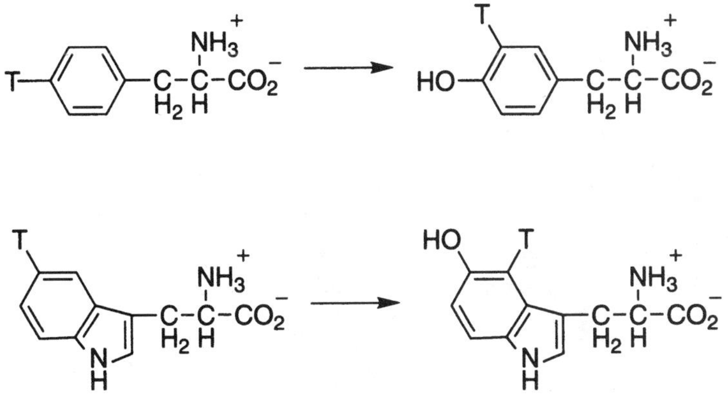

Figure 14. The NIH shift in the phenylalanine and tryptophan hydroxylase reactions.

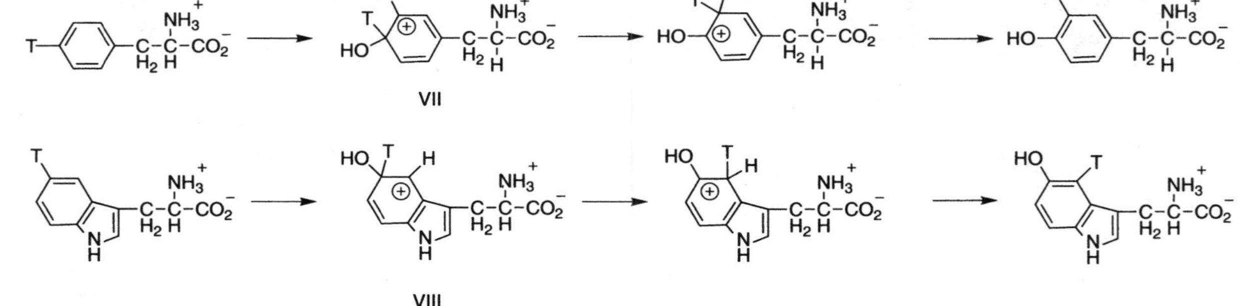

Figure 15. Proposed cationic intermediates (Guroff et al., 1967) involved in the NIH shift in the phenylalanine and tryptophan hydroxylase reactions.

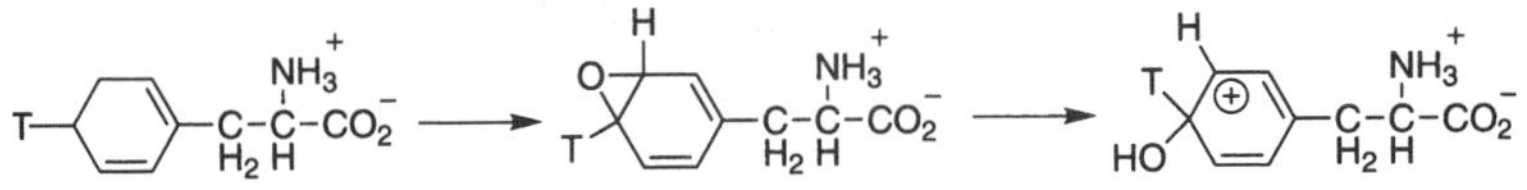

Figure 16. Proposed arene oxide intermediate in the phenylalanine hydroxylase reaction.

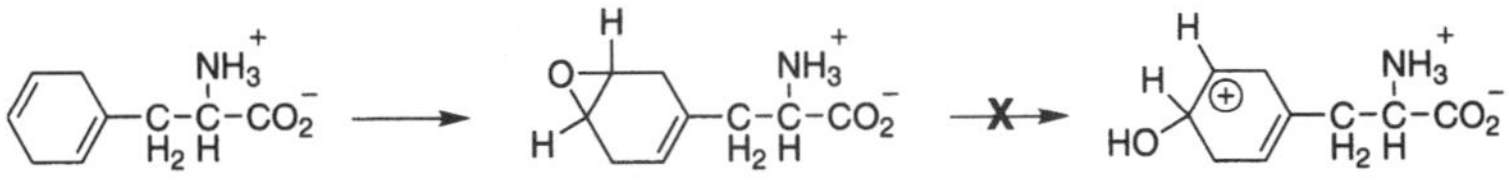

Figure 17. Reaction of [2,5-H_2]-phenylalanine with phenylalanine hydroxylase.

this epoxide by PAH would occur in a similar fashion to the formation of the phenylalanine epoxide of Figure 16. In contrast to the case with the physiological substrate, PAH would not catalyze the rearrangement of the epoxide and the attendant NIH shift.

C. RECENT MECHANISTIC STUDIES OF TYROSINE AND TRYPTOPHAN HYDROXYLASE

Essentially all of the mechanistic studies described so far were carried out with PAH. This was due to the relative ease of obtaining sufficient amounts of purified PAH for study rather than any intrinsic advantage of this enzyme over the others. Indeed, the complex regulatory properties of PAH (vide infra) have significantly complicated mechanistic studies of that enzyme. Still, the studies with PAH have identified the pterin product and provided evidence for a ferryl oxygenating species that reacts with the amino acid substrate to form an arene oxide. The availability of recombinant enzymes has made mechanistic studies of TYH and, very recently, TRH possible. This section will focus on more recent progress toward elucidating the catalytic mechanisms of the pterin-dependent hydroxylases, focusing on these two enzymes.

One of the first steps in any mechanistic examination of an enzyme is determining the kinetic mechanism. Knowledge of the kinetic mechanism establishes the protein-substrate complexes that are along the catalytic pathway, placing limits on the possible chemical mechanisms that must be considered. In the case of the pterin-dependent enzymes, there are three substrates to consider, complicating the kinetic analyses. The analysis can be simplified if the concentration of one substrate is maintained as a constant molar proportion of another (Rudolph and Fromm, 1979). Under such conditions, squared terms can be introduced into the rate equation if an individual term contains the concentrations of more than one substrate, allowing one to determine more readily the relevant steady-state kinetic equation. Such an analysis was carried out for tyrosine hydroxylase using 6-methyltetrahydropterin (Fitzpatrick, 1991a) and for bacterial PAH using 6,7–dimethyltetrahydropterin (Pember et al., 1989). The results with both enzymes were interpreted in favor of ter-bi sequential mechanisms, but the proposed mechanisms differed in the order of addition of the three substrates (Fig. 18). In the case of TYH, catalysis required ordered binding of tetrahydropterin, then oxygen, then tyrosine to form the quaternary complex. Oxygen was proposed to bind in rapid equilibrium since the data would be fit without assuming a K_m for oxygen. Tyrosine could bind to the free enzyme producing the dead-end complex responsible for the substrate inhibition seen at high concentrations of tyrosine. With bacterial PAH, oxygen was proposed to bind first, followed by random binding of tetrahydropterin and phenylalanine. However, it was noted by Pember et al. (1989) that there may be a preferred pathway with tetrahydropterin binding before phenylalanine, because substrate inhibition was seen with the amino acid. If this is the case, the principle difference between the two mechanisms is in the order of binding of oxygen and the tetrahydropterin. In both mechanisms all three substrates must be bound before any chemistry occurs, so that no partial reactions occur in the absence of one or more substrates. Consistent with this model, no enzyme-catalyzed oxidation of tetrahydropterin by TYH can be detected in the absence of an amino acid (Fitzpatrick, 1991b).

The kinetic mechanisms of PAH and TRH were determined with different tetrahydropterins. Recent results with TRH suggest that the identity of the tetrahydropterin does affect whether random or ordered mechanisms occur (Moran and Fitzpatrick, 2000). With tetrahydrobiopterin as substrate, significant substrate inhibition is seen with tryptophan as the varied substrate. In contrast, no inhibition is seen when 6-methyltetrahydropterin is used. The data for both tetrahydropterins could be fit by a model in which either the amino acid or the tetrahydropterin can bind to the free enzyme (Figure 18). However, binding of tetrahydropterin to the enzyme-tryptophan binary complex is hindered or prevented when tetrahydrobiopterin is used,

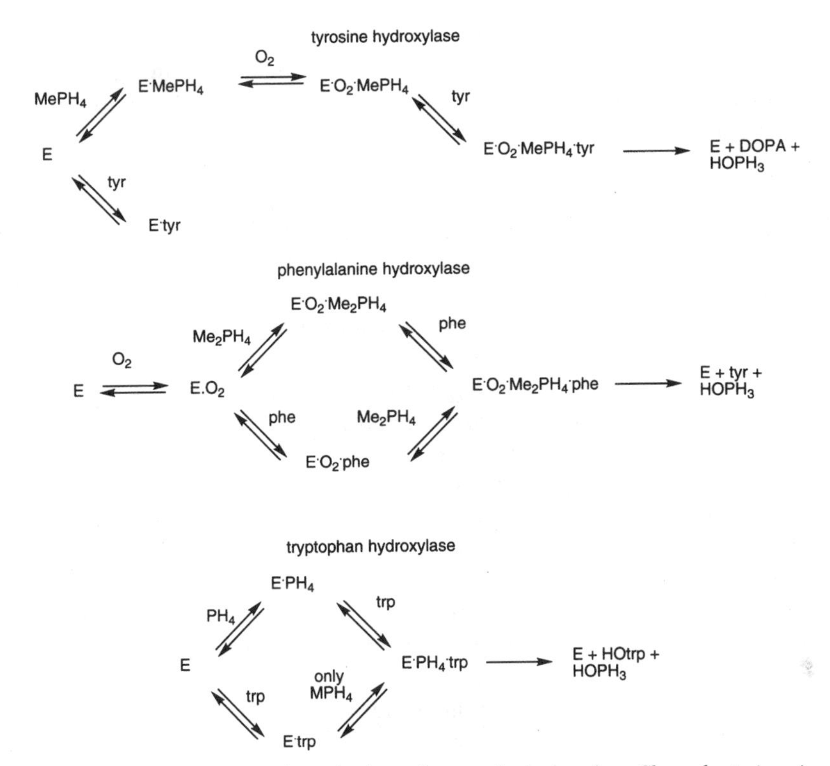

Figure 18. Steady-state kinetic mechanisms of rat tyrosine hydroxylase, *Chromobacterium violaceum* phenylalanine hydroxylase, and rabbit tryptophan hydroxylase.

but can occur when 6-methyltetrahydropterin is the substrate. The binding of oxygen was not addressed in these studies.

The kinetic mechanisms of Figure 18 combine catalysis and product release into a single step. To determine whether catalysis or product release limits turnover by TYH or TRH, the rates of formation of DOPA (Fitzpatrick, 1991a) or hydroxytryptophan (Moran et al., 2000a), respectively, during the first two to three turnovers have been determined. In both cases, the rate of hydroxylated amino production increases linearly with time rather than showing a burst; this establishes that all of the steps after amino acid hydroxylation, such as release of products from the enzyme, must be relatively rapid. These results are consistent with chemical steps being rate-limiting.

As a probe of the relative rates of formation of the hydroxylating intermediate and of its subsequent reaction with the amino acid in the overall TYH reaction, the kinetic parameters were determined with amino acids other than tyrosine as substrates. Since there is often an excess of tetrahydropterin consumed over amino acid hydroxylated when nonphysiological substrates are used as substrates for the pterin-dependent hydroxylases, it is necessary to measure the rate of tetrahydropterin consumption to ensure that the total rate of catalysis is measured. When that is done, the V_{max} value changes less than twofold with a series of *p*-substituted phenylalanines as substrates (Fitzpatrick, 1991b). The most straightforward interpretation of this result is that formation of the hydroxylating intermediate is much slower than the subsequent transfer of oxygen to the amino acid in the case of TYH.

The model studies of the autoxidation reaction of tetrahydropterins described earlier provide an attractive model for the initial events in the formation of the hydroxylating intermediate. Two series of experiments have been carried out to address more directly the mechanism of the reaction of oxygen with TYH. Solvent kinetic isotope effects were measured to determine if an exchangeable proton is in flight during this rate-limiting step in catalysis (Fitzpatrick, 1991b). There is no change in either the V_{max} value or the V/K value for tyrosine when assays are carried out in D_2O instead of H_2O (Fitzpatrick, 1991b). This rules out the possibility that an exchangeable proton is in flight in the transition state for formation of the hydroxylating intermediate. Such a result is consistent with the mechanism of Figure 10, in which electron transfer is rate-limiting.

As an alternative probe of the initial events in the reaction of oxygen with TYH, ^{18}O kinetic isotope effects have been determined (Francisco et al., 1998). Experimentally, this involves determining the isotopic composition of the remaining O_2 as the reaction proceeds. In the case of TYH, the analysis was carried out with *p*-methoxyphenylalanine rather than tyrosine to avoid inhibition by DOPA. Because the analysis involves a competition between $^{16}O_2$ and $^{16}O^{18}O$ present naturally in oxygen, the effect that is measured is on the V/K value for oxygen. Since oxygen binds in rapid equilibrium, the isotope effect will include the chemical step if the concentration of the next substrate, the amino acid, is not too high. With *p*-methoxyphenylalanine as substrate for rat TYH, the $^{18}V/K_{O2}$ was 1.0175 ± 0.0019 (Francisco et al., 1998). This value was unaffected by varying the concentration of the amino acid severalfold above or below its K_m value, by changing the pH from 6.5 to 8, or by using tetrahydropterin instead of 6-methyltetrahydropterin. The observation of a significant ^{18}O isotope effect rigorously establishes that there is a change in the bond order to oxygen during an early step in the reaction of oxygen with the enzyme. To gain insight into the nature of the change in the bonds to oxygen that gives rise to the isotope effect, it is necessary to compare the value with those expected for possible reactions on the enzyme surface. The equilibrium isotope effect, which sets an upper limit for the kinetic isotope effect in this case, has been calculated for a number of reactions involving oxygen (Tian and Klinman, 1993). Formation of an oxygen species in which there is no change in the bond order, such as hydrogen peroxide or protonated superoxide, would not result in an isotope effect, so that these species can be ruled out as products of the initial reaction with oxygen. The equilibrium isotope effects for formation of superoxide anion, peroxide anion, and peroxide dianion are 1.033, 1.034, and 1.0496, respectively. The lack of a solvent isotope effect would appear to rule out formation of peroxide anion, while formation of superoxide anion would seem more likely than peroxide dianion. A reasonable conclusion from both the ^{18}O and the solvent isotope effects is that the initial steps in the formation of the hydroxylating intermediate do indeed resemble Figure 10, leading to the formation of an enzyme-bound peroxypterin.

An important question is whether a peroxypterin is the hydroxylating intermediate or whether it is simply along the catalytic pathway to formation of the actual intermediate. The ability of the pterin-dependent enzymes to catalyze hydroxylation of benzylic and aliphatic carbons would seem to require a more powerful hydroxylating species than a peroxypterin (vide infra), although the peroxypterin-iron species in Figure 12 may be sufficiently reactive. If either form of the peroxypterin is the hydroxylating intermediate, cleavage of the oxygen-oxygen bond must be concerted with oxygen insertion into the amino acid substrate. Heterolytic cleavage to donate oxygen to the amino acid will leave the 4a-hydroxypterin, so that the amount of hydroxypterin produced should equal the amount of amino acid hydroxylated. As discussed above, excess hydroxypterin is reported when tyrosine is substrate for PAH but not with *p*-chlorophenylalanine (Dix and Benkovic, 1985; Davis and Kaufman, 1989).

An alternative approach to the use of nonphysiological substrates has been taken to this question with TYH. As described earlier, site-directed mutagenesis has been used to probe the role of individual amino acid residues in the active site of TYH. A mutant protein that is germane to the identity of the hydroxylating intermediate arises from mutagenesis of Ser395. In the active site of TYH, Ser395 is appropriately placed to form a hydrogen bond to His331, a ligand to the iron. Ser395 has been mutated to alanine and the resulting protein characterized (Ellis et al., 2000a). The K_m values for both tyrosine and tetrahydrobiopterin are relatively unaffected by the mutation, suggesting that this residue is not involved in binding of either substrate. In contrast, the V_{max} value for the hydroxylation of tyrosine to DOPA is decreased two orders of magnitude, establishing that this residue is required for hydroxylation of the amino acid. However, the V_{max} value for tetrahydropterin consumption by the mutant protein is comparable to that for the wild-type enzyme. More critically, this enzyme still catalyzes the formation of the hydroxypterin at a rate similar to that seen with the wild type enzyme. Thus, in this mutant protein, formation of the hydroxypterin can occur independent of amino acid hydroxylation. This is not the result predicted if cleavage of the peroxypterin oxygen-oxygen bond is concerted with amino acid hydroxylation, as required if the peroxypterin is the hydroxylating intermediate. Instead, amino acid hydroxylation must involve another intermediate formed subsequent to heterolytic cleavage of the peroxypterin oxygen-oxygen bond. A reasonable candidate for the subsequent species is a ferryl oxo species. Presumably Ser395 is required to properly position His331 as a ligand to the iron. Loss of the hydrogen bond to Ser395 must destabilize the ferryl oxo intermediate so that it breaks down more rapidly than it can react with the amino acid.

Studies with TRH also support the involvement of a hydroxylating intermediate distinct from a peroxypterin. TRH will catalyze the formation of DOPA from tyrosine, but the rate is 5000–fold slower than hydroxylation of tryptophan (Moran et al., 1998). The rate of DOPA production is only 1% the rate of tetrahydropterin consumption. However, about one-fifth of the tetrahydropterin consumed forms the 4a-hydroxypterin (Moran et al, 2000a). This result establishes that cleavage of the oxygen-oxygen bond occurs in a separate step from hydroxylation of the amino acid in the case of TRH also.

The second half of the catalytic reaction is the hydroxylation of the amino acid. Because the formation of the hydroxylating intermediate is rate limiting with TYH, simple measurement of rates of turnover with different amino acids is not likely to provide much insight into the hydroxylation mechanism. However, if multiple products are formed from specific amino acids, the partitioning among the different products can be informative as to the nature of the chemical steps at which the partitioning to form the different products occurs.

When phenylalanine is used as a substrate for TYH, the major product is tyrosine (Daly et al., 1968), but *m*-hydroxyphenylalanine is also formed at low levels (Fukami et al., 1990). This is the result expected if an arene oxide is formed initially (Fig. 19). While opening of the epoxide ring to form tyrosine would be expected to be the predominant path based upon the relative stabilities of the two possible carbocations, a small amount of *m*-hydroxyphenylalanine could be formed if the epoxide ring opened in the other direction. The effects of deuteration of specific carbons in the aromatic ring of phenylalanine have been used as a probe for the intermediacy of the arene oxide (Fitzpatrick, 1994). The NIH shift involves breaking of a carbon-hydrogen bond, a step whose rate will be decreased significantly by replacement with deuterium of the hydrogen undergoing the 1,2 shift. If both hydroxylated amino acid products originate from a single intermediate such as an arene oxide and if the partitioning of that intermediate is altered by selective deuteration, the relative amounts of the two products will be altered accordingly. This change in product ratios can be detected even if hydroxylation is not rate-limiting. When 4-^{2}H-phenylalanine is used as a substrate for rat TYH, the amount of tyrosine formed decreases by a factor of 1.2, while the amount of 3–hydroxyphenylalanine is unchanged. With 3,5-$^{2}H_2$-phenylalanine, the amount of 3–hydroxyphenylalanine decreases by a factor of 1.7, but the amount of tyrosine formed is unchanged. In both cases there is a decrease in the total amount of hydroxylated amino acids formed and an increase in the amount of unproductive tetrahydropterin oxidation. The observation that deuteration of the aromatic ring only alters the amount of hydroxylation at the site of deuteration rules out the partitioning of a common intermediate that already contains oxygen, such as an arene oxide. The observation of a normal isotope effect upon deuteration at either site is also not consistent with model studies of arene oxides, which show that no isotope effect is seen on the rate of ring opening (Bruice and Bruice, 1976). The observed isotope effects suggest that the hydrogen is already undergoing the 1,2-shift in the transition state for oxygen addition.

Key to the interpretation of the isotope effects on product ratios is the fact that a significant amount of the tetrahydropterin is oxidized without concomitant amino acid hydroxylation when phenylalanine is substrate. If tetrahydropterin oxidation were tightly coupled to amino acid hydroxylation, any decrease in the amount of one product would be reflected in an equal increase in the amount of other products. This phenomenon was termed metabolic switching when it was observed with cytochrome P_{450} (Harada et al., 1984). In the case of hydroxylation of phenylalanine by TYH, the rate of unproductive decay of the hydroxylating intermediate must be comparable to the rate of oxygen addition to the amino acid. A decrease in the rate of

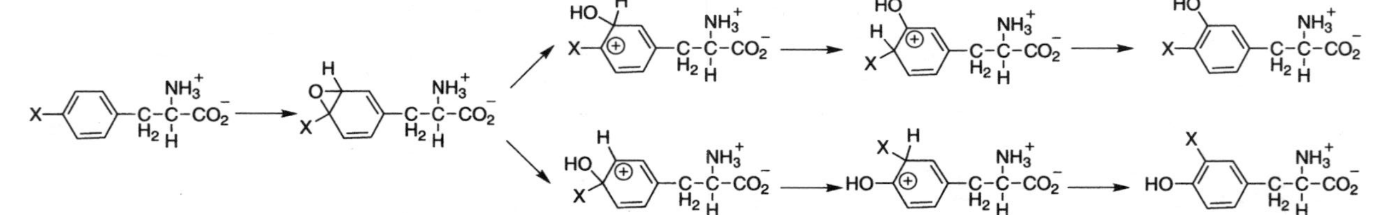

Figure 19. Formation of tyrosine and *m*-hydroxyphenylalanine from phenylalanine by tyrosine hydroxylase via an arene oxide intermediate.

oxygen attack due to deuteration of a specific carbon is consequently reflected in increased unproductive decay of the hydroxylating intermediate.

As a probe of the hydroxylation mechanism, a number of *p*-substituted phenylalanines have been characterized as substrates for rat TYH by Hillas and Fitzpatrick (1996). The results are summarized in Table 1. As is the case with phenylalanine, hydroxylation can occur at both the *meta* and the *para* position of the aromatic ring when *p*-methyl, *p*-Cl-, or *p*-Br-phenylalanine is a substrate for TYH. In contrast, *p*-F-phenylalanine is only hydroxylated on the *para*-carbon and *p*-methoxyphenylalanine is only hydroxylated on the *meta*-carbon. There is in fact a relatively good correlation between the size of the substituent at the *para*-position of the substituted phenylalanines and the site of hydroxylation. With small substituents such as hydrogen or F, hydroxylation occurs overwhelmingly at the *para*-position. With the largest substituent, the methoxy group, hydroxylation occurs only at the unsubstituted *meta*-position. When the substituent is intermediate in size between these extremes, hydroxylation can occur at both positions. This suggests that the predominant determinant of the site of hydroxylation by TYH of these amino acids is sterics. Larger substituents alter the position of the aromatic

TABLE 1
Products from Alternate Substrates of Tyrosine Hydroxylase

Substrate X	Product Y	Z	% Formation	Amino Acid Hydroxylated/ Tetrahydropterin Consumed	Amount of NIH Shift	*Para* versus *Meta* Addition
H	H	OH	4	0.27	0.84	24:1
	HO	H	96			
CH_3	CH_3	HO	51	0.71	0.41	0.41:1
	HO	CH_3	21			
	$HOCH_2$	H	28			
CH_3O	CH_3O	HO	100	0.88	0	<1:100
F	HO	H	100	0.34	0	>99:1
Cl	Cl	HO	51	0.10	0.22	0.96:1
	HO	Cl	11			
	HO	H	38			
Br	Br	HO	40	0.02	0.61	1.5:1
	HO	Br	35			
	HO	H	25			

Source: From Hillas and Fitzpatrick (1996).

ring of the amino acid in the active site, resulting in a change in the location of hydroxylation.

Table 1 also shows that the relative amount of tetrahydrobiopterin consumed per amino acid hydroxylated is altered by the substituent on the phenylalanine ring. Quantitative interpretation of the effect requires reconsideration of the kinetics of TYH. At high concentrations of substrates, only the chemical steps need be considered. As discussed earlier, these can be divided into the irreversible formation of the hydroxylating intermediate and the transfer of oxygen from the hydroxylating intermediate to the amino acid. Oxidation of tetrahydrobiopterin without amino acid hydroxylation must be due to breakdown of the hydroxylating intermediate. This basic scheme is depicted in Figure 20. The partitioning between amino acid hydroxylation and unproductive tetrahydrobiopterin oxidation will be determined by the relative values of the rate constants k_{hyd} and k_{un}. This partitioning is described by equation 1, which can be rearranged to yield equation 2. Given the relative independence of the total turnover to the identity of the amino acid, it is a reasonable assumption that k_{un} is relatively insensitive to the identity of the amino acid, at least for the series used in this study. If that is the case, then equation 2 gives an explicit relationship between the relative stoichiometry of tetrahydrobiopterin consumption and amino acid hydroxylation and the rate of attack of the hydroxylating intermediate on the amino acid. The values of $-\log$ ([BH_4 consumed/aa-OH produced] – 1) calculated from the data in Table 1 can be used to construct a Hammett plot. The best correlation is obtained when σ^+ is used as a measure of the electron-donating ability of the substituent on the aromatic ring of

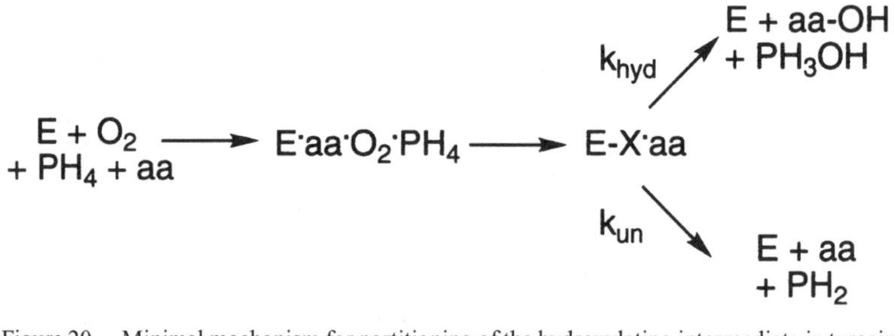

$$E + O_2 + PH_4 + aa \longrightarrow E{\cdot}aa{\cdot}O_2{\cdot}PH_4 \longrightarrow E\text{-}X{\cdot}aa$$

$$E\text{-}X{\cdot}aa \xrightarrow{k_{hyd}} E + aa\text{-}OH + PH_3OH$$

$$E\text{-}X{\cdot}aa \xrightarrow{k_{un}} E + aa + PH_2$$

Figure 20. Minimal mechanism for partitioning of the hydroxylating intermediate in tyrosine hydroxylase between amino acid hydroxylation and unproductive decay.

phenylalanine, for a ρ value of –4.6. A negative ρ value of this magnitude is most consistent with an extremely electron deficient transition state. A reasonable conclusion is that hydroxylation involves attack of an electrophilic oxygenating intermediate on the amino acid substrate to form a cation.

$$\frac{\text{aa-OH produced}}{\text{BH}_4\ \text{consumed}} = \frac{k_{\text{hyd}}}{k_{\text{hyd}} + k_{\text{un}}} \tag{1}$$

$$-\log\left(\frac{\text{BH}_4\ \text{consumed}}{\text{aa-OH produced}} - 1\right) = \log k_{\text{hyd}} - \log k_{\text{un}} \tag{2}$$

The propensities of the different substituents to undergo an NIH shift provides further support for the involvement of a cationic intermediate. The relative amount of the individual product formed via an NIH shift varies with the substituent, in the order Br > CH_3 > Cl >> F ~ CH_3O. Previous studies with phenylalanine as substrate had established that 84% of the tyrosine formed from 4-^{2}H-phenylalanine still contained deuterium (Daly et al., 1968), so that hydrogen is even more likely than Br to undergo a 1,2-shift. This ranking of substituents agrees with their ability to form the positively charged three-centered transition state involved in the NIH shift.

The product distributions seen with this series of substituted phenylalanines as substrates for TYH are consistent with the mechanism in Figure 21. The electrophilic oxygen in the hydroxylating intermediate can add to either the *m*- or the *p*-position of the aromatic ring; for substrates other than tyrosine the position of attack is determined by whether the size of the substituent at the *p*-position allows approach to that carbon. Attack at the *p*-carbon is followed by an NIH shift of the substituent at that position if the resulting cation is stable. Rearomatization then generates the final hydroxylated amino acid. The similarity of this mechanism to that in Figure 15 proposed by Guroff et al. (1967) over thirty years ago is obvious.

The mechanism of Figure 21 predicts that substitution of deuterium for hydrogen at the site of hydroxylation will affect the rates of several steps. The initial oxygen addition should show an inverse isotope effect due to the change in hybridization of the carbon to which oxygen adds from sp^2 to sp^3. The subsequent NIH shift and rearomatization would be expected to show normal isotope effects. When 3,5-$^2H_2$-tyrosine is used as a substrate for TYH, there is no significant effect on the rate of any kinetic parameter, since cleavage of the amino acid carbon-hydrogen bond does not occur in a rate-limiting

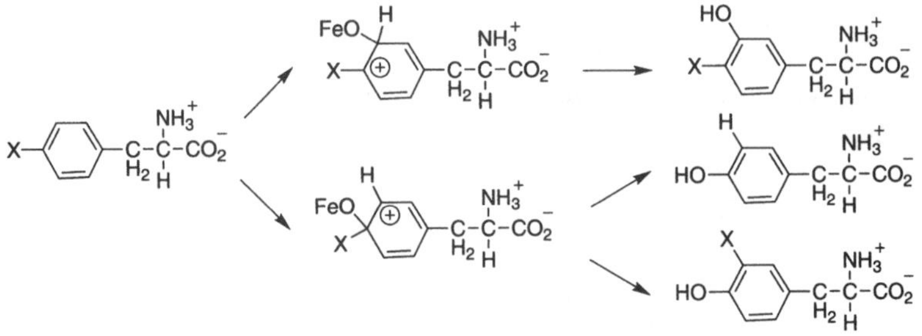

Figure 21. Mechanism proposed by Hillas and Fitzpatrick (1996) for hydroxylation by tyrosine based on partitioning of substituted phenylalanines.

step (Fitzpatrick, 1991a). When deuterated phenylalanine is used as a substrate for either TYH or PAH, a small normal isotope effect is found on the V_{max} value if 6-methyltetrahydropterin is used; no effect is seen with PAH using tetrahydrobiopterin (Abita et al., 1984). The normal isotope effect can be attributed to partial rate limitation by one of the latter steps. In the case of TRH, the isotope effect on the V_{max} value with [ring-$^2H_5$]-tryptophan as substrate depends on the tetrahydropterin used. With tetrahydrobiopterin, a small normal effect is seen, as was the case with PAH; in contrast, an effect of 0.81 is seen with 6–methyltetrahydropterin (Moran et al., 2000a). The observation that the magnitude of the isotope varies from normal to inverse depending upon the structure of the tetrahydropterin suggests that the rates of the individual steps in Figure 21 are similar and that their relative magnitude can be perturbed by small changes in the orientation of the tetrahydropterin in the active site. The inverse effect seen with 6–methyltetrahydropterin is consistent with a rate-limiting attack of the hydroxylating intermediate on the amino acid substrate to form a cation similar to VIII in Figure 15. Measurements with 5-^{2}H- and 4-^{2}H-tryptophan as substrates establish that the inverse isotope effect arises from changes at carbon 5 (Moran et al., 2000a), providing further evidence for this mechanism.

Based upon the individual mechanistic studies described here, it is possible to propose the relatively detailed mechanism for the aromatic amino acid hydroxylases shown in Figure 22. The initial reaction of the tetrahydropterin would be a single electron transfer to form superoxide and the pterin cation radical, as indicated by the model studies and the ^{18}O isotope effects. The two radicals would then collapse to form a peroxypterin. The data with metal-free bacterial PAH suggest that the iron is not required for this reaction, although it may be required for the eukaryotic enzymes. If the iron is not required for tetrahydropterin oxidation, the peroxypterin must be a discrete intermediate. While there is no direct evidence for the iron peroxypterin species shown as the next intermediate, it is similar to peroxy metal species proposed for a number of monooxygenases (Klinman, 1996; Que and Ho, 1996; Sono et al., 1996). The subsequent heterolytic cleavage of the oxygen-oxygen bond would be facilitated by formation of an iron-oxygen bond. The resulting ferryl oxygen species should be sufficiently electrophilic for aromatic hydroxylation. Attack of the electrophilic ferryl oxygen on the aromatic ring of the amino acid substrate would generate the previously proposed cation. Cleavage of the iron-oxygen bond would regenerate the ferrous iron, while loss of a proton from the cationic amino acid followed by rearomatization would generate the hydroxylated product.

The ferryl intermediate proposed in Figure 22 would be expected to be capable of more difficult reactions than aromatic hydroxylation. Consistent with such an expectation, the pterin-dependent enzymes have been reported to carry out benzylic hydroxylation, in that both PAH and TYH can catalyze the hydroxylation of *p*-methylphenylalanine to *p*-hydroxymethylphenylalanine (Daly and Guroff, 1968; Hillas and Fitzpatrick, 1996) and TRH can catalyze formation of 5-hydroxymethyltryptophan from 5-methyltryptophan

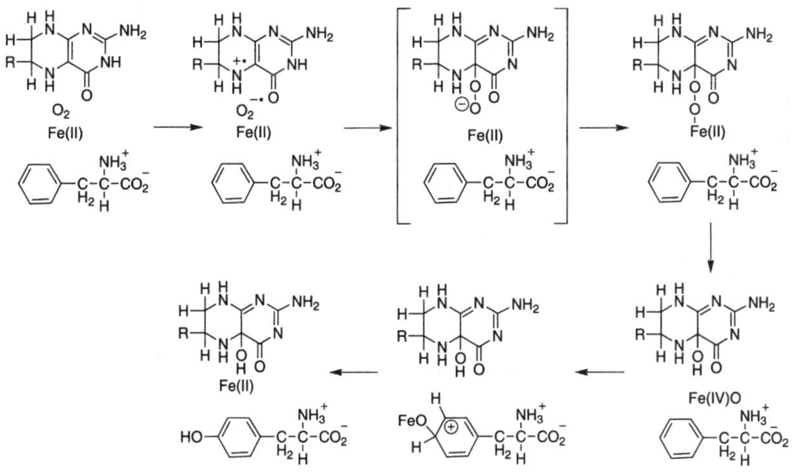

Figure 22. Possible mechanism for hydroxylation of phenylalanine by the aromatic amino acid hydroxylases.

(Moran et al., 1999). Rat liver PAH is reported to be able to hydroxylate norleucine (Kaufman and Mason, 1982), while the bacterial enzyme is reported to catalyze the hydroxylation of cyclohexylalanine (Carr et al., 1995); these are both examples of aliphatic hydroxylation. This range of reactions resembles that of cytochrome P_{450}-catalyzed reactions, consistent with the predicted properties of a ferryl species.

VI. Regulation

A. PHENYLALANINE HYDROXYLASE

Despite more than two decades of study, the complex regulatory properties of PAH are only partly understood. The activity of this enzyme is determined by the competing allosteric effects of the amino acid and tetrahydropterin substrates, as modulated by reversible phosphorylation of Ser16 in the regulatory domain. The rate of tyrosine formation shows a sigmoidal dependence on the concentration of phenylalanine (Fisher and Kaufman, 1973a). Phosphorylation of the enzyme by cAMP-dependent protein kinase increases the maximum rate of tyrosine formation about threefold, but the substrate dependence is still sigmoidal rather than hyperbolic (Abita et al., 1976). Several other protein kinases will also phosphorylate PAH on the same residue (Doskeland et al., 1984b). Sigmoidal kinetics are seen only when tetrahydrobiopterin is used; the kinetic pattern is hyperbolic with 6,7-dimethyltetrahydropterin. Shiman and coworkers showed that the sigmoidal kinetics were abolished if the enzyme was preincubated with phenylalanine beforehand; preincubation with phenylalanine abolishes a long lag in the formation of tyrosine seen with enzyme that had not been treated with the amino acid (Shiman et al., 1979). Similar activation is seen if the enzyme is treated with lysolecithin (Fisher and Kaufman, 1973a; Shiman and Gray, 1980). The activation can be successfully modeled by assuming that the native form of the enzyme has very little activity and that binding of phenylalanine converts the native form to a highly active form (Shiman and Gray, 1980). Several other amino acids are also able to activate PAH in a similar manner (Kaufman and Mason, 1982). Activation by phenylalanine or lysolecithin exposes a sulfhydryl group on the enzyme (Fisher and Kaufman, 1973a) and converts it to a form that absorbs to hydrophobic surfaces (Shiman et al., 1979), consistent with a conformational change accompanying activation. Conversely, treatment with sulfhydryl reagents activates the enzyme (Parniak and Kaufman, 1981); the modified thiol has been identified as Cys236 in the rat enzyme (Gibbs and Benkovic, 1991). Treatment with chymotrypsin converts the enzyme to a form that no longer requires pretreatment with phenylalanine; the change in molecular weight upon proteolysis is consistent with removal of the regulatory domain (Fisher and Kaufman, 1973a). Removal of the regulatory domain of rat or human PAH by site-directed mutagenesis similarly results in enzyme that is fully active without preincubation (Knappskog et al., 1996; Daubner et al., 1997a, 1997b).

Shiman and coworkers have carried out an extensive series of kinetic analyses of the effects of both phenylalanine and tetrahydrobiopterin on the activity of rat PAH (Shiman et al., 1990, 1994a, 1994b; Xia et al., 1994). Based upon the results they proposed a more extensive model for the allosteric effects of the two substrates. In this model, tetrahydrobiopterin can bind to a regulatory site on the resting enzyme, forming an inactive binary complex that phenylalanine cannot activate. Other pterins, such as 6-methyltetrahydropterin, bind much more weakly to the pterin regulatory site. Phenylalanine binds to a separate regulatory site on the resting enzyme to activate it. Binding to either regulatory site is not in rapid equilibrium, so that time-dependent activation and inhibition is seen. Phosphorylation would result in an increase in the affinity of the regulatory site for phenylalanine, with no other changes in kinetic parameters (Xia et al., 1994). Binding of phenylalanine increases the rate of phosphorylation of PAH, while binding of tetrahydrobiopterin inhibits phosphorylation (Doskeland et al., 1984a).

The model developed by Shiman is fully consistent with the three-dimensional structure of rat PAH recently described by Kobe et al. (1999). The structure of the regulatory domain allows identification of putative regulatory sites for both phenylalanine and tetrahydrobiopterin. In the absence of either ligand, the N-terminus physically blocks the active site, as shown in Figure 9. The kinetic model suggests that phenylalanine binds to a conformation in which the active site is not blocked. Tetrahydropterin binding would stabilize the conformation in which the regulatory domain hinders access to the active site. Although no density was seen for the N-terminal 18 amino acid residues, including the phosphorylation site, the structures do not show any difference upon phosphorylation. This suggests that phosphorylation makes displacement of the regulatory domain from the active site easier without causing a complete conformational change. A possible hinge for a rotation of the N-terminus has been identified at Gly33, very close to the proposed tetrahydropterin binding site. This would provide a possible structural basis for the inhibitory effects of this substrate.

Inhibition by steric restriction of the active site rather than a modification of the active site is consistent with spectral studies of the effects of activation of PAH. No changes in the ligand field of the iron upon phenylalanine activation have been detected by any of a variety of spectroscopic methods (Kappock et al., 1995; Loev et al., 1997). However, changes in the overall shape consistent with a more open conformation due to movement of the N-terminal inhibitory peptide can be detected by gel filtration (Kappock et al., 1995).

B. TYROSINE HYDROXYLASE

As is the case with PAH, tyrosine hydroxylase is regulated by reversible phosphorylation. More than two decades ago, several laboratories reported that the enzyme is a substrate for cAMP-dependent protein kinase in vitro (Ames et al., 1978; Joh et al., 1978; Yamauchi and Fujisawa, 1979). Since then the purified enzyme has been shown to be a substrate for several other kinases (Zigmond et al., 1989; Haycock et al., 1992; Sutherland et al., 1993). Campbell et al. (1986) initially identified three different serine residues in the intact rat enzyme as phosphorylation sites in vitro using purified protein kinases. cAMP-dependent protein kinase catalyzed phosphorylation of Ser40, calmodulin-dependent protein kinase catalyzed phosphorylation of both Ser40 and Ser19, and an unidentified protein kinase that copurified with TYH catalyzed phosphorylation of Ser8. More recently, Ser31 was identified as the target of the MAP kinases ERK1 and ERK2 (Haycock et al., 1992). The situation with human TYH is somewhat more complicated. Due to differential splicing of the mRNA, there are four different isoforms of human TYH in cells (Grima et al., 1987; Kaneda et al., 1987; O'Malley et al., 1987). The human form 1 (hTH1) corresponds to the rat enzyme. Isoform hTH2 contains a four amino acid insert after Met30, hTH3 contains a different 27 amino acid insert after Met30, and hTH4 contains a 31 amino acid insert consisting of the insert in hTH2 followed by that in hTH3. Isoforms hTH1, hTH2, and hTH3 are phosphorylated by cAMP-dependent protein kinase on the residue that corresponds to Ser40 in hTH1 and the rat enzyme (Le Bourdellès et al., 1991; Alterio et al., 1998). MAP kinase (ERK) phosphorylates hTH1, hTH3, and hTH4 on the residue corresponding to Ser31; hTH2 is also phosphorylated but very slowly (Sutherland et al., 1993). Calmodulin-dependent protein kinase II phosphorylates hTH1–3 on both Ser19 and the residue corresponding to Ser40 (Le Bourdellès et al., 1991; Alterio et al., 1988). In addition, hTH2 is phosphorylated on the residue corresponding to Ser31 by this kinase; this change in specificity has been attributed to a change in the recognition sequence surrounding this residue as a result of the introduction of the four additional amino acids immediately preceding this residue (Sutherland et al., 1993).

An important question is which of these phosphorylation sites are physiologically significant. The most extensive studies have been carried out with bovine chromaffin granules and rat PC12 pheochromocytoma cells. It was initially reported that agents such as adenosine or 8-bromo-cAMP that increase the activity of cAMP-dependent protein kinase in cells increased conversion of tyrosine to DOPA, increased formation of catecholamines, and increased the level of phosphorylation of TYH at a single site (Vaccaro et al., 1980; Erny et al., 1981; Haycock et al., 1982a, 1982b; Meligeni et al., 1982). Susequent studies by several laboratories showed that multiple sites could be phosphorylated under different conditions (Haycock et al., 1982a, 1982b; McTigue et al., 1985; Tachikawa et al., 1986; Griffith and Schulman, 1988; Waymire et al., 1988). The specific residues that are phosphorylated when cells are subjected to various treatments have been identified by Haycock (1990). Increased calcium influx leads predominantly to phosphorylation of Ser19; treatment of cells with nerve growth factor or phorbol esters results primarily in phosphorylation of Ser31, also described by Mitchell et al. (1998); Ser8 phosphorylation is seen upon treatment with a phosphatase inhibitor; and increased cAMP levels leads to phosphorylation of Ser40. Similar results have been obtained from bovine chromaffin cells: calcium influx leads to phosphorylation of Ser19 initially and of Ser40 and Ser31 after longer times; treatment with nerve growth factor, angiotensin II, endothelin III, prostaglandin E1, or γ-aminobutyric acid results in phosphorylation of Ser31; bradykinin or histamine treatment leads to rapid phosphorylation of Ser19 followed by slower phosphorylation of Ser31 and Ser40; and increased cAMP levels result in phosphorylation of Ser40. A number of pharmacological agents have been identified that cause phosphorylation of all three of these sites. A direct involvement of an ERK kinase in phosphorylation of TYH in bovine chromaffin cells has been demonstrated by Halloran and Vulliet (1994).

The combined results of experiments in cells and with purified proteins present a consistent picture. Agents that increase the level of cAMP regulate TYH by increasing the level of phosphorylation of Ser40, agents that increase cellular levels of calcium increase the level of phosphorylation of

Ser19, and agents that act through the phosphoinositol/protein kinase C pathway increase the level of phosphorylation of Ser31 (Haycock, 1993). However, a simple model in which a single kinase phosphorylates a single site on TYH is not fully consistent with either the in vivo or the in vitro data. In many cases more than one serine residue is phosphorylated when a pure protein is exposed to a given kinase or cells are treated with a specific compound. In the case of the pure proteins, it may be that the individual kinases are not completely specific for a given residue. The extent of phosphorylation obtained by different investigators ranges from less than 10% to 100%. Experiments that use high levels of kinases to obtain stoichiometric phosphorylation will mask differences in rates of phosphorylation at different residues. A notable exception to such qualitative approaches examined the rates of phosphorylation by cAMP-dependent protein kinase of peptides corresponding to phosphorylation sites in rat TYH; this study showed that Ser40 was preferred over Ser19 by three orders of magnitude (Roskoski and Ritchie, 1991). In the case of experiments with cells, there have only been a few instances where the time course of phosphorylation at different sites was determined (Waymire et al., 1988; Haycock, 1993; Haycock et al., 1998). In those cases, the rates of phosphorylation of different residues are clearly different. For example, when PC12 cells are treated with KCl, Ser19 is phosphorylated rapidly, followed by much slower phosphorylation of Ser40; the time course for the increase in TYH activity is closer to that for Ser40 phosphorylation, although the amount of the increase is quite small (Haycock et al., 1998). In addition, the actual level of maximal phosphorylation of Ser19 is much higher than that of Ser40, with stoichiometries of 0.7 and 0.1 per subunit, respectively. The different time courses raise the possibility that multiple phosphorylation cascades become involved at longer times.

Site-directed mutagenesis of phosphorylation sites in TYH to nonphosphorylatable residues has been used to determine the relative contribution of individual sites to regulation in vivo. Serines 8, 19, 31, and 40 in the rat enzyme have been individually mutated to leucine and the recombinant enzymes expressed in AT20 cells, a mouse anterior pituitary cell line capable of dopamine production (Wu et al., 1992). The S40L protein behaves as if it were phosphorylated on Ser40, with higher V_{max} and dopamine K_i values, a decreased K_m value for tetrahydropterin (Wu et al., 1992), and increased DOPA production in cells containing the mutant protein (Harada et al., 1996). When the cells were treated with forskolin to increase cAMP levels, cells expressing wild-type TYH showed an increase in DOPA production of about twofold, while cells expressing the S40L protein showed no change in the already high levels of DOPA production (Harada et al., 1996). The wild-type protein was phosphorylated on four serine residues under these conditions, although the level of phosphorylation of Ser8 was relatively low, while the mutant protein was phosphorylated on the three remaining serines. Thus, no further activation of the already activated S40L protein occurred upon phosphorylation of the additional residues. In contrast, treatment of cells with either high levels of KCl, which would be expected to result in phosphorylation of Ser19, Ser31, and Ser40 (McTigue et al., 1985; Haycock, 1993), or the phosphatase inhibitor okadoic acid increased DOPA production both from cells expressing wild-type TYH and from those expressing the S40L protein (Harada et al., 1996). Unfortunately, the relative levels of phosphorylation of the individual serine residues were not determined, making it difficult to rationalize the different results obtained upon stimulation by forskolin, okadoic acid, and KCl. A subsequent study showed that treatment of AT20 cells with KCl results in comparable increases in DOPA production if cells are transfected with wild-type TYH or the S19L protein (Haycock et al., 1998).

There have been a number of reports describing the effects of phosphorylation on TYH activity and on individual kinetic parameters. While it is possible to build up a qualitative picture of the effects of phosphorylation from these, it is more difficult to reach a consensus on the quantitative effects. A number of factors have contributed to variability in the effects described by different laboratories. Typically, the activity of TYH is determined using end point assays; crude preparations of enzyme have been reported to show lags in the rate of DOPA formation (Miller and Lovenberg, 1985; Bailey et al., 1989), so that the activity detected in an assay may depend upon the assay period used. There has been a wide variation in the extent of phosphorylation upon treatment with protein kinases; few studies carried out before 1990 report stoichiometries of phosphorylation. In addition, enzymes purified from eukaryotic sources contain endogenous levels of phosphate at different levels. For example, different preparations of enzyme purified from bovine adrenal medulla have been reported to contain 0.07 and 0.7 moles of covalent phosphate per monomer (Nelson and Kaufman, 1987; Haavik et al., 1988), while the phosphate content of enzyme from unactivated PC12 cells is reported to be 0.05 or less at Ser40 and 0.15–0.2 at Ser19 (Haycock et al., 1998).

Many of the initial experiments determined the effects on activity at pH 6 because that is the pH optimum for crude enzyme from nonrecombinant sources. At pH 6, phosphorylation with cAMP-dependent protein kinase is generally reported to result in a increase in the V_{max} value of two- to threefold and a decrease in the K_m value for either tetrahydrobiopterin or 6-methyltetrahydropterin at phosphorylation stoichiometries of 0.25–0.75 (Joh et al., 1978; Markey et al., 1980; Vulliet et al., 1980, 1984; Lazar et al., 1982). In addition, the K_i value for dopamine increases severalfold (Markey et al., 1980; Lazar et al., 1982). The K_m value for tyrosine has consistently been reported to be unaffected by phosphorylation. More recently, analyses have measured the activity at pH 7 because the effects are larger at this pH (Pollock et al., 1981; Richtand et al., 1985). In addition, changes in activity at pH 7 are more likely to reflect physiologically relevant differences. Qualitatively, the effects are the same as at pH 6. The K_m values for tetrahydropterins have been reported to decrease from two- to sevenfold; there is no correlation between reported degree of phosphorylation and extent of change in the K_m value when results from different laboratories are compared (Markey et al., 1980; Lazar et al., 1982; Le Bourdellès et al., 1991; Almas et al., 1992; Daubner et al., 1992; Sutherland et al., 1993; Alterio et al., 1998). Recombinant enzymes expressed in *E. coli* are unphosphorylated when purified, unlike the nonrecombinant enzymes (Almas et al., 1992; Daubner et al., 1992), allowing the fully unphosphorylated enzyme to be used for comparison. At pH 7, the V_{max} value and the K_m value for tyrosine are unaffected when the recombinant rat enzyme is phosphorylated with cAMP-dependent protein kinase, while the K_m value for tetrahydrobiopterin decreases twofold (Daubner et al., 1992). The effects on activity described for the human enzyme vary among the different isoforms. Alterio et al. (1998) reported that phosphorylation of Ser40 alone by cAMP-dependent protein kinase activates isoforms hTH1, 2, and 4, but not hTH3, but Sutherland et al. (1993) reported that phosphorylation of Ser40 by MAPKAP kinase 1 activates all four isoforms about threefold. In the case of hTH1 and hTH2, the activation was due to a decrease in the K_m for tetrahydrobiopterin of about twofold and an increase in the K_i value for dopamine of two- to threefold.

The effects of phosphorylation of residues other than Ser40 have been less studied. Treatment with calmodulin-dependent protein kinase II, which phosphorylates both Ser19 and Ser40, is reported to have no effect on the activity of the rat enzyme in the absence of an activator protein (Yamauchi et al., 1981; Albert et al., 1984; Vulliet et al., 1984; Atkinson et al., 1987), even in cases where phosphorylation of Ser40 alone results in increased catalytic activity (Funakoshi et al., 1991). In the presence of the activator protein an increase in activity of up to twofold has been reported upon phosphorylation by calmodulin-dependent protein kinase II (Yamauchi et al., 1981; Atkinson et al., 1987). The activator protein has been identified as a member of the 14-3-3 family of proteins (Ichimura et al., 1987); these are small proteins involved in a number of signal transduction pathways (Aitken, 1996). Incorporation into Ser40 of 0.43 phosphates per subunit catalyzed by protein kinase C has been reported to have no effect on activity while incorporation of 0.78 phosphates into the same residue by cAMP-dependent protein kinase yields a threefold increase in activity (Funakoshi et al., 1991). The reported effects on the human enzymes are again more complex. Phosphorylation of both Ser19 and Ser40 by MAPKAP kinase 2 is reported to increase the activity of all four isoforms about threefold, but the combination of MAPKAP kinases 1 and 2 gives the same degree of activation as treatment by MAPKAP kinase 1 alone (Sutherland et al., 1993). In contrast, phosphorylation of Ser19 and Ser40 by calmodulin-dependent protein kinase II is reported both to have no effect on the activity of hTH1 and 3 (Le Bourdellès et al., 1991; Alterio et al., 1998) and to increase the activity of hTH1 by 50% (Sutherland et al., 1993). MAP kinase phosphorylates Ser31 or the equivalent residue in hTH1, 3, and 4. With hTH1 and 3 this results in an increase in activity of up to twofold. However, enzyme that has been phosphorylated on Ser40 by MAPKAP kinase 1 shows no further increase upon further phosphorylation by MAP kinase (Sutherland et al., 1993). Finally, treatment of hTH2 with calmodulin-dependent protein kinase phosphorylates Ser31 in addition to Ser19 and Ser40, but no change in activity can be detected (Le Bourdellès et al., 1991).

There is general agreement among the many reports of the effects of phosphorylation that the activity of TYH increases when Ser40 alone is phosphorylated, although there is a report that phosphorylation of this residue in nonrecombinant enzyme by protein kinase C has no effect (Funakoshi et al., 1991). The data are also consistent with phosphorylation of Ser40 leading to maximal activation, in that subsequent phosphorylation of additional sites causes no further increase in activity (Sutherland et al., 1993). The effects of phosphorylation of Ser19 and Ser31 are much less consistent. In all cases in which phosphorylation of these residues has been described, Ser40 was also phosphorylated. Consequently, it is not possible to rule out the possibility that the increased activity results from phosphorylation of the latter residue.

Even in the case of phosphorylation of Ser40, the changes in steady-state kinetic parameters that have been reported are generally quite modest. It seems unlikely that the activity of TYH could be effectively regulated by changing the K_m for tetrahydrobiopterin by no more than twofold. The first indication that the activity of TYH was influenced by factors in addition to the phosphorylation state came when Andersson et al. (1988) reported that enzyme isolated from bovine adrenal gland contained a substoichiometric amount of epinephrine and norepinephrine bound to the iron. The enzyme from PC12 cells was similarly shown to have a mixture of dopamine and norepinephrine bound (Andersson et al., 1992). As noted earlier, the iron in the enzyme as isolated is in the ferric form. Catechols such as the catecholamines will form tight ferric complexes; in essence, the PC12 and bovine enzymes are significantly inhibited when isolated. Evidence that this inhibition is related to the regulation came with the observation that reduction by 6-methyltetrahydropterin of the iron in the bovine enzyme to the active ferrous state occurs much more readily once the enzyme has been phosphorylated by cAMP-dependent protein kinase (Andersson et al., 1989). A reasonable rationale for this observation is that phosphorylation results in weaker binding of the catecholamines; Haavik et al. (1990) subsequently reported that the dissociation rate for added norepinephrine increased severalfold upon phosphorylation. These results suggested that the observed effects of phosphorylation would be dependent upon the catecholamine content of the enzyme. By using recombinant rat TYH, which lacks both catecholamine and phosphorylated serines, Daubner et al. (1992) were able to separate the contribution of these two factors to the total activity. The purified recombinant enzyme is significantly more active than enzyme isolated from PC12 cells. The only effect of phosphorylation on the steady-state kinetic parameters is a twofold decrease in the K_m value for tetrahydrobiopterin. Dopamine binds tightly to the recombinant enzyme, resulting in a species with significantly lower activity. The kinetic parameters of this complex match those of the nonrecombinant enzyme. Furthermore, phosphorylation of dopamine-bound recombinant enzyme results in a large increase in its activity that agrees quantitatively with the effect seen with the nonrecombinant enzyme, a twofold decrease in the K_m value for tetrahydrobiopterin and a 10-fold increase in the V_{max} value. A reasonable conclusion from these results is that the primary result of phosphorylation of Ser40 is to relieve TYH from inhibition by bound catecholamine.

As noted above, when TYH is purified the iron is ferric, whether catecholamines are bound or not (Haavik et al., 1988; Ramsey et al., 1996), even though the enzyme requires ferrous iron for activity. Catechols bind much more tightly to ferric than to ferrous iron. However, K_i values measured in initial rate assays necessarily measure binding to the ferrous form. Effects of phosphorylation on the rates of interconversion of the ferrous and ferric form and on binding to the ferric form would not be measured in such assays. The effects of phosphorylation on binding of catecholamines to the different redox states of TYH were recently determined directly (Ramsey and Fitzpatrick, 1998). Binding to the ferrous form was determined by measuring inhibition constants. These analyses were done using a continuous assay rather than the end point assay more typically used for TYH. Use of a continuous assay makes it apparent that the rate of catalysis is not constant for the several minutes required for an end point assay, but instead decreases rather rapidly from the initial rate due to formation of the ferric enzyme-DOPA complex during turnover (Ramsey et al., 1996). As a consequence of this decrease in activity, estimates of initial rates based on end point assays of several minutes duration can yield inaccurate estimates of the binding to the active ferrous enzyme. In contrast to previous reports that the K_i values for catecholamines increase significantly upon phosphorylation of Ser40 by TYH (Le Bourdellès et al., 1991; Lazar et al., 1982), the K_i values measured using initial rates were unaffected by phosphorylation (Ramsey and Fitzpatrick, 1998). In addition, the K_i values for dopamine and DOPA were significantly greater than those previously reported, suggesting that previous measurements were indeed affected by nonlinear rates. In contrast, phosphorylation has large effects on the binding of catecholamines to the ferric enzyme. The K_d value for DOPA increases from 1 to 18 μM, while that for dopamine increases from 0.7 nM to 0.2 μM. The decreases in affinity are almost completely due to increases in the rates of dissociation of bound catecholamines. Based upon these results, the effect of phosphorylation of Ser40 is primarily to increase the rate of dissociation of bound catecholamines from the inactive ferric enzyme, thereby allowing the iron to be reduced to the active form. Whether the rate of reduction of the ferric enzyme is also unaffected has not been determined. In addition, similar analyses have not been carried out to determine the effects of phosphorylation of Ser19 and Ser31.

A model in which phosphorylation of TYH alters the balance between two inactive forms of the enzyme resembles the regulatory model for PAH proposed by Xia et al. (1994). Figure 23 depicts both models in order to make the similarities more apparent. The resting form of PAH has low activity; binding of tetrahydrobiopterin to a regulatory site traps the enzyme in an inhibited complex, while binding of phenylalanine to a regulatory site converts the enzyme to a more active form. In the presence of oxygen, TYH is oxidized to the inactive ferric enzyme. Binding of a catecholamine to this

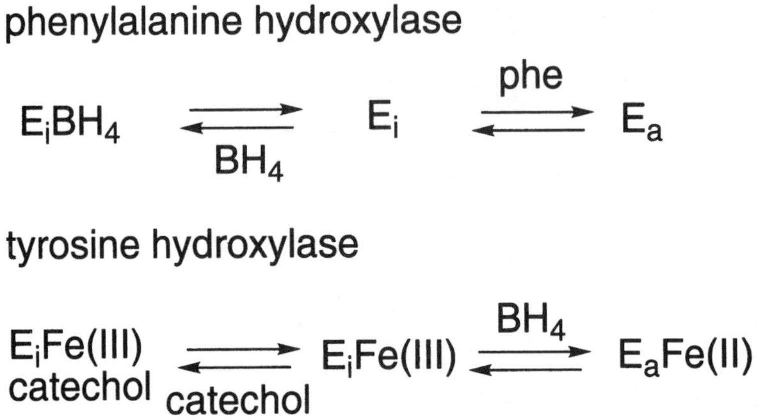

phenylalanine hydroxylase

$$E_iBH_4 \underset{BH_4}{\rightleftharpoons} E_i \overset{phe}{\rightleftharpoons} E_a$$

tyrosine hydroxylase

$$\begin{array}{c}E_iFe(III)\\ catechol\end{array} \underset{catechol}{\rightleftharpoons} E_iFe(III) \overset{BH_4}{\rightleftharpoons} E_aFe(II)$$

Figure 23. Comparison of the regulatory models for phenylalanine and tyrosine hydroxylase.

species traps the enzyme in an inactive catecholamine complex. In the case of PAH, phosphorylation facilitates the binding of phenylalanine to the regulatory site. In the case of TYH, phosphorylation facilitates release of catecholamines from the inhibited complex.

In light of the analogies between the two regulatory mechanisms, one might ask whether there are also structural analogies and specifically whether the regulatory domain of TYH physically interacts with the active site as is the case with PAH. The regulatory domains of TYH and PAH show no sequence similarities, and the structure of the regulatory domain of TYH has not been reported. However, there is evidence that the regulatory domain of TYH interacts with the active site and that this interaction is affected by phosphorylation. TYH is quite sensitive to proteases, with the initial cleavages occurring in the regulatory domain (Abate and Joh, 1991). Treatment of the unmodified enzyme with very low concentrations of trypsin results in rapid decrease in the molecular weight of about 4000 due to cleavage of four bonds between residue 33 and 51 (McCulloch and Fitzpatrick, 1999). Phosphorylation of Ser40 increases the rate of proteolysis of this region by an order of magnitude, while binding of dopamine decreases the rate by a comparable amount. These results are consistent with this region of the regulatory domain interacting with dopamine bound in the active site. Phosphorylation of Ser40 would disrupt this interaction, decreasing the affinity of the enzyme for dopamine. Such a direct interaction of a portion of the regulatory domain with the active site is analogous to the situation in PAH (Kobe et al., 1999).

C. TRYPTOPHAN HYDROXYLASE

The present understanding of the regulatory properties of tryptophan hydroxylase is far behind that of the other two pterin-dependent hydroxylases. Just as was the case for PAH and TYH, TRH is a substrate for several protein kinases. Initially, TRH activity in crude lysates was reported to increase about twofold under conditions that would activate calmodulin-dependent protein kinase; the activation was due to an increase in the V_{max} value of two and a decrease in the K_m value for 6-methyltetrahydropterin of 50% (Edmondson et al., 1987). Subsequent purification of the reaction components showed that the activation by cAMP-dependent protein kinase required a 14-3-3 protein, the same protein proposed as an activator of phosphorylated TYH (Yamauchi et al., 1981; Ichimura et al., 1987). The activator protein forms a complex with the phosphorylated TRH, but not with the unphosphorylated enzyme (Furukawa et al., 1993). A similar effect of phosphorylation by cAMP-dependent protein on TRH activity has also been demonstrated. The extent of the increase in activity is the same for both kinases; activation by cAMP-dependent protein kinase is reported to require a 14-3-3 protein in addition to phosphorylation (Makita et al., 1990). The site of phosphorylation by cAMP-dependent protein kinase has been identified by site directed mutagenesis; mutation of Ser58 to either alanine or arginine prevents phosphorylation (Kuhn et al., 1997; Kumer et al., 1997).

VII. Conclusion

The aromatic amino acid hydroxylases constitute a small family of physiologically critical and chemically interesting enzymes. Clearly, significant progress has been made in recent years in elucidating their catalytic and regulatory mechanisms. Still, a number of outstanding issues remain unsettled. While it is possible to propose an overall catalytic mechanism, none of the proposed intermediates have actually been directly detected. The roles of individual amino acid residues in catalysis and specificity are only beginning to be determined. The structural changes responsible for the regulatory properties remain unknown, although the recent structure of the regulatory and catalytic domains of phenylalanine should allow more rapid progress with this enzyme. The structural changes associated with the multiple phosphorylation sites on tyrosine hydroxylase will clearly require structures of the intact protein. The regulatory effects of phosphorylation at Ser19 and Ser31 of tyrosine hydroxylase remain obscure. Study of the regulatory properties of tryptophan hydroxylase has barely begun.

Acknowledgments

The author wishes to thank present and past colleagues who are responsible for the results from his laboratory described here. Special thanks is due to Dr. Colette Daubner for helpful comments during the preparation of this review. The preparation of this review was supported in part by a grant from the NIH (GM47291).

References

Abate C, Joh tH (1991): Limited proteolysis of rat brain tyrosine hydroxylase defines an N-terminal region required for regulation of cofactor binding and directing substrate specificity. J Mol Neurosci 2:203–215.

Abita J-P, Milstien S, Chang N, Kaufman S (1976): *In vitro* activation of rat liver phenylalanine hydroxylase by phosphorylation. J Biol Chem 251:5310–5314.

Abita J-P, Parniak M, Kaufman S (1984): The activation of rat liver phenylalanine hydroxylase by limited proteolysis, lysolecithin, and tocopherol phosphate. Changes in conformation and catalytic properties. J Biol Chem 259:14560–14566.

Aitken A (1996): 14–3–3 and its possible role in co-ordinating multiple signalling pathways. Trends Cell Biol 6: 341–347.

Albert KA, Helmer-Matyjek E, Nairn AC, Muller TH, Haycock JW, Greene LA, Goldstein M, Greengard P (1984): Calcium/phospholipid-dependent protein kinase (protein kinase C) phosphorylates and activates tyrosine hydroxylase. Proc Natl Acad Sci U S A 81: 7713–7717.

Almas B, Le Bourdelles B, Flatmark T, Mallet J, Haavik J (1992): Regulation of recombinant human tyrosine hydroxylase isozymes by catecholamine binding and phosphorylation structure/activity studies and mechanistic implications. Eur J Biochem 209: 249–255.

Alterio J, Ravassard P, Haavik J, Le Caer J, Biguet N, Waksman G, Mallet J (1998): Human tyrosine hydroxylase isoforms. J Biol Chem 273: 10196–10201.

Ames MM, Lerner P, Lovenberg W (1978): Tyrosine hydroxylase. Activation by protein phosphorylation and end product inhibition. J Biol Chem 253: 27–31.

Andersson KK, Cox DD, Que L Jr, Flatmark T, Haavik J (1988): Resonance Raman studies on the blue-green-colored bovine adrenal tyrosine 3–monooxygenase (tyrosine hydroxylase). Evidence that the feedback inhibitors adrenaline and noradrenaline are coordinated to iron. J Biol Chem 263: 18621–18626.

Andersson KK, Haavik J, Martinez A, Flatmark T, Petersson L (1989): Evidence from EPR spectroscopy that phosphorylation of Ser-40 in bovine adrenal tyrosine hydroxylase facilitates the reduction of high-spin Fe(III) under turnover conditions. FEBS Letts 258: 9–12.

Andersson KK, Vassort C, Brennan BA, Que L Jr, Haavik J, Flatmark T, Gros F, Thibault J (1992): Purification and characterization of the blue-green rat phaeochromocytoma (PC12) tyrosine hydroxylase with a dopamine-Fe(III) complex reversal of the endogenous feedback inhibition by phosphorylation of serine-40. Biochem J 284: 687–695.

Archer MC, Scrimgeour KG (1970): Rearrangement of quinonoid dihydropteridines to 7,8-dihydropteridines. Can J Biochem 48: 278–287.

Atkinson J, Richtand N, Schworer C, Kuczenski R, Soderling T (1987): Phosphorylation of purified rat striatal tyrosine hydroxylase by Ca^{2+}/calmodulin-dependent protein kinase II: effect of an activator protein. J Neurochem 49: 1241–1249.

Bailey SW, Ayling JE (1978): Pyrimidines as cofactors for phenylalanine hydroxylase. Biochem Biophys Res Commun 85: 1614–1621.

Bailey SW, Ayling JE (1980): Cleavage of the 5–amino substituent of pyrimidine cofactors by phenylalanine hydroxylase. J Biol Chem 255: 7774–7781.

Bailey SW, Weintraub ST, Hamilton SM, Ayling JE (1982): Incorporation of molecular oxygen into pyrimidine cofactors by phenylalanine hydroxylase. J Biol Chem 257: 8253–8260.

Bailey SW, Dillard SB, Thomas KB, Ayling JE (1989): Changes in the cofactor binding domain of bovine striatal tyrosine hydroxylase at physiological pH upon cAMP-dependent phosphorylation mapped with tetrahydrobiopterin analogues. Biochemistry 28: 494–504.

Bailey SW, Rebrin I, Boerth SR, Ayling JE (1995): Synthesis of 4a-hydroxytetrahydropterins and the mechanism of their nonenzymatic dehydration to quinoid dihydropterins. J Am Chem Soc 117: 10203–10211.

Balasubramanian S, Carr RT, Bender CJ, Peisach J, Benkovic SJ (1994): Identification of metal ligands in Cu(II)-inhibited *Chromobacterium violaceum* phenylalanine hydroxylase by electron spin echo envelope modulation analysis of histidine to serine mutations. Biochemistry 33: 8532–8537.

Benkovic S, Wallick D, Bloom L, Gaffney BJ, Domanico P, Dix T, Pember S (1985a): On the mechanism of action of phenylalanine hydroxylase. Biochem Soc Trans 13: 436–438.

Benkovic SJ, Sammons D, Armarego WLF, Waring P, Inners R (1985b): Tautomeric nature of quinonoid 6,7-dimethyl-7,8-dihydro-6*H*-pterin in aqueous solution: A ^{15}N NMR study. J Am Chem Soc 107: 3706–3712.

Blackburn NJ, Strange RW, Carr RT, Benkovic SJ (1992): X-ray absorption studies of the Cu-dependent phenylalanine hydroxylase from *Chromobacterium violaceum.* Comparison of the copper coordination in oxidized and dithionite-reduced enzymes. Biochemistry 31: 5298–5303.

Blair JA, Pearson AJ (1974): Kinetics and mechanism of the autoxidation of the 2-amino-4-hydroxy-5,6,7,8-tetrahydropteridines. J C S Perkin II 80–88.

Bloom LM, Benkovic SJ, Gaffney BJ (1986): Characterization of phenylalanine hydroxylase. Biochemistry 25: 4204–4210.

Boadle-Biber MC (1993): Regulation of serotonin synthesis. Prog Biophys Molec Biol 60: 1–15.

Bruice TC, Bruice PY (1976): Solution chemistry of arene oxides. Acc Chem Res 9: 378–384.

Bruice TC, Noar JB, Ball SS, Venkatara UV (1983): Monooxygen donation potential of 4a-hydroperoxyflavins as compared with those of a percarboxylic acid and other hydroperoxides. Monooxygen donation to olefin, tertiary amine, alkyl sulfide, and iodide ion. J Am Chem Soc 105: 2452–2465.

Brunner HG, Nelen M, Breakefield XO, Ropers HH, van Oost BA (1993): Abnormal behavior associated with point mutation in the structural gene for monoamine oxidase A. Science 262: 578–580.

Byerley W, Plaetke R, Hoff M, Jensen S, Holik J, Reimherr F, Mellon C, Wender P, O'Connell P, Leppert M (1992): Tyrosine hydroxylase gene not linked to manic-depression in seven of eight pedigrees. Hum Hered 42: 259–263.

Campbell DG, Hardie DG, Vulliet PR (1986): Identification of four phosphorylation sites in the N-terminal region of tyrosine hydroxylase. J Biol Chem 261: 10489–10492.

Carr RT, Benkovic SJ (1993): An examination of the copper requirement of phenylalanine hydroxylase from *Chromobacterium violaceum*. Biochemistry 32: 14132–14138.

Carr RT, Balasubramanian S, Hawkins PCD, Benkovic SJ (1995): Mechanism of metal-independent hydroxylation by *Chromobacterium violaceum* phenylalanine hydroxylase. Biochemistry 34: 7525–7532.

Chen D, Frey P (1998): Phenylalanine hydroxylase from *Chromobacterium violaceum*. Uncoupled oxidation of tetrahydropterin and the role of iron in hydroxylation. J Biol Chem 273: 25594–25601.

Cox DD, Benkovic SJ, Bloom LM, Bradley FC, Nelson MJ, Que L Jr, Wallick DE (1988): Catecholate LMCT bands as probes for the active sites of nonheme iron oxygenases. J Am Chem Soc 110: 2026–2032.

Daly J, Guroff G (1968): Production of *m*-methyltyrosine and *p*-hydroxymethylphenylalanine from *p*-methylphenylalanine by phenylalanine hydroxylase. Arch Biochem Biophys 125: 136–141.

Daly J, Levitt M, Guroff G, Udenfriend S (1968): Isotope studies on the mechanism of action of adrenal tyrosine hydroxylase. Arch Biochem Biophys 126: 593–598.

Daly JW, Jerina DM, Witkop B (1972): Arene oxides and the NIH shift: the metabolism, toxicity and carcinogenicity of aromatic compounds. Experientia 28: 1129–1264.

Daubner SC, Fitzpatrick PF (1998): Mutation to phenylalanine of tyrosine 371 in tyrosine hydroxylase increases the affinity for phenylalanine. Biochemistry 37: 16440–16444.

Daubner SC, Fitzpatrick PF (1999): Site-directed mutants of charged residues in the active site of tyrosine hydroxylase. Biochemistry 38: 4448–4454.

Daubner SC, Lauriano C, Haycock JW, Fitzpatrick PF (1992): Site-directed mutagenesis of Serine 40 of rat tyrosine hydroxylase. Effects of dopamine and cAMP-dependent phosphorylation on enzyme activity. J Biol Chem 267: 12639–12646.

Daubner SC, Hillas PJ, Fitzpatrick PF (1997a): Characterization of chimeric pterin dependent hydroxylases: contributions of the regulatory domains of tyrosine and phenylalanine hydroxylase to substrate specificity. Biochemistry 36: 11574–11582.

Daubner SC, Hillas PJ, Fitzpatrick PF (1997b): Expression and characterization of the catalytic domain of human phenylalanine hydroxylase. Arch Biochem Biophys 348: 295–302.

Davis MD, Kaufman S (1989): Evidence for the formation of the 4a-carbinolamine during the tyrosine-dependent oxidation of tetrahydrobiopterin by rat liver phenylalanine hydroxylase. J Biol Chem 264: 8585–8596.

Davis MD, Kaufman S (1991): Studies on the partially uncoupled oxidation of tetrahydropterins by phenylalanine hydroxylase. Neurochem Res 16: 813–819.

De bruyn A, Mendelbaum K, Sandkuijl LA, Delvenne V, Hirsch D, Staner L, Mendlewicz J, Van Broeckhoven C (1994): Nonlinkage of bipolar illness to tyrosine hydroxylase, tyrosine, and D_2 and D_4 dopamine receptor genes on chromosome 11. Am J Psychiatry 151: 102–106.

Dickson PW, Jennings IG, Cotton RGH (1994): Delineation of the catalytic core of phenylalanine hydroxylase and identification of glutamate 286 as a critical residue for pterin function. J Biol Chem 269: 20369–20375.

Dix TA, Benkovic SJ (1985): Mechanism of "uncoupled" tetrahydropterin oxidation by phenylalanine hydroxylase. Biochemistry 24: 5839–5846.

Dix TA, Bollag G, Domanico PL, Benkovic SJ (1985): Phenylalanine hydroxylase: absolute configuration and source of oxygen of the 4a-hydroxytetrahydropterin species. Biochemistry 24: 2955–2958.

Dix TA, Kuhn DM, Benkovic SJ (1987): Mechanism of oxygen activation by tyrosine hydroxylase. Biochemistry 26: 3354–3361.

Doskeland AP, Doskeland SO, Ogreid D, Flatmark T (1984a): The effect of ligands of phenylalanine 4-monooxygenase on the cAMP-dependent phosphorylation of the enzyme. J Biol Chem 259: 11242–11248.

Doskeland AP, Schworer CM, Doskeland SO, Chrisman TD, Soderling TR, Corbin JD, Flatmark T (1984b): Some aspects of the phosphorylation of phenylalanine 4-monooxygenase by a calcium-dependent and calmodulin-dependent protein kinase. Eur J Biochem 145: 31–37.

Eberlein G, Bruice TC (1983): The chemistry of a 1,5-diblocked flavin. 2. Proton and electron transfer steps in the reaction of dihydroflavins with oxygen. J Am Chem Soc 105: 6685–6697.

Eberlein GA, Bruice TC, Lazarus RA, Henrie R, Benkovic SJ (1984): The interconversion of the 5,6,7,8-tetrahydro-, 7,8-dihydro-, and radical forms of 6,6,7,7-tetramethyldihydropterin. A model for the biopterin center of aromatic amino acid mixed function oxidases. J Am Chem Soc 106: 7916–7924.

Edmondson DE, Hazzard JT, Tollin G (1987): Laser flash photolysis studies of intramolecular electron transfer between the FAD and iron-sulfur II centers in xanthine oxidase. In Edmondson DE, McCormick DB (eds): "Flavins and Flavoproteins." Hawthorne, N.Y.: Walter de Gruyter, pp. 403–408.

Ellis HR, Daubner SC, McCulloch RI, Fitzpatrick PF (1999): Conserved phenylalanine residues in the active-site of tyrosine hydroxylase: mutagenesis of Phe300 and Phe309 to alanine and metal ion-catalyzed hydroxylation of Phe 300. Biochemistry 38: 10909–10914.

Ellis HR, Daubner SC, Fitzpatrick PF: Mutation of serine 395 of tyrosine hydroxylase decouples oxygen-oxygen bond cleavage and tyrosine hydroxylation. (Submitted for publication).

Endrizzi JA, Cronk JD, Wang W, Crabtree GR, Alber T (1995): Crystal structure of DCoH, a bifunctional, protein-binding transcriptional coactivator. Science 268: 556–559.

Entsch B, Ballou DP, Massey V (1976): Flavin-oxygen derivatives involved in hydroxylation by *p*-hydroxybenzoate hydroxylase. J Biol Chem 251: 2550–2563.

Erlandsen H, Fusetti F, Martinez A, Hough E, Flatmark T, Stevens RC (1997): Crystal structure of the catalytic domain of human phenylalanine hydroxylase reveals the structural basis for phenylketonuria. Nat Struct Biol 4: 995–1000.

Erlandsen H, Flatmark T, Stevens RC, Hough E (1998): Crystallographic analysis of the human phenylalanine hydroxylase catalytic domain with bound catechol inhibitors at 2.0 Å resolution. Biochemistry 37: 15638–15646.

Erny RE, Berezo MW, Perlman RL (1981): Activation of tyrosine 3-monooxygenase in pheochromocytoma cells by adenosine. J Biol Chem 256: 1335–1339.

Fisher DB, Kaufman S (1973a): The stimulation of rat liver phenylalanine hydroxylase by lysolecithin and α-chymotrypsin. J Biol Chem 248: 4345–4353.

Fisher DB, Kaufman S (1973b): Tetrahydropterin oxidation without hydroxylation catalyzed by rat liver phenylalanine hydroxylase. J Biol Chem 248: 4300–4304.

Fitzpatrick PF (1989): The metal requirement of rat tyrosine hydroxylase. Biochem Biophys Res Commun 161: 211–215.

Fitzpatrick PF (1991a): The steady state kinetic mechanism of rat tyrosine hydroxylase. Biochemistry 30: 3658–3662.

Fitzpatrick PF (1991b): Studies of the rate-limiting step in the tyrosine hydroxylase reaction: alternate substrates, solvent isotope effects, and transition state analogs. Biochemistry 30: 6386–6391.

Fitzpatrick PF (1994): Kinetic isotope effects on hydroxylation of ring-deuterated phenylalanines by tyrosine hydroxylase provide evidence against partitioning of an arene oxide intermediate. J Am Chem Soc 116: 1133–1134.

Francisco WA, Tian G, Fitzpatrick PF, Klinman JP (1998): Oxygen-18 kinetic isotope effect studies of the tyrosine hydroxylase reaction: evidence of rate limiting oxygen activation. J Am Chem Soc 120: 4057–4062.

Fukami MH, Haavik J, Flatmark T (1990): Phenylalanine as substrate for tyrosine hydroxylase in bovine adrenal chromaffin cells. Biochem J 268: 525–528.

Funakoshi H, Okuno S, Fujisawa H (1991): Different effects on activity caused by phosphorylation of tyrosine hydroxylase at serine 40 by three multifunctional protein kinases. J Biol Chem 266: 15614–15620.

Furukawa Y, Ikuta N, Omata S, Yamauchi T, Isobe T, Ichimura T (1993): Demonstration of the phosphorylation-dependent interaction of tryptophan hydroxylase with the 14-3-3 protein. Biochem Biophys Res Commun 194: 144–149.

Fusetti F, Erlandsen H, Flatmark T, Stevens RC (1998): Structure of tetrametic human phenylalanine hydroxylase and its implications for phenylketonuria. J Biol Chem 273: 16962–16967.

Gibbs BS, Benkovic SJ (1991): Affinity labeling of the active site and the reactive sulfhydryl associated with activation of rat liver phenylalanine hydroxylase. Biochemistry 30: 6795–6802.

Gibbs BS, Wojchowski D, Benkovic SJ (1993): Expression of rat liver phenylalanine hydroxylase in insect cells and site-directed mutagenesis of putative non-heme iron-binding sites. J Biol Chem 268: 8046–8052.

Goodwill KE, Sabatier C, Marks C, Raag R, Fitzpatrick PF, Stevens RC (1997): Crystal structure of tyrosine hydroxylase at 2.3 Å and its implications for inherited diseases. Nat Struct Biol 4: 578–585.

Goodwill KE, Sabatier C, Stevens RC (1998): Crystal structure of tyrosine hydroxylase with bound cofactor analogue and iron at 2.3 Å resolution: self-hydroxylation of phe300 and the pterin-binding site. Biochemistry 37: 13437–13445.

Grahame-Smith DG (1992): An overview of serotonin and psychiatry. In Stahl SM (ed): "Serotonin 1A Receptors in Depression and Anxiety." New York: Raven Press, Ltd., pp. 1–21.

Grenett HE, Ledley FD, Reed LL, Woo SLC (1987): Full-length cDNA for rabbit tryptophan hydroxylase: functional domains and evolution of aromatic amino acid hydroxylases. Proc Natl Acad Sci U S A 84: 5530–5534.

Griffith LC, Schulman H (1988): The multifunctional Ca^{2+}/calmodulin-dependent protein kinase mediates Ca^{2+}-dependent phosphorylation of tyrosine hydroxylase. J Biol Chem 263: 9542–9549.

Grima B, Lamouroux A, Blanot F, Biguet NF, Mallet J (1985): Complete coding sequence of rat tyrosine hydroxylase mRNA. Proc Natl Acad Sci U S A 82: 617–621.

Grima B, Lamouroux A, Boni C, Julien J-F, Javoy-Agid F, Mallet J (1987): A single human gene encoding multiple tyrosine hydroxylases with different predicted functional characteristics. Nature 326: 707–711.

Guengerich FP, MacDonald TL (1984): Chemical mechanisms of catalysis by cytochromes P-450: a unified view. Acc Chem Res 17: 9–16.

Guroff G, Levitt M, Daly J, Udenfriend S (1966): The production of meta-tritiotyrosine from *p*-tritio-phenylalanine by phenylalanine hydroxylase. Biochem Biophys Res Commun 25: 253–259.

Guroff G, Daly JW, Jerina DM, Renson J, Witkop B, Udenfriend S (1967): Hydroxylation-induced migration: the NIH shift. Science 157: 1524–1530.

Haavik J, Flatmark T (1987): Isolation and characterization of tetrahydropterin oxidation products generated in the tyrosine 3-monooxygenase (tyrosine hydroxylase) reaction. Eur J Biochem 168: 21–26.

Haavik J, Andersson KK, Petersson L, Flatmark T (1988): Soluble tyrosine hydroxylase (tyrosine 3-monooxygenase) from bovine adrenal medulla: large-scale purification and physicochemical properties. Biochim Biophys Acta 953: 142–156.

Haavik J, Martinez A, Flatmark T (1990): pH-dependent release of catecholamines from tyrosine hydroxylase and the effect of phosphorylation of Ser-40. FEBS Lett 262: 363–365.

Halloran SM, Vulliet PR (1994): Microtubule-associated protein kinase-2 phosphorylates and activates tyrosine hydroxylase following depolarization of bovine adrenal chromaffin cells. J Biol Chem 269: 30960–30965.

Hamilton GA (1971): The proton in biological redox reactions. In Kaiser ET, Kezdy FJ (eds): "Progress in Biorganic Chemistry, Vol. 1." New York: John Wiley & Sons, Inc., pp. 83–157.

Harada K, Wu J, Haycock JW, Goldstein M (1996): Regulation of L-DOPA biosynthesis by site-specific phosphorylation of tyrosine hydroxylase in AtT-20 cells expressing wild-type and serine 40-substituted enzyme. J Neurochem 67: 629–635.

Harada N, Miwa GT, Walsh JS, Lu AYH (1984): Kinetic isotope effects on cytochrome P-450-catalyzed oxidation reactions. J Biol Chem 259: 3005–3010.

Haycock JW (1990): Phosphorylation of tyrosine hydroxylase *in situ* at serine 8, 19, 31, and 40. J Biol Chem 265: 11682–11691.

Haycock JW (1993): Multiple signaling pathways in bovine chromaffin cells regulate tyrosine hydroxylase phosphorylation at Ser^{19}, Ser^{31}, and Ser^{40}. Neurochem Res 18: 15–26.

Haycock JW, Bennett WF, George RJ, Waymire JC (1982a): Multiple site phosphorylation of tyrosine hydroxylase. Differential regulation *in situ* by 8-bromo-cAMP and acetylcholine. J Biol Chem 257: 13699–13703.

Haycock JW, Meligeni JA, Bennett WF, Waymire JC (1982b): Phosphorylation and activation of tyrosine hydroxylase mediate the acetylcholine-induced increase in catecholamine biosynthesis in adrenal chromaffin cells. J Biol Chem 257: 12641–12648.

Haycock JW, Ahn NG, Cobb MH, Krebs EG (1992): ERK1 and ERK2, two microtubule-associated protein 2 kinases, mediate the phosphorylation of tyrosine hydroxylase at serine-31 *in situ*. Proc Natl Acad Sci U S A 89: 2365–2369.

Haycock JW, Lew JY, Garcia-Espana A, Lee KY, Harada K, Meller E, Goldstein M (1998): Role of serine-19 phosphorylation in regulating tyrosine hydroxylase studied with site- and phosphospecific antibodies and site-directed mutagenesis. J Neurochem 71: 1670–1675.

Hegg EL, Que L Jr (1997): The 2–His-1-carboxylate facial triad. An emerging structural motif in mononuclear non-heme iron(II) enzymes. Eur J Biochem 250: 625–629.

Hillas PJ, Fitzpatrick PF (1996): A mechanism for hydroxylation by tyrosine hydroxylase based on partitioning of substituted phenylalanines. Biochemistry 35: 6969–6975.

Ichimura T, Isobe T, Okuyama T, Yamauchi T, Fujisawa H (1987): Brain 14–3–3 protein is an activator protein that activates tryptophan 5–monooxygenase and tyrosine 3–monooxygenase in the presence of Ca^{2+}, calmodulin-dependent protein kinase II. FEBS Letts. 219: 79–82.

Joh TH, Park DH, Reis DJ (1978): Direct phosphorylation of brain tyrosine hydroxylase by cyclic AMP-dependent protein kinase: mechanism of enzyme activation. Proc Natl Acad Sci U S A 75: 4744–4748.

Kaneda N, Kobayashi K, Ichinose H, Kishi F, Nakazawa A, Kurosawa Y, Fujita K, Nagatsu T (1987): Isolation of a novel cDNA clone for human tyrosine hydroxylase: Alternative RNA splicing produces four kinds of mRNA from a single gene. Biochem Biophys Res Commun 147: 971–975.

Kappock TJ, Caradonna JP (1996): Pterin-dependent amino acid hydroxylases. Chem Rev 96: 2659–2756.

Kappock TJ, Harkins PC, Friedenberg S, Caradonna JP (1995): Spectroscopic and kinetic properties of unphosphorylated rat hepatic phenylalanine hydroxylase expressed in *Escherichia coli.* Comparison of resting and activated states. J Biol Chem 270: 30532–30544.

Kaufman S (1963): The structure of the phenylalanine-hydroxylation cofactor. Biochemistry 50: 1085–1093.

Kaufman S (1979): The activity of 2,4,5-triamino-6-hydroxypyrimidine in the phenylalanine hydroxylase system. J Biol Chem 254: 5150–5154.

Kaufman S (1985): Aromatic amino acid hydroxylases. Biochem Soc Trans 13: 433–436.

Kaufman S (1995): Tyrosine hydroxylase. In Meister A (ed): "Advances in Enzymology and Related Areas of Molecular Biology." Vol. 70. John Wiley & Sons, Inc., pp. 103–220.

Kaufman S, Mason K (1982): Specificity of amino acids as activators and substrates for phenylalanine hydroxylase. J Biol Chem 257: 14667–14678.

Kaufman S, Bridgers WF, Eisenberg F, Friedman S (1962): The source of oxygen in the phenylalanine hydroxylase and the dopamine-β-hydroxylase catalyzed reactions. Biochem Biophys Res Commun 9: 497–502.

Kemsley JN, Mitic N, Zaleski KL, Caradonna JP, Solomon EI (1999): Circular dichroism and magnetic circular dichroism spectroscopy of the catalytically competent ferrous active site of phenylalanine hydroxylase and its interaction with pterin cofactor. J Am Chem Soc 121: 1528–1536.

Klinman JP (1996): Mechanisms whereby mononuclear copper proteins functionalize organic substrates. Chem Rev 96: 2541–2562.

Knappskog PM, Flatmark T, Mallet J, Ludecke B, Bartholome K (1995): Recessively inherited L-DOPA-responsive dystonia caused by a point mutation (Q381K) in the tyrosine hydroxylase gene. Hum Mol Genet 4: 1209–1212.

Knappskog PM, Flatmark T, Aarden JM, Haavik J, Martinez A (1996): Structure/function relationships in human phenylalanine hydroxylase. Effect of terminal deletions on the oligomerization, activation and cooperativity of substrate binding to the enzyme. Eur J Biochem 242: 813–821.

Kobayashi K, Morita S, Sawada H, Mizuguchi T, Yamada K, Nagatsu I, Hata T, Watanabe Y, Fujita K, Nagatsu T (1995): Targeted disruption of the tyrosine hydroxylase locus results in severe catecholamine depletion and perinatal lethality in mice. J Biol Chem 270: 27235–27243.

Kobe B, Jennings IG, House CM, Michell BJ, Goodwill KE, Santarsiero BD, Stevens RC, Cotton RGH, Kemp BE (1999): Structural basis of intrasteric and allosteric controls of phenylalanine hydroxylase. Nat Struct Biol 6: 442–448.

Kobe G, Jennings IG, House CM, Feil SC, Michell BJ, Tiganis T, Parker MW, Cotton RGH, Kemp BE (1997): Regulation and crystallization of phosphorylated and dephosphorylated forms of truncated dimeric phenylalanine hydroxylase. Prot Sci 6: 1352–1357.

Köster S, Stier G, Ficner R, Hölzer M, Curtius H-C, Suck D, Ghisla S (1996): Location of the active site and proposed catalytic mechanism of pterin-4a-carbinolamine dehydratase. Eur J Biochem 241: 858–864.

Kuhn DM, Arthur R,Jr., States JC (1997): Phosphorylation and activation of brain tryptophan hydroxylase: identification of serine-58 as a substrate site for protein kinase A. J Neurochem 68: 2220–2223.

Kumer SC, Mockus SM, Rucker PJ, Vrana KE (1997): Amino-terminal analysis of tryptophan hydroxylase: protein kinase phosphorylation occurs at serine-58. J Neurochem 69: 1738–1745.

Lazar MA, Lockfeld AJ, Truscott RJW, Barchas JD (1982): Tyrosine hydroxylase from bovine striatum: catalytic properties of the phosphorylated and nonphosphorylated forms of the purified enzyme. J Neurochem 39: 409–422.

Lazarus RA, Dietrich RF, Wallick DE, Benkovic SJ (1981): On the mechanism of action of phenylalanine hydroxylase. Biochemistry 20: 6834–6841.

Lazarus RA, DeBrosse CW, Benkovic SJ (1982): Phenylalanine hydroxylase: structural determination of the tetrahydropterin intermediates by ^{13}C NMR spectroscopy. J Am Chem Soc 104: 6869–6871.

Le Bourdellès B, Horellou P, Le Caer J-P, Denèfle P, Latta M, Haavik J, Guibert B, Mayaux J-F, Mallet J (1991): Phosphorylation of human recombinant tyrosine hydroxylase isoforms 1 and 2: an additional phosphorylated residue in isoform 2, generated through alternative splicing. J Biol Chem 266: 17124–17130.

Ledley FD, DiLella AG, Kwok SCM, Woo SLC (1985): Homology between phenylalanine and tyrosine hydroxylases reveals common structural and functional domains. Biochemistry 24: 3389–3394.

Letendre CH, Dickens G, Guroff G (1974): The tryptophan hydroxylase of *Chromobacterium violaceum*. J Biol Chem 249: 7186–7191.

Loev KE, Westre TE, Kappock TJ, Mitic N, Glasfeld E, Caradonna JP, Hedman B, Hodgson KO, Solomon EI (1997): Spectroscopic characterization of the catalytically competent ferrous site of the resting, activated, and substrate-bound forms of phenylalanine hydroxylase. J Am Chem Soc 119: 1901–1915.

Lohse DL, Fitzpatrick PF (1993): Identification of the intersubunit binding region in rat tyrosine hydroxylase. Biochem Biophys Res Commun 197: 1543–1548.

Lovenberg W, Jequier E, Sjoerdsma A (1967): Tryptophan hydroxylation: measurement in pineal gland, brainstem, and carcinoid tumor. Science 155: 217–219.

Ludecke B, Dworniczak B, Bartholome K (1995): A point mutation in the tyrosine hydroxylase gene associated with Segawa's syndrome. Hum Genet 95: 123–125.

Ludecke B, Knappskog PM, Clayton PT, Surtees RAH, Clelland JD, Heales SJR, Brand MP, Bartholome K, Flatmark T (1996): Recessively inherited L-DOPA-responsive parkinsonism in infancy caused by a point mutation (L205) in the tyrosine hydroxylase gene. Hum Mol Genet 5: 1023–1028.

Makita Y, Okuno S, Fujisawa H (1990): Involvement of activator protein in the activation of tryptophan hydroxylase by cAMP-dependent protein kinase. FEBS Lett 268: 185–188.

Markey KA, Kondo S, Shenkman L, Goldstein M (1980): Purification and characterization of tyrosine hydroxylase from a clonal pheochromocytoma cell line. Molec Pharmacol 17: 79–85.

Marota JJA, Shiman R (1984): Stoichiometric reduction of phenylalanine hydroxylase by its cofactor: a requirement for enzymatic activity. Biochemistry 23: 1303–1311.

Massey V (1994): Activation of molecular oxygen by flavins and flavoproteins. J Biol Chem 269: 22459–22462.

Matsuura S, Sugimoto T, Murata S, Sugawara Y, Iwasaki H (1985): Stereochemistry of biopterin cofactor and facile methods for the determination of the stereochemistry of a biologically active 5,6,7,8-tetrahydropterin. J.Biochem. 98: 1341–1348.

McCulloch RI, Fitzpatrick PF (1999): Limited proteolysis of tyrosine hydroxylase identifies residues 33–50 as conformationally sensitive to phosphorylation state and dopamine binding. Arch Biochem Biophys 367: 143–145.

McTigue M, Cremins J, Halegoua S (1985): Nerve growth factor and other agents mediate phosphorylation and activation of tyrosine hydroxylase. A convergence of multiple kinase activities. J Biol Chem 260: 9047–9056.

Meligeni JA, Haycock JW, Bennett WF, Waymire JC (1982): Phosphorylation and activation of tyrosine hydroxylase mediate the cAMP-induced increase in catecholamine biosynthesis in adrenal chromaffin cells. J Biol Chem 257: 12632–12640.

Meyer-Klaucke W, Winkler H, Schünemann V, Trautwein AX, Nolting H-F, Haavik J (1996): Mössbauer, electron-paramagnetic-resonance and X-ray-absorption fine-structure studies of the iron environment in recombinant human tyrosine hydroxylase. Eur J Biochem 241: 432–439.

Michaud-Soret I, Andersson KK, Que L Jr, Haavik J (1995): Resonance Raman studies of catecholate and phenolate complexes of recombinant human tyrosine hydroxylase. Biochemistry 34: 5504–5510.

Miller LP, Lovenberg W (1985): The use of the natural cofactor (6R)-L-erythrotetrahydrobiopterin in the analysis of nonphosphorylated and phosphorylated rat striatal tyrosine hydroxylase at pH 7.0. Biochem Int 7: 689–697.

Miller RJ, Benkovic SJ (1988): L-[2,5-H_2]Phenylalanine, an alternate substrate for rat liver phenylalanine hydroxylase. Biochemistry 27: 3658–3663.

Mitchell JP, Hardie DG, Vulliet PR (1998): Site-specific phosphorylation of tyrosine hydroxylase after KCl depolarization and nerve growth factor treatment of PC12 cells. J Biol Chem 265: 22358–22364.

Moad G, Luthy CL, Benkovic PA, Benkovic SJ (1979): Studies on 6-methyl-5-deazatetrahydropterin and its 4a adducts. J Am Chem Soc 101: 6068–6076.

Moran GR, Derecskei-Kovacs, Fitzpatrick PF (2000): On the catalytic mechanism of tryptophan hydroxylase. (Submitted for publication).

Moran GR, Fitzpatrick PF (2000): Investigations of the kinetic mechanism of tryptophan hydroxylase. (Unpublished manuscript).

Moran GR, Daubner SC, Fitzpatrick PF (1998): Expression and characterization of the catalytic core of tryptophan hydroxylase. J Biol Chem 273: 12259–12266.

Moran GR, Phillips RMI, Fitzpatrick PF (1999): The influence of steric bulk and electrostatics on the hydroxylation regiospecificity of tryptophan hydroxylase: characterization of methyltryptophans and azatryptophans as substrates. Biochemistry 38: 16283–16289.

Nagatsu T, Levitt M, Udenfriend S (1964): Tyrosine hydroxylase the initial step in norepinephrine biosynthesis. J Biol Chem 239: 2910–2917.

Nakata H, Fujisawa H (1982): Tryptophan 5–monooxygenase from mouse mastocytoma P815 a simple purification and general properties. Eur J Biochem 124: 595–601.

Nakata H, Yamauchi T, Fujisawa H (1979): Phenylalanine hydroxylase from *Chromobacterium violaceum* purification and characterization. J Biol Chem 254: 1829–1833.

Nelson TJ, Kaufman S (1987): Interaction of tyrosine hydroxylase with ribonucleic acid and purification with DNA-cellulose or poly(A)-sepharose affinity chromatography. Arch Biochem Biophys 257: 69–84.

Nielsen DA, Goldman D, Virkkunen M, Tokola R, Rawlings R, Linnoila M (1994): Suicidality and 5–hydroxyindoleacetic acid concentration associated with a tryptophan hydroxylase polymorphism. Arch Gen Psychiatry 51: 34–38.

Okuno S, Fujisawa H (1982): Purification and some properties of tyrosine 3–monooxygenase from rat adrenal. Eur J Biochem 122: 49–55.

O'Malley KL, Anhalt MJ, Martin BM, Kelsoe JR, Winfield SL, Ginns EI (1987): Isolation and characterization of the human tyrosine hydroxylase gene: Identification of 5′ alternative splice sites responsible for multiple mRNAs. Biochemistry 26: 6910–6914.

Parniak MA, Kaufman S (1981): Rat liver phenylalanine hydroxylase. Activation by sulfhydryl modification. J Biol Chem 256: 6876–6882.

Pember SO, Villafranca JJ, Benkovic SJ (1986): Phenylalanine hydroxylase from *Chromobacterium violaceum* is a copper-containing monooxygenase. Kinetics of the reductive activation of the enzyme. Biochemistry 25: 6611–6619.

Pember SO, Johnson KA, Villafranca JJ, Benkovic SJ (1989): Mechanistic studies on phenylalanine hydroxylase from *Chromobacterium violaceum*. Evidence for the formation of an enzyme-oxygen complex. Biochemistry 28: 2124–2130.

Pollock RJ, Kapatos G, Kaufman S (1981): Effect of cyclic AMP-dependent protein phosphorylating conditions on the pH-dependent activity of tyrosine hydroxylase from beef and rat striata. J Neurochem 37: 855–860.

Que L Jr, Ho RYN (1996): Dioxygen activation by enzymes with mononuclear non-heme iron active sites. Chem Rev 96: 2607–2524.

Quinsey NS, Luong AQ, Dickson PW (1998): Mutational analysis of substrate inhibition in tyrosine hydroxylase. J Neurochem 71: 2132–2138.

Ramsey AJ, Fitzpatrick PF (1998): Effects of phosphorylation of serine 40 of tyrosine hydroxylase on binding of catecholamines: evidence for a novel regulatory mechanism. Biochemistry 37: 8980–8986.

Ramsey AJ, Daubner SC, Ehrlich JI, Fitzpatrick PF (1995): Identification of iron ligands in tyrosine hydroxylase by mutagenesis of conserved histidinyl residues. Prot Sci 4: 2082–2086.

Ramsey AJ, Hillas PJ, Fitzpatrick PF (1996): Characterization of the active site iron in tyrosine hydroxylase: redox states of the iron. J Biol Chem 271: 24395–24400.

Renson JD, Daly J, Weissbach H, Witkop B, Udenfriend S (1966): Enzymatic conversion of 5–tritiotryptophan to 4–tritio-5–hydroxytryptophan. Biochem Biophys Res Commun 25: 504–513.

Ribeiro P, Wang Y, Citron BA, Kaufman S (1993): Deletion mutagenesis of rat PC12 tyrosine hydroxylase regulatory and catalytic domains. J Molec Neurosci 4: 125–139.

Richtand NM, Inagami T, Misono K, Kuczenski R (1985): Purification and characterization of rat striatal tyrosine hydroxylase. Comparison of the activation by cyclic AMP-dependent phosphorylation and by other effectors. J Biol Chem 260: 8465–8473.

Roskoski R Jr, Ritchie P (1991): Phosphorylation of rat tyrosine hydroxylase and its model peptides *in vitro* by cyclic AMP-dependent protein kinase. J Neurochem 56: 1019–1023.

Roy A, DeJong J, Linnoila M (1989): Cerebrospinal fluid monoamine metabolites and suicidal behavior in depressed patients. Arch Gen Psychiatry 46: 609–612.

Rudolph FB, Fromm HJ (1979): Plotting methods for analyzing enzyme rate data. Methods Enzymol 63: 138–159.

Schuller DJ, Grant GA, Banaszak LJ (1995): The allosteric ligand site in the V_{max}-type cooperative enzyme phosphoglycerate dehydrogenase. Nat Struct Biol 2: 69–76.

Shamoo Y, Ghosaini LR, Keating KM, Williams KR, Sturtevant JM, Konigsberg WH (1989): Site-specific mutagenesis of T4 gene 32: the role of tyrosine residues in protein-nucleic acid interactions. Biochemistry 28: 7409–7417.

Shiman R (1985): Phenylalanine hydroxylase and dihydropterin reductase. In Blakley RL, Benkovic SJ (eds): "Folates and Pterins, Vol. 2." New York: John Wiley & Sons, pp. 179–249.

Shiman R, Gray DW (1980): Substrate activation of phenylalanine hydroxylase. A kinetic characterizaion. J Biol Chem 255: 4793–4800.

Shiman R, Gray DW, Pater A (1979): A simple purification of phenylalanine hydroxylase by substrate-induced hydrophobic chromatography. J Biol Chem 254: 11300–11306.

Shiman R, Jones SH, Gray DW (1990): Mechanism of phenylalanine regulation of phenylalanine hydroxylase. J Biol Chem 265: 11633–11642.

Shiman R, Gray DW, Hill MA (1994a): Regulation of rat liver phenylalanine hydroxylase. I. Kinetic properties of the enzyme's iron and enzyme reduction site. J Biol Chem 269: 24637–24646.

Shiman R, Xia T, Hill MA, Gray DW (1994b): Regulation of rat liver phenylalanine hydroxylase. II. Substrate binding and the role of activation in the control of enzymatic activity. J Biol Chem 269: 24647–24656.

Sono M, Roach MP, Coulter ED, Dawson JH (1996): Heme-containing oxygenases. Chem Rev 96: 2841–2888.

Stadtman ER (1993): Oxidation of free amino acids and amino acid residues in proteins by radiolysis and by metal-catalyzed reactions. Ann Rev Biochem 62: 797–821.

Sutherland C, Alterio J, Campbell DG, Le Bourdelles B, Mallet J, Haavik J, Cohen P (1993): Phosphorylation and activation of human tyrosine hydroxylase *in vitro* by mitogen-activated protein (MAP) kinase and MAP-kinase-activated kinases 1 and 2. Eur J Biochem 217: 715–722.

Tachikawa E, Tank AW, Yanagihara N, Mosimann W, Weiner N (1986): Phosphorylation of tyrosine hydroxylase on at least three sites in rat pheochromocytoma PC12 cells treated with 56 mM K^+: Determination of the sites on tyrosine hydroxylase phosphorylated by cyclic AMP-dependent and calcium/calmodulin-dependent protein kinases. Molec Pharmacol 30: 476–485.

Tian G, Klinman JP (1993): Discrimination between ^{16}O and ^{18}O in oxygen binding to the reversible oxygen carriers hemoglobin, myoglobin, hemerythrin, and hemocyanin: a new probe for oxygen binding and reductive activation by proteins. J Am Chem Soc 115: 8891–8897.

Udenfriend S, Cooper JR (1952): The enzymatic conversion of phenylalanine to tyrosine. J Biol Chem 194: 503–511.

Vaccaro KK, Liang BT, Perelle BA, Perlman RL (1980): Tyrosine 3-monooxygenase regulates catecholamine synthesis in pheochromocytoma cells. J Biol Chem 255: 6539–6541.

van den Heuvel LPWJ, Luiten B, Smeitink JAM, deRijk-van Andel JF, Hyland K, Steenbergen-Spanjers GCH, Janssen RJT, Wevers RA (1998): A common point mutation in the tyrosine hydroxylase gene in autosomal recessive L-DOPA-responsive dystonia in the Dutch population. Hum Genet 102: 644–646.

Vulliet PR, Langan TA, Weiner N (1980): Tyrosine hydroxylase: a substrate of cyclic AMP-dependent protein kinase. Proc Natl Acad Sci U S A 77: 92–96.

Vulliet PR, Woodgett JR, Cohen P (1984): Phosphorylation of tyrosine hydroxylase by calmodulin-dependent multiprotein kinase. J Biol Chem 259: 13680–13683.

Wallick DE, Bloom LM, Gaffney BJ, Benkovic SJ (1984): Reductive activation of phenylalanine hydroxylase and its effect on the redox state of the non-heme iron. Biochemistry 23: 1295–1302.

Waymire JC, Johnston JP, Hummer-Lickteig K, Lloyd A, Vigny A, Craviso GL (1988): Phosphorylation of bovine adrenal chromaffin cell tyrosine hydroxylase. Temporal correlation of acetylcholine's effect on site phosphorylation, enzyme activation, and catecholamine synthesis. J Biol Chem 263: 12439–12447.

Wu J, Filer D, Friedhoff AJ, Goldstein M (1992): Site-directed mutagenesis of tyrosine hydroxylase role of serine 40 in catalysis. J Biol Chem 267: 25754–25758.

Xia T, Gray DW, Shiman R (1994): Regulation of rat liver phenylalanine hydroxylase. III. Control of catalysis by (6*R*)-tetrahydrobiopterin and phenylalanine. J Biol Chem 269: 24657–24665.

Yamauchi T, Fujisawa H (1979): *In vitro* phosphorylation of bovine adrenal tyrosine hydroxylase by adenosine 3′: 5′-monophosphate-dependent protein kinase. J Biol Chem 254: 503–507.

Yamauchi T, Nakata H, Fujisawa H (1981): A new activator protein that activates tryptophan 5-monooxygenase and tyrosine 3-monooxygenase in the presence of Ca^{2+}-, calmodulin-dependent protein kinase. Purification and characterization. J Biol Chem 256: 5404–5409.

Zhao G, Xia T, Song J, Jensen RA (1994): *Pseudomonas aeruginosa* possesses homologues of mammalian phenylalanine hydroxylase and 4α-carbinolamine dehydratase/DCoH as part of a three-component gene cluster. Proc Natl Acad Sci U S A 91: 1366–1370.

Zhou Q-Y, Quaife CJ, Palmiter RD (1995): Targeted disruption of the tyrosine hydroxylase gene reveals that catecholamines are required for mouse fetal development. Nature 374: 640–643.

Zigmond RE, Schwarzschild MA, Rittenhouse AR (1989): Acute regulation of tyrosine hydroxylase by nerve activity and by neurotransmitters via phosphorylation. Ann Rev Neurosci 12: 415–461.

L-ASPARTASE: NEW TRICKS FROM AN OLD ENZYME

By RONALD E. VIOLA, *Department of Chemistry, University of Akron, Akron, Ohio 44325-3601*

CONTENTS

Advances in Enzymology and Related Areas of Molecular Biology, Volume 74: Mechanism of Enzyme Action, Part B, Edited by Daniel L. Purich
ISBN 0-471-34921-6

I. Introduction

The enzyme L-aspartate ammonia-lyase (aspartase) catalyzes a conceptually straightforward reaction, the reversible deamination of the amino acid L-aspartic acid to produce fumaric acid and ammonium ion (Scheme 1). Numerous studies have reported on various properties of this enzyme, and several earlier reviews have been written (e.g., Virtanen and Ellfolk, 1955; Williams and Lartigue, 1969; Tokushige, 1985). Despite this large body of work, spanning nearly 100 years, a number of more recent studies have revealed some interesting and unexpected new aspects of this reasonably well-characterized enzyme. The nonlinear kinetics that are seen under certain conditions have been shown to be the result of a separate regulatory site at which the substrate, aspartic acid, can also play the role of an activator. Truncation of the carboxyl terminus at specific positions causes an enhancement of the catalytic activity of aspartase. Quite recently this enzyme has been shown to have a separate, nonenzymatic property in certain organisms as an effector that contributes to the activation of the precursor plasminogen to the clot-dissolving enzyme plasmin. All of these unusual properties have been discovered in an enzyme that, for nearly 30 years, was thought to be a well-characterized, classical enzyme. Before examining some of the more extraordinary aspects of this enzyme it will be helpful to consider the background of aspartase, including the early characterization of this enzyme, a wide range of mechanistic studies, and the recent structural determination of aspartase.

II. Early History

The ability of bacteria to reduce aspartic acid to succinic acid was first reported at the beginning of the twentieth century (Harden, 1901). In 1926 researchers showed that resting bacteria can accelerate the establishment of an equilibrium between L-aspartic acid, fumaric acid, and ammonia (Quastel and Woolf, 1926). Subsequent work established that this enzyme was a

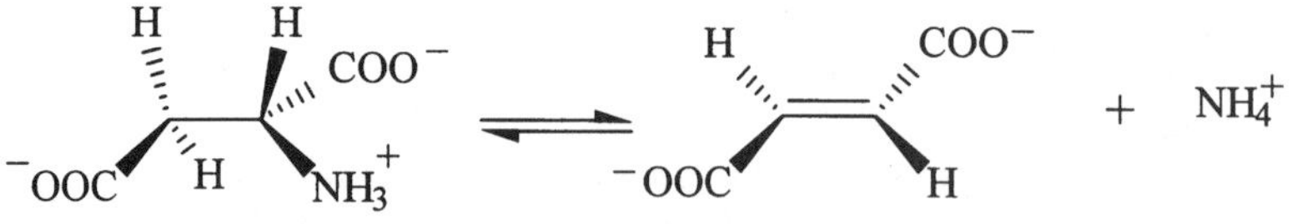

Scheme 1. The reversible deamination of L-aspartic acid catalyzed by aspartase.

deaminase and not an oxidase, and was given the name aspartase (Woolf, 1929). Additional studies through the following decade examined the rate of ammonia and fumarate production under a range of conditions (Virtanen and Tarnanen, 1932) and quantitated the equilibrium constant for this new catalyst (Jacobsohn and Pereira, 1936). Significantly, solvent extraction of *Pseudomonas fluorescens* to release aspartase produced the first cell-free sample that was able to catalyze the synthesis of an amino acid (Virtanen and Tarnanen, 1932). After these early investigations on aspartase, a number of years passed during which the concepts of biological catalysis were refined, and the nature of these new biological entities was debated and clarified. Once these principles were firmly established, the stage was set to begin a detailed examination of the catalytic mechanism of enzymes such as aspartase. This pursuit has continued to define the field of enzymology to this day.

III. Purification and Characterization

Aspartase has now been identified and purified from a variety of microorganisms. Very early work (Cook and Woolf, 1928) detected aspartase activity in several different facultative aerobes. The initial definitive work on aspartase was a series of papers published from 1953 to 1954 by Ellfolk that reported the partial purification of the enzyme (Ellfolk, 1953a), metal ion and chemical modification studies (Ellfolk, 1953b), and an examination of the substrate specificity of aspartase (Ellfolk, 1954). Aspartase was initially purified by alcohol extraction, followed by either ammonium sulfate or acetone fractionation (Ellfolk, 1953a). Treatment of crude extracts from *Hafnia alvei* (formerly known as *Bacterium cadaveris*) by starch electrophoresis led to the separation of aspartase from other interfering activities (Wilkinson and Williams, 1961). The aspartase from *Escherichia coli* was purified in 20% overall yield from a seven-step procedure (Rudolph and Fromm, 1971), and in slightly higher yield from a procedure that utilized four separate chromatography steps (Suzuki et al., 1973). A further improvement on this procedure utilized the published low pH/heat and ammonium sulfate fractionation steps from these earlier purification schemes, followed by ion-exchange and dye-ligand chromatography as the key steps to obtain homogeneous enzyme. This improved scheme resulted in over 800-fold purification of aspartase in 50% yield from wild-type *E. coli* (Karsten et al., 1985). The *aspA* gene, encoding for aspartase, had been mapped in the *E. coli* genome (Spencer et al., 1976). Subsequent incorporation of this gene into an efficient vector has resulted in a dramatic overexpression of aspartase in *E. coli*, frequently reaching as high as 25% of the soluble cellular proteins. These higher levels of aspartase have allowed the elimination of the early steps in the purification scheme. Aspartase can now be purified to homogeneity from these overproducing strains in about 2 to 3 days, typically yielding at least 100 mg of enzyme from about 10 g of cells.

The enzyme from *E. coli* is a tetramer composed of four identical subunits of molecular weight 52,200 Da (Takagi et al., 1985). Purified aspartase from *P. fluorescens* is also a homotetramer, with a subunit of 50,900 Da (Takagi et al., 1986a). Denaturation of each of these enzymes by the addition of high concentrations of guanidine.HCl leads to reversible dissociation and reassembly of the tetrameric structure (Tokushige et al., 1977). The resulting monomers, and the intermediate dimers and trimers that are formed during reactivation, were reported to be devoid of catalytic activity (Imaishi et al., 1989), although later studies observed the presence of an active dimeric form of aspartase under appropriate denaturation conditions (Murase et al., 1993b). Despite the high sequence homology between the aspartases from *E. coli* and *P. fluorescens* (56% sequence identity), the dissociated subunits do not form hybrid tetramers when they are mixed and allowed to renature (Yumoto et al., 1992).

IV. Metabolic Significance

Despite many years of investigation, the precise metabolic role of aspartase has been difficult to resolve. Aspartase activity is found predominantly in bacterial species. There have been isolated reports of aspartase activity detected in some plants and aquatic species, however aspartase has not generally been found in higher plants and mammals. The work of Halpern and Umbarger suggested that aspartase functions primarily in a degradative role, releasing ammonia and producing fumarate that enters into the tricarboxylic acid cycle (Halpern and Umbarger, 1960). Examination of *E. coli* mutants in which the enzyme glutamate dehydrogenase is absent found that these organisms are still capable of growth on glutamic acid as a carbon source. Under these growth conditions the levels of aspartase are elevated, suggesting that aspartase can catalyze amino acid formation via the amination of fumarate in vivo (Vender and Rickenberg, 1964). In mutants that are lacking glutamate decarboxylase activity, transamination with oxaloacetate was proposed to yield α-ketoglutarate and aspartate. α-Ketoglutarate can enter directly into the tricarboxylic acid cycle. Deamination of aspartate would allow its entry by way of fumarate, and there are enhanced aspartase levels seen in these mutants (Vender et al., 1965). The identification of additional glutamate-metabolizing enzymes, glutamine synthetase and glutamate synthase, along with glutamate dehydrogenase eliminated the need to propose a biosynthetic role for aspartase (Berberich, 1972).

V. Commercial Applications

The excellent catalytic properties of enzymes has led to ever-increasing commercial applications of these biological catalysts. Aspartase has been one of the early success stories in these applications, with the high specificity and catalytic activity of aspartase being exploited for the commercial production of L-aspartic acid. *Escherichia coli* cells have been immobilized in a polyacrylamide gel lattice (Tosa et al., 1974), in polyurethane (Fusee, 1987), or with carrageenan (Nishida et al., 1979). The aspartase activity in these immobilized cells shows a significantly higher stability against detergents and thermal denaturation (Tosa et al., 1977). The immobilized and stabilized cells are used in flow systems for the continuous enzymatic conversion of fumaric acid and ammonium salts to L-aspartic acid. Further advances in the commercial applications of aspartase have included the immobilization of engineered strains of *E. coli* that overproduce aspartase (Chibata et al., 1986), and patented processes for the production of heat-stable aspartase (Kimura et al., 1983) and for the coupling of maleate isomerase and aspartase to allow the conversion of ammonium maleate to L-aspartic acid (Sakano et al., 1996).

VI. Substrate Specificity

Aspartase is among the most specific enzymes known, with extensive studies over the years failing to identify any alternative amino acid substrates that can replace L-aspartic acid. Aspartase has no activity with either D-aspartic acid or crotonic acid (Virtanen and Ellfolk, 1955), and earlier work showed that glycine, alanine, glutamine, maleic acid, and glutaconic acid were not substrates (Quastel and Woolf, 1926). Also, L-cysteic acid, diaminosuccinic acid, leucine, mesaconic acid, aconitic acid, sorbic acid, and the diamine or mono- or diethyl esters of fumaric acid have all failed to show turnover with this highly specific enzyme (Virtanen and Tarnanen, 1932; Ellfolk, 1954). While the specificity for L-aspartic acid had been clearly demonstrated, hydroxylamine was later found to be an alternative substrate for ammonia leading to the production of *N*-hydroxyaspartic acid (Emery, 1963).

A more extensive investigation of structural analogs of aspartic acid also failed to uncover any alternative substrates for aspartase (Falzone et al., 1988). However, a wide variety of inhibitors of the enzyme were identified (Table 1), with inhibition constants ranging from an order of magnitude lower to an order of magnitude higher than the Michaelis constant for L-aspartic acid. Binding of substrate analogs at the active site requires the presence of a carboxylate or a similarly charged functional group at each end of the four-carbon chain. Substitution with a phospho-, phosphono-, or nitro-group can fulfill the electrostatic requirement for binding. Greater structural variation is permitted at the α-amino group. Decreased affinity is observed when the amino group is missing (succinate), however substitution with a phosphate

TABLE 1
Inhibitors of Aspartase

	Structure[a]		
Substrate Analog	R_1	R_2	K_i (mM)
O-phospho-D-serine	NH_3^+	OPO_3^{2-}	0.20
D-malate	OH	COO^-	0.66
DL-2-amino-3- phosphonopropionate	NH_3^+	PO_3^{2-}	0.66
3-nitropropionate	H	NO_2	0.83
2,3-diphosphoglycerate	OPO_3^{2-}	OPO_3^{2-}	1.1
mercaptosuccinate	SH	COO^-	1.6
L-2-chlorosuccinate	Cl	COO^-	1.7
DL-2-bromosuccinate	Br	COO^-	2.3
DL-2-amino-4-phosphonobutyrate	NH_3^+	$CH_2PO_3^{2-}$	2.4
D-2-methylmalate[b]	OH	COO^-	3.7
2-hydroxy-3-nitropropionate	OH	NO_2	6.7
N-acetyl-L-aspartate	$NHC(=O)CH_3$	COO^-	6.7
β-aspartylhydrazine	NH_3^+	$C(=O)NHNH_2$	18
methylsuccinate	CH_3	COO^-	23
phosphoglycolate	H	OPO_3^{2-}	23
succinate	H	COO^-	24
L-malate	OH	COO^-	31
(3-aminopropyl)phosphonate[c]	NH_3^+	$CH_2PO_3^{2-}$	36

[a]Parent structure: $R_2CH_2CH(R_1)COO^-$

[b]α-Proton replaced with an α-methyl group.

[c]α-Carboxyl replaced with a proton.

group (2,3-diphosphoglycerate), a halide (bromo- or chlorosuccinate), or a thiol (mercaptosuccinate) still allows binding with substantial affinity (Table 1). The wide variety of competitive inhibitors identified for aspartase demonstrates that the high substrate specificity of this enzyme is not dictated solely by active site binding discrimination. Although all of the compounds listed are competitive inhibitors of aspartase, subtle changes in the binding orientation of these inhibitors must alter the positioning of these analogs with respect to the active site catalytic groups, thus influencing the conformational changes that are required for a productive catalytic cycle.

VII. Mechanism of Action

The family of ammonia-lyases, of which aspartase is a member, seem to have very little in common with each other besides the overall deamination reaction that they catalyze. No sequence homology has been identified across this diverse family except that seen between the histidine and phenylalanine ammonia-lyases (Taylor et al., 1990). This subfamily of ammonia-lyases also seems to use a common mechanism for deamination, utilizing an electrophilic dehydroalanine (Hanson and Havir, 1970) that is generated by a posttranslational dehydration of an active site serine (Hernandez et al., 1993; Langer et al., 1994). Other members of this ammonia-lyase family appear to use different mechanisms to catalyze the deamination reaction. Ethanolamine ammonia-lyase contains an active site cobalamin that is essential for catalysis (O'Brien et al., 1985). Even methylaspartase, which catalyzes the closely related deamination of methylaspartic acid to produce methylfumarate (mesaconitate), has no sequence identity to aspartase (Goda et al., 1992), has different metal ions requirements (Bright, 1967), and is proposed to function by a different mechanism (Botting and Gani, 1989). Based on the work that has been carried out to date with these enzymes, the ammonia-lyases appear to be an enzyme family separated by their common mechanistic differences.

A. METAL ION REQUIREMENTS OF ASPARTASE

The addition of magnesium was shown to affect the equilibrium constant for the aspartase-catalyzed deamination (Jacobsohn and Pereira, 1936), and the presence of certain divalent metal ions was reported to contribute to the stability of the enzyme (Ichihara et al., 1955). In contrast to the very high substrate specificity, aspartase is quite nonspecific for metal ion activators. Studies with purified aspartase from *H. alvei* reported activation of the en-

TABLE 2
Activation of Aspartase by Divalent Metal Ions[a]

Metal Ion	K_a (μM)
Zn^{2+}	0.21
Cd^{2+}	0.7
Mn^{2+}	0.8
Co^{2+}	2.0
Mg^{2+}	4.9
Ca^{2+}	19

[a]Assay conditions: 100 mM Hepes, pH 7.0, and 30 mM L-aspartic acid.

zyme upon addition of Mn^{2+}, Mg^{2+}, or Zn^{2+} (Wilkinson and Williams, 1961). Later work indicated only minimal activation by either Zn^{2+} or Co^{2+}, and no activation with Ca^{2+} (Nuiry et al., 1984). At higher pH the *E. coli* enzyme has an absolute requirement for a divalent metal ion (Rudolph and Fromm, 1971), but at lower pH the enzyme possesses some activity in the absence of added metal ions (Suzuki et al., 1973). An extensive study of *E. coli* aspartase (Falzone et al., 1988) demonstrated a range of activating divalent metal ions, with Zn^{2+} (K_a = 0.2 μM) being the most potent (Table 2).

There has been some disagreement concerning the role of divalent metal ions in the catalytic activity of aspartase. Kinetic studies of the enzyme from *H. alvei* led to a proposed role for a divalent metal ion interacting with the β-carboxyl group of the substrate (Nuiry et al., 1984), similar to the role postulated in β-methylaspartase (Bright, 1965). However, nuclear magnetic resonance (NMR) relaxation studies (Falzone et al., 1988) have shown that the divalent metal ion required for activity at high pH in the *E. coli* enzyme is bound at an activator site and not at the active site of the enzyme (Bright, 1965). It now appears that many of these divergent and conflicting results can be ascribed to the different sources of the enzyme.

B. MECHANISM OF ASPARTASE

The results obtained from kinetic and isotope effect studies of aspartase are consistent with the rapid equilibrium-ordered addition of the divalent metal ion activator prior to aspartate binding followed, after catalysis, by the random release of products (Nuiry et al., 1984). However, the kinetic mechanism of aspartase is complicated by the nonlinear kinetics that are observed above neutral pH (Williams and Lartigue, 1967; Rudolph and Fromm,

1971). The sigmoidal kinetics have been suggested to be evidence for cooperativity in substrate binding and, based on this evidence, the aspartase from *H. alvei* was proposed to be a regulatory enzyme (Williams and Lartigue, 1967). Later studies (see Section IX) have detected the presence of an activator site that completely explains the nonlinear kinetics of aspartase.

The enzyme-catalyzed amination of fumarate was shown to yield monodeutero-L-aspartic acid when the reaction was conducted in D_2O (Englard, 1958). These results support a stereospecific addition of ammonia across the double bond of fumarate. These authors further concluded, arguing by analogy with the earlier work on the mechanism of fumarase (Farrar et al., 1957), and their work on succinate dehydrogenase (Englard and Colowick, 1956) and aconitase (Englard and Colowick, 1957), that the aspartase-catalyzed deamination is a *cis*-elimination that proceeds by initial elimination of ammonia to produce a carbocation intermediate (Englard, 1958). However Gawron demonstrated, from NMR characterization of the deuterium-labeled products, that the mechanisms of both aspartase and fumarase involve a *trans*-addition of ammonia and water, respectively, to fumaric acid (Gawron and Fondy, 1959).

The absence of a primary deuterium isotope effect for the deamination of 3-deutero-L-aspartate was presented as evidence for a rate-limiting C–N bond cleavage, and therefore for development of carbocation character at carbon-2 in the intermediate (Dougherty et al., 1972). In contrast, Hanson has provided evidence (Hanson and Havir, 1972) in support of a 3-carbanion intermediate, followed by rate-limiting elimination of ammonium ion. To address this controversy, a transition state analog of the proposed carbanion intermediate was examined. 2-Amino-3-nitro propionate is an aspartic acid analog in which the β-carboxyl group has been replaced with a nitro group. The electron-withdrawing nitro group supports the ionization of the proton at carbon-3, with resonance stabilization of the resulting carbanion (Scheme 2) lowering this pK value to below 8 (Porter and Bright, 1980). In contrast to the rapid ionization of oxyacids, the ionization of carbon acids involves a considerable energy barrier, which makes the attainment of equilibrium quite slow. Thus it takes several minutes for the ionization of 2-amino-3-nitropropionate (Scheme 2) after the pH is raised above the pK value of the C-3 proton. This slow ionization provides an opportunity to determine which ionization form is binding to the active site of aspartase with high affinity. When this nitro analog is added, at low concentrations as the conjugate acid (left side of Scheme 2), to a reaction mixture with aspartase at pH 8.5 there is a lag before substantial inhibition is observed (Porter and Bright,

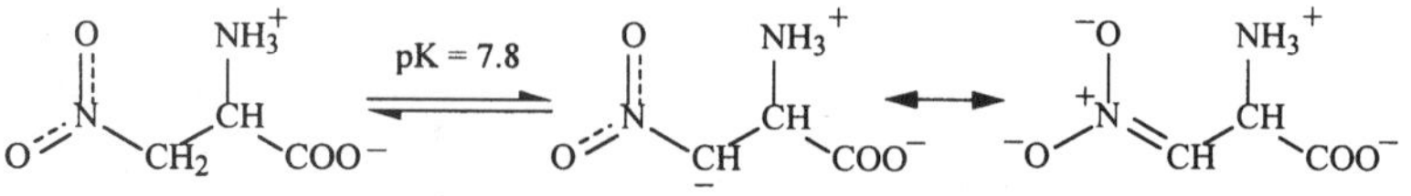

$$O=N(=O)-CH_2-CH(NH_3^+)-COO^- \underset{}{\overset{pK = 7.8}{\rightleftharpoons}} O=N(=O)-\overline{C}H-CH(NH_3^+)-COO^- \longleftrightarrow {}^-O-N^+(O^-)=CH-CH(NH_3^+)-COO^-$$

Scheme 2. Ionization of 2-amino-3-nitropropionate.

1980). This lag corresponds to the rate of ionization at carbon-3 and demonstrates that 2-amino-3-nitropropionate binds very tightly in the ionized (carbanion) form. The inhibition constant for the carbanion is some 1600-fold tighter than the Michaelis constant for aspartic acid, suggesting that this nitro analog is binding as a transition state or intermediate analog. The high affinity for the ionized form of 2-amino-3-nitropropionate provides strong evidence in support of a carbanion mechanism for aspartase.

C. IDENTITY OF ACTIVE SITE RESIDUES

Numerous studies of aspartase have been conducted with the aim of identifying the active site residues that are involved in substrate binding and in catalysis. These studies have used specific reagents to examine the sensitivity of aspartase to inactivation (chemical modification studies) or have examined the effect of changes in ionization states on aspartase activity (pH studies). These approaches have led to the identification of a number of different amino acid functional groups that are suggested to play critical roles in the activity of aspartase.

1. Chemical Modification

Chemical modification studies of the enzyme have indicated possible roles for sulfhydryl and imidazole groups in the aspartase-catalyzed reaction. Treatment of *E. coli* aspartase with either *N*-ethylmaleimide (NEM) or 5,5′-dithiobis-(2-nitrobenzoate) (DTNB) results in the modification of two sulfhydryl groups per subunit, and led to the complete loss of catalytic activity (Mizuta and Tokushige, 1975). The presence of aspartic acid, or various aspartic acid derivatives, provides substantial protection against inactivation by these reagents. Similar results were obtained with the *H. alvei* enzyme, and the pH dependence of the inactivation rate suggested that an amino acid with a p*K* value of 8.3 was being modified (Shim et al., 1997). Treatment of *E. coli* aspartase with the fluorescent reagent *N*-(7-dimethylamino-4-methylcoumarynyl)maleimide (DACM), followed by HPLC analysis of a tryptic digest, revealed the selective modification of two cysteinyl residues, Cys-141[1] and Cys-431 (Ida and Tokushige, 1985). Protection against modification of these residues in the presence of aspartic acid led to the suggestion that these cysteines are located in the substrate binding site of aspartase (Ida and Tokushige, 1985). Similarly, the selective modification of a single histidine by diethylpyrocarbonate supported the suggestion of an essential histidine at the active site of the enzyme (Ida and Tokushige, 1984).

When aspartase was incubated with *o*-phthalaldehyde (OPA), a linear relationship was observed between the loss of enzymatic activity and changes in both the absorbance and the fluorescence spectra of the enzyme. This loss of activity is correlated with an increase in absorbance at 377 nm, and the appearance of a peak in the fluorescence emission spectrum at 420 nm when this new absorbance band is excited. These absorbance and fluorescence changes are characteristic of isoindole ring formation resulting from the modification of adjacent lysine and cysteine residues (Wong et al., 1985). Complete protection against both inactivation and spectral modification by OPA was observed in the presence of saturating levels of a substrate (fumarate), an activator (α-methylaspartate) and a divalent metal ion (Mg^{2+}), while protection was lost if any one of these components was omitted. The results of these experiments suggest the juxtaposition of a lysyl residue near one of the previously identified cysteines.

2. *pH studies*

An initial pH study of aspartase from *H. alvei* suggested catalytic roles for sulfhydryl and histidine residues based on the observed p*K* values in a bell-shaped pH profile (Lartigue, 1965). We have conducted a more extensive study of the pH dependence of the kinetic parameters of the aspartase-catalyzed reaction in both the amination and deamination directions. The enzyme isolated from *E. coli* exists in a pH-dependent equilibrium between a high pH form that has an absolute requirement for substrate activation and a divalent metal ion, and a low pH form that does not require activation by either substrate or metal ions (Fig. 1). The interconversion between these enzyme forms is observed near neutral pH in both the maximum velocity (V_{max}) and V_{max}/K_m pH profiles examined for the reaction in either direction

[1]The original amino acid numbering scheme has been increased by one since the publication of the high-resolution structure (Shi et al., 1997) to account for the *N*-terminal methionine that is present in the mature enzyme.

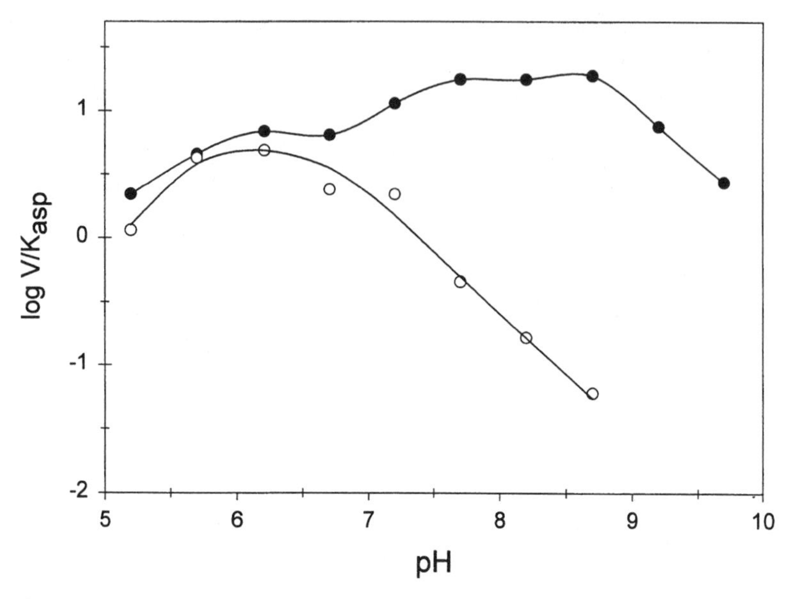

Figure 1. pH dependence ($V_{max}/K_{aspartase}$) of the deamination reaction of aspartase. (●) Enzyme activity in the presence of 10 mM Mg^{2+}; (○) Enzyme activity in the absence of divalent metal ions (Karsten and Viola, 1991).

(Karsten and Viola, 1991). A similar pH-dependent activation by divalent metal ions is observed at high pH for *H. alvei* aspartase. However, significant metal ion activation is also seen for this enzyme below pH 7 (Yoon et al., 1995). Loss of activity is observed with both enzymes in the presence of metal ions at high pH, with a p*K* value near 9. The pH profiles of competitive inhibitors such as 3–nitropropionic acid and succinic acid have shown that the enzyme group responsible for this activity loss is required to be protonated for substrate binding at the active site. Based on these results, and on the structure of the substrate for this reaction, a lysyl residue is the most likely candidate. An enzymic group has also been identified that must be protonated in the amination reaction, with a p*K* value near 6.5, and deprotonated in the deamination reaction. This group, tentatively assigned as a histidyl residue, fulfills the criteria for the acid-base catalyst at the active site of aspartase (Karsten and Viola, 1991).

The most significant difference between the enzyme from these organisms is found in the ionization state of ammonia utilized by aspartase in the amination reaction. The pH profile for ammonia ($V_{max}/K_{ammonia}$) decreases at low pH with the *E. coli* enzyme (Karsten and Viola, 1991), requiring neutral ammonia to be the substrate, while with the *H. alvei* enzyme the decrease in activity is observed at high pH (Yoon et al., 1995) suggesting protonated ammonium ion as the active species. Clearly, the nature of the acid-base catalytic mechanism in the amination reaction of aspartase must accommodate the differences between these enzymes.

D. MECHANISM-BASED INACTIVATION

Treatment of an enzyme with a group-specific reagent can potentially lead to the covalent modification of any accessible and reactive amino acid of the appropriate type. Higher modification specificity can be achieved if a highly reactive reagent is generated only upon the catalytic action of the enzyme on a suitably designed substrate analog. The substrate analog L-aspartate-β-semialdehyde (ASA) has been identified as a mechanism-based inactivator of aspartase from *E. coli* (Higashi et al., 1988; Schindler and Viola, 1994). Incubation of aspartase with ASA results in the irreversible inactivation of the enzyme (Fig. 2). However, subsequent work has shown that ASA itself is not the inactivating species. Aspartase catalyzes the deamination of ASA to yield fumaric acid semialdehyde (FAA) and ammonia, thus identifying the first alternative amino acid substrate for this enzyme (Schindler and Viola, 1994). It is the FAA product that then reacts to inactivate aspartase. Once this had been established the questions that remained to be addressed were: (1) where is the site (or sites) of modification that result in enzyme inactivation, and (2) what is the mechanism of inactivation?

1. *Identification of the Inactivation Sites*

After treatment of aspartase with ASA the inactivated enzyme was subjected to proteolytic digestion. When these peptide fractions were compared to those generated from identical treatment of the unmodified enzyme, an alteration was observed in the retention time of a single peptide from an HPLC tryptic map (Giorgianni et al., 1995). Further proteolytic digestion and mass spectral analysis has lead to the identification of this tryptic peptide as encompassing amino acids 253 to 285 of the aspartase sequence (Fig. 3). A V8 protease fragment of this tryptic peptide, containing the amino acids Ala-270 to Lys-285 (m/z = 1681), undergoes a mass shift upon FAA derivatization. A chymotrypsin fragment of the same peptide, containing amino acids 253 to 277, also undergoes a similar mass shift. These results

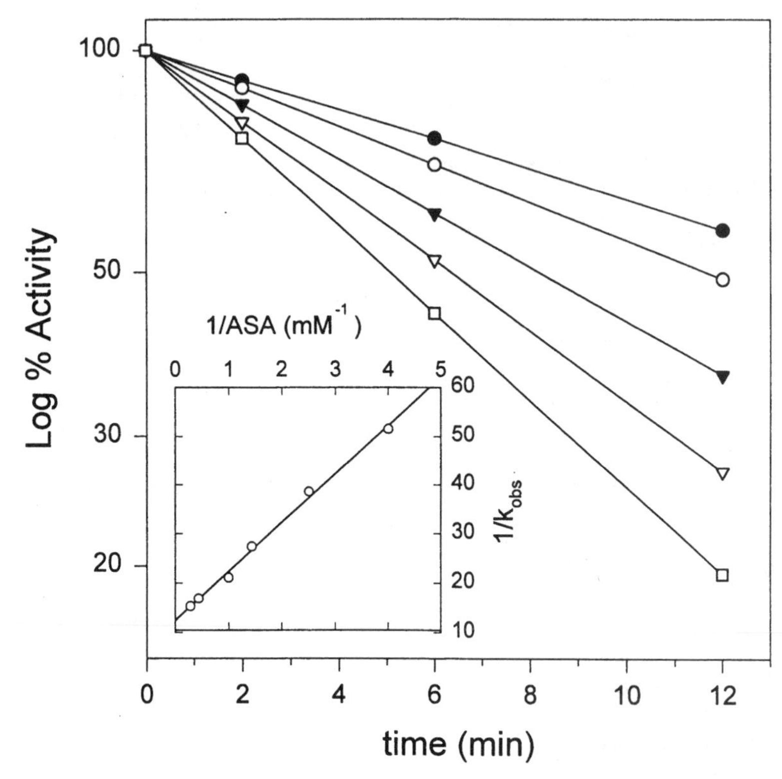

Figure 2. Time course for the inactivation of aspartase at pH 8.0. Aspartase (0.5 μM) was incubated at 25°C with increasing concentrations of ASA. [ASA] = 0.25 mM (●); 0.40 mM (○); 0.70 mM (▼); 1.00 mM (▽); and 2.29 mM (□). *Inset:* reciprocal plot of the observed rate constant (k_{obs}) for ASA inactivation of aspartase as a function of ASA concentration (Schindler and Viola, 1994).

localize the modification of aspartase by FAA within the amino acid sequence between residues 270 and 277, the region of overlap between the mass shifted V8 protease and chymotrypsin fragments:

^{270}ala-thr-ser-asp-CYS-gly-ala-tyr^{277}

This fragment contains Cys-274, one of the candidates for the nucleophile that is being modified. While the enzyme from *E. coli* is highly sensitive to this mechanism-based FAA inactivation, studies with aspartase from *P. flu-*

orescens have reported that this enzyme, which is 77% identical in amino acid sequence to the *E. coli* enzyme, is completely insensitive to treatment with ASA (Takagi et al., 1984). Significantly, the cysteine that would be at the corresponding position in the *Pseudomonas* enzyme has been replaced by a methionine. Replacement of Cys-274 in the *E. coli* enzyme with either serine or alanine has no observable effect on the catalytic activity of aspartase (Giorgianni et al., 1995). This establishes that this residue plays no direct role in the mechanism of the deamination of aspartic acid. However, while catalysis is not affected, substitution of the cysteine side chain with either the hydroxymethyl group of serine or with the methyl group of alanine substantially decreases, but does not eliminate, the sensitivity of this enzyme to inactivation by the alternative product FAA (Fig. 4). This residual reactivity requires the presence of additional modification site(s) that can lead to the slower inactivation of aspartase.

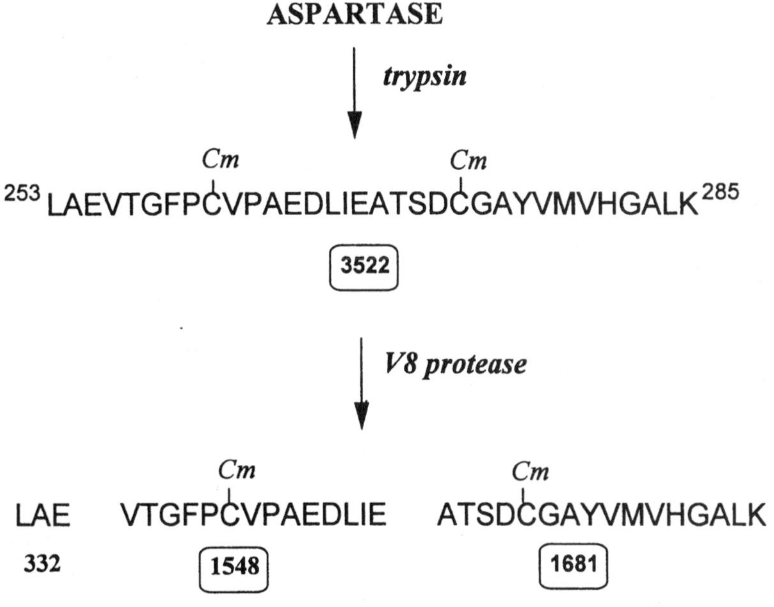

Figure 3. Proteolytic digestion of aspartase by trypsin and *Staphylococcus aureus* V8 protease. The calculated mass of each fragment is shown, with the mass in boxes indicated those fragments that have been directly observed by mass spectroscopy. The sites of carboxymethylation (*Cm*) are indicated (Giorgianni et al., 1995).

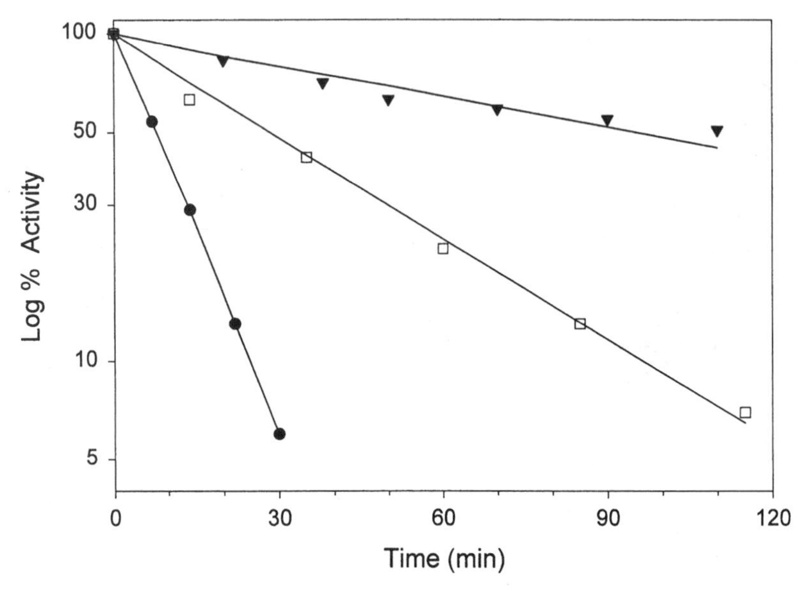

Figure 4. Sensitivity of aspartase mutants to inactivation by aspartate semialdehyde (ASA). Aspartase samples (3 μM) was incubated in 0.5 M Taps buffer, pH 8.5, at 30°C with 5 mM ASA. (●) Wild-type enzyme; (□) C274A mutant; (▼) C141S/C274A double mutant (Giorgianni et al., 1997).

Earlier kinetic and chemical modification studies on aspartase (discussed below in Section VII.C) provided evidence for the possible involvement of a lysine residue in substrate binding at the active site of the enzyme (Karsten and Viola, 1991). To test the hypothesis that a lysyl Schiff base is involved in the inactivation by FAA an extra step, consisting of reduction of the inactivated aspartase with sodium cyanoborohydride, was added to the enzyme modification protocol. To increase the yield of modified enzyme these mapping studies were carried out using the ethyl ester of FAA. This compound is more stable to hydrolysis, is a more reactive inactivator than its free acid precursor, and its inactivation specificity toward aspartase has already been established (Schindler and Viola, 1994). Peptide mapping of FAA ethyl ester-modified aspartase shows the presence of the previously characterized peptide containing cysteine-274, along with an additional modified peptide (Giorgianni et al., 1997). The fact that both ASA and FAA treatment of aspartase yield a common modified residue supports the postulated mechanism-based inactivation by ASA. This additional modification site was identified, by proteolysis and mass spectral analysis, as occurring in the peptide from amino acids 129 to 146. This peptide contains the Cys-141 that has been previously modified by treatment with DACM (Ida and Tokushige, 1985), and the adjacent Lys-140. The failure of trypsin to cleave at this lysine upon proteolytic treatment indicates the involvement of this functional group in the modification by FAA.

2. *Generation of Inactivation Resistant Mutants*

Site-directed mutagenesis experiments were carried out to confirm the identity and to elucidate the role of the amino acids residues that had been identified by chemical modification and enzyme mapping studies. Substitution of cysteine-141 with a serine had no effects on the kinetic parameters for the natural substrate. This suggests that while cysteine-141 is a very reactive residue that is located in an accessible position for modification by substrate analog electrophiles, it is not essential for catalysis. Lysine-140, the other amino acid residue identified by the modification studies, represents a good candidate for a residue that can be involved in binding to a substrate that contains two negatively charged carboxylate groups. The kinetic parameters for the K140I mutant show a K_m value for L-aspartic acid that is 10-fold higher than that of the wild-type enzyme, and a comparable increase in the K_i for competitive inhibitors (Giorgianni et al., 1997). However, replacement of lysine-140 has no effect on the rate of inactivation of aspartase by either ASA or FAA ethyl ester. Since Cys-274 has been previously identified as a site of modification, the double mutant C141S/C274A was generated to test if replacement of the two reactive thiols would further reduce the rate of enzyme inactivation toward mechanism-based inactivator by FAA and its derivatives. The results in Figure 4 show that the sensitivity of this mutant aspartase to inactivation has been almost completely eliminated, confirming these thiols as the primary sites of inactivation.

3. *Inactivation Mechanisms*

The inactivation of aspartase by FAA involves interactions at two distinct sites. At the first site attack of the thiolate of Cys-274 at the α-carbon of FAA yields a stable Michael-type enzyme adduct (Fig. 5). Subsequent formation of a hydrazone upon treatment of the enzyme adduct with 2,4-dinitrophenylhydrazine confirms this mechanism by verifying the presence of the unreacted aldehydic group of FAA. This proposed mechanism of in-

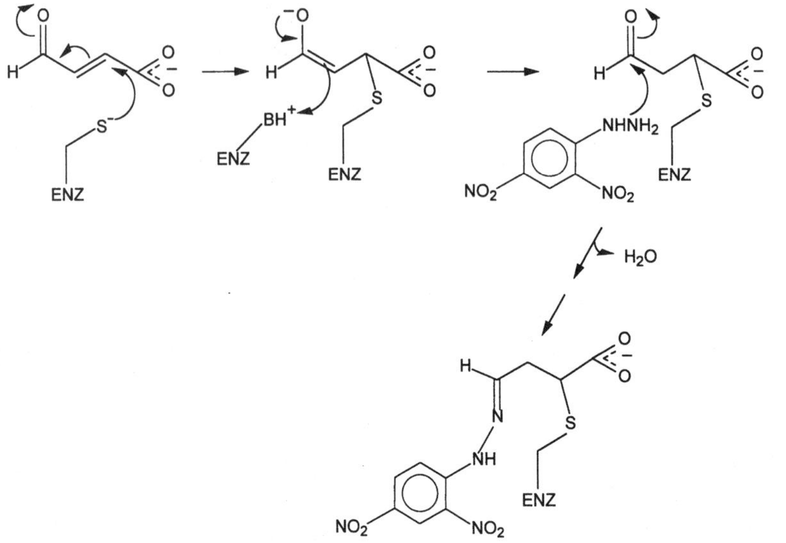

Figure 5. Proposed reaction of an active site nucleophile (S^-) with the double bond of FAA. Subsequent derivatization with 2,4-dinitrophenylhydrazine results in an enzyme-bound hydrazone adduct (Schindler and Viola, 1994).

activation of aspartase by aldehydic product analogs predicts that the rate of enzyme inactivation would be enhanced by substituents that make carbon-2 more electron deficient. This prediction was tested by examining several fumarate analogs with different electron withdrawing substituents. The methyl ester of FAA leads to the inactivation of aspartase with a rate of inactivation that is nearly five orders of magnitude greater than that of FAA, indicating that as greater electron withdrawing groups (with $COOCH_3$ >> COOH > CH_3) are attached to these fumarate analogs, the rate of inactivation increases (Baer and Urbas, 1970). For substituents that are located at the position β to the proposed site of nucleophilic attack, resonance stabilization will contribute to making the α-position more electron deficient. The methyl ester of β-nitroacrylate is a very potent inactivator of aspartase, with a rate of inactivation that is more than 10 orders of magnitude greater than that of FAA. The enhanced reactivity with aspartase is due to the presence of the highly electron withdrawing ester and the additional stabilization of the Michael adduct that is provided by the β- nitro group (Schindler and Viola,

1994). Thus, Cys-274 has been identified as a nucleophile that, while not directly involved in catalysis in aspartase, is poised to attack an activated double bond in an enzyme-bound product analog.

For the reaction of FAA at the second site in aspartase removal of Lys-140 by site-directed mutagenesis does not affect the rate of inactivation. This suggests that the initial modification at this site involves derivatization of Cys-141. Subsequent attack of the adjacent lysyl residue on the aldehyde of bound FAA would lead to a cross-linked adduct (Fig. 6) that would be stabilized by reduction. This adduct is resistant to tryptic digestion and would lead to the peptide that is isolated.

This investigation has added new information about the mechanism of modification of the enzyme aspartase from *E. coli* by identifying the labeled residues and characterizing the chemical structure of the labeling groups. Site-directed mutagenesis studies have ruled out the possibility that either of these reactive cysteine residues are directly involved in the catalytic activity

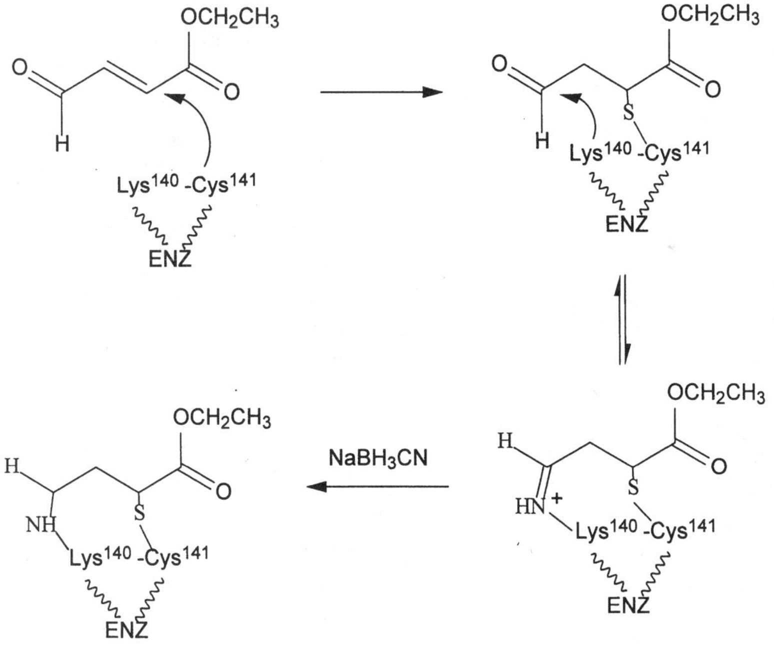

Figure 6. Proposed mechanism of inactivation of aspartase by FAA ethyl ester at Cys-141 and Lys-140. Reduction with cyanoborohydride results in a stable adduct (Schindler and Viola, 1994).

of aspartase, as had previously been postulated. A previously unidentified nucleophile, Lys-140, also takes part in enzyme inactivation, and mutagenesis studies indicate that this residue can have a direct role in enzyme catalysis by contributing to substrate binding. These results also have broader implications in the interpretation of mechanism-based inactivation studies. The tacit assumption in these types of studies has been that the reactive species that is generated by the catalytic action of the enzyme is constrained to react with a proximal active site functional group. However, as has now been shown with FAA, in cases where the species that is produced is somewhat stable, significant migration may occur before subsequent reaction. Thus, the assignment of a direct catalytic role to an amino acid functional group that has been modified, even by a highly specific mechanism-based inactivator, must be interpreted with caution.

E. SEQUENCE HOMOLOGY

The *asp*A gene encoding aspartase has been cloned and sequenced from several different bacterial sources (Guest et al., 1984; Takagi and Kisumi, 1985; Takagi et al., 1986a; Sun and Setlow, 1991). The *asp*A gene in *E. coli* encodes a 478 amino acid protein with a subunit molecular weight of 52 kDa (Takagi et al., 1985), and the catalytically active enzyme is a tetramer composed of identical subunits (Williams and Lartigue, 1967). There is good homology among the aspartases that have been sequenced, including 40–50% sequence identity among three different bacterial species. This homology extends to functionally related enzymes such as the class II fumarases, with 72–76% identity among seven species ranging from *E. coli* to human (Acuna et al., 1991), the argininosuccinate lyases with 51% identity between two species (Takagi et al., 1986b; Woods et al., 1986, 1988), and a prokaryotic and eukaryotic adenylosuccinate lyase with 27% sequence identity (Aimi et al., 1990). Additional studies have shown that δ-crystallin II (Matsubasa et al., 1989), a putative argininosuccinate lyase, and 3-carboxymuconate lactonizing enzyme (Williams et al., 1992) are also members of this homologous fumarase-aspartase family. While the overall sequence homology in this family is less than 20%, alignment of the amino acid sequences of representative members of each enzyme shows the presence of some highly homologous regions throughout the extended family. In particular there is a highly conserved sequence centered around an active site lysine (Fig. 7) that serves as the signature sequence to identify members of the fumarase-aspartase family.

Species/enzyme											327									
EcAspA	E	L	Q	A	G	S	S	I	M	P	A	**K**	V	N	P	V	V	P	E	V
PfAspA	A	R	Q	P	G	S	S	I	M	P	G	**K**	V	N	P	V	I	P	E	A
BsAspA	A	R	Q	P	G	S	S	I	M	P	G	**K**	V	N	P	V	M	A	E	L
EcFumC	E	N	E	P	G	S	S	I	M	P	G	**K**	V	N	P	T	Q	C	E	A
RatFum	E	N	E	P	G	S	S	I	M	P	G	**K**	V	N	P	T	Q	C	E	A
HumFum	E	N	E	P	G	S	S	I	M	P	G	**K**	V	N	P	T	Q	C	E	A
YeaArs	A	Y	S	T	G	S	S	L	M	P	Q	**K**	K	N	A	D	S	L	E	L
HumArs	A	Y	S	T	G	S	S	L	M	P	Q	**K**	K	N	P	D	S	L	E	L
DkDcII	A	Y	S	T	G	S	S	L	M	P	Q	**K**	K	N	P	D	S	L	E	L
CkDcII	A	Y	S	T	G	S	S	L	L	P	Q	**K**	K	N	P	D	S	L	E	L
BsAdl	K	G	Q	K	G	S	S	A	M	P	H	**K**	R	N	P	I	G	S	E	N
HumAdl	K	D	Q	I	G	S	S	A	M	P	Y	**K**	R	N	P	M	R	S	E	R
PpPcaB	P	G	K	G	G	S	S	T	M	P	H	**K**	R	N	P	V	G	A	A	V

Figure 7. A highly conserved region among the fumarase-aspartase enzyme family. Species: *Ec, Escherichia coli; Pf, Pseudomonas fluorescens; Bs, Bacillus subtilis; Hum*, human; *Yea*, yeast; *Dk*, duck; *Ck*, chicken; *Pp, Pseudomonas putida*. Enzymes: *AspA*, aspartase; *Fum*, fumarase; *Ars*, argininosuccinate lyase; *DcII*, δ-crystallin II; *Adl*, adenylosuccinate lyase; *PcaB*, 3-carboxymuconate lactonizing enzyme. The *dark shaded region* encompasses those residues that are identical between the *E. coli* aspartase and the other members of this enzyme family. The *light shaded areas* enclose residues that are identical among the functionally related subfamilies, the fumarases, the argininosuccinate lyases, and the δ-crystallins. The active site lysine is shown in *bold*.

F. MUTAGENESIS

In the absence of a high-resolution structure of an enzyme, homology between similar enzymes can aid in the selection of suitable functional amino acid targets for examination by site-specific mutagenesis. Previous modification and pH studies had identified cysteinyl, histidinyl, and lysyl residues as playing essential roles in the catalytic activity of aspartase. However, substitution of the most highly conserved cysteine with either serine or alanine, or the most highly conserved histidine with leucine, has no significant effect on the activity of aspartase (Table 3). In addition the sensitivity of these

TABLE 3
Kinetic Parameters of the Aspartase Mutants[a]

Enzyme	V_{max}/E_t (min^{-1})	Percent Activity	k_{cat}/K_m M^{-1} s^{-1}	K_m (mM) (L-aspartate)
Native	61.4	88	2.4×10^2	4.3
C390S	48.2	69	2.7×10^2	3.0
C390A	66.2	95	2.1×10^2	5.2
H124L	50.7	72	4.0×10^2	2.1
K327R	0.20	0.3	1.5×10^{-1}	23.8
K55R	—	0.0	—	—

[a]Assay conditions: 30 mM Hepes buffer, pH 7.0, 10 mM Mg acetate, and varied concentrations of L-aspartic acid at 30°C. The data were fitted to the Michaelis-Menten equation to obtain the kinetic parameters.

[b]Relative to the consensus V_{max}/E_t value of 70 min^{-1} for the freshly purified native enzyme under these assay conditions.

mutated aspartases to cysteine and histidine specific modifying reagents is not dramatically affected. But alteration of each of the two most conserved lysines to arginine does cause drastic changes in the catalytic properties of the enzyme.

1. *Cysteine Residues*

N-ethylmaleimide modification of aspartase, and protection by aspartic acid, has suggested the involvement of a cysteine in the catalytic mechanism of aspartase (Mizuta and Tokushige, 1975). Among the 11 cysteines that are present in the *E. coli* enzyme, only three are conserved among the bacterial aspartases that have been sequenced. One of these conserved cysteines, cysteine-390, is the only cysteine that is also conserved among all of the members of this homologous fumarase-aspartase enzyme family. The cysteine at this position was replaced with a serine to test the role of this sulfhydryl group in the catalytic activity of aspartase from *E. coli* (Saribas et al., 1994). No significant decrease was observed in the kinetic parameters of this C390S mutant (Table 3). Next, the cysteine at this position was replaced with an alanine. This nonconservative substitution also failed to alter the kinetic properties of the enzyme. To test the reactivity of this cysteine, the C390A mutant was incubated with varying amounts of NEM. A time-dependent loss of activity was observed. When examined in parallel experiments, the rate constant for activity loss was found to be identical to that measured for the native enzyme. The addition of saturating levels of fumarate provided complete protection against NEM inactivation for both the native and the C390A enzymes. Titration of aspartase with DTNB revealed the presence of 1.7 reactive sulfhydryl groups per subunit for the native enzyme. The C390A enzyme was found to have 1.9 reactive cysteinyl residues under the same conditions. When these titrations were repeated in the presence of saturating levels of fumarate, α-methylaspartate, and Mg^{2+}, the stoichiometry of reactive sulfhydryl groups dropped to between 0.7 and 0.9 per subunit for both the native enzyme and for the C390A mutant.

These results show that the highly conserved Cys-390 is not the reactive sulfhydryl group that is modified. Systematic replacement of each of the 11 cysteines in *E. coli* aspartase by a serine led to a family of mutants in which only minor changes in the kinetic parameters are observed (Chen et al., 1996). These results clearly demonstrate that, despite the suggestions from earlier chemical modification studies, the cysteinyl residues of aspartase are not directly involved in substrate binding or in catalysis for this enzyme.

2. *Histidine Residues*

The putative role of a histidine in the activity of aspartase was also examined by site-directed mutagenesis. Chemical modification studies with diethylpyrocarbonate (DEPC) (Ida and Tokushige, 1984) and pH-dependent activity changes (Karsten and Viola, 1991) have suggested the presence of an essential histidine in aspartase. However, none of the eight histidines in *E. coli* aspartase are conserved throughout the entire fumarase-aspartase family of enzymes, and only one, His-124, is found in the corresponding position in each of the aspartases. Mutation of His-124 to leucine was carried out to yield an active enzyme that behaved similarly to the native enzyme during purification. The V_{max} for this mutant has decreased slightly from that of the native enzyme, while the k_{cat}/K_m value increased by a factor of two (Table 3). Treatment of the H124L mutant with DEPC showed that this enzyme has the same sensitivity to this reagent as was observed for the native enzyme. Analogous replacement of each of the eight histidines of aspartase with a glutamine also had little effect on catalysis. Only two of these mutants showed as much as a several-fold decrease in activity (Chen et al., 1997).

3. *Lysine Residues*

Chemical modification and pH-dependent activity studies have also suggested the importance of a lysyl residue in aspartase (Karsten and Viola, 1991). There are 27 lysines in the enzyme, however only two of these, Lys-

55 and Lys-327, are completely conserved within the fumarase-aspartase enzyme family. Lys-327 is also located in the highly conserved signature sequence that is found throughout this family of related enzymes (Fig. 7). Changing Lys-327 to an arginine results in an enzyme that is expressed to high levels in aspartase-deficient cells (Guest et al., 1984), but which has very low catalytic activity. This K327R mutant was also found to be resistant to inactivation by OPA, even in the absence of added substrates or activators, and no spectral changes were observed even after prolonged incubation of this mutated aspartase with this reagent. Purification of the K327R mutant was accomplished on a Red A affinity column, to which the enzyme showed similar high binding affinity to that observed for the native enzyme. However, in contrast to the native enzyme, no significant elution from this column was obtained with L-aspartic acid, even when added at 10 times the usual eluant levels. This enzyme was eluted, with a high degree of purity, by using a potassium chloride ionic strength gradient. The purified mutant enzyme has a V_{max} that is only 0.3% that of the native enzyme, and a k_{cat}/K_m that has decreased by five orders of magnitude compared to the wild-type aspartase (Table 3). The failure of L-aspartic acid to efficiently elute this enzyme during purification is probably due to the decreased affinity of K327R for its substrate, since the K_m for L-aspartic acid has increased by a factor of five in this mutant. Mutation of this lysyl residue also causes a change in the divalent metal ion requirements of the enzyme. This mutant now has an absolute requirement for a divalent metal ion across the entire pH range, and also shows no lag in the reverse reaction even at high pH (Saribas et al., 1994). These changes are probably not the result of any direct metal ion interactions with this lysine, but instead are due to a change in communication between the active site and the activator site. Based on these results we can conclude that Lys-327 plays a role in the enzymatic activity of aspartase, and is most likely involved in substrate binding.

Lys-55 is also absolutely conserved throughout this enzyme family, and was therefore examined by site-directed mutagenesis. The K55R mutant that was constructed is overproduced in aspartase-deficient cells, however, no catalytic activity was observed in this cell line. A large portion of this protein was not solubilized into the supernatant fraction when these cells were initially disrupted by sonication. The soluble protein fraction containing a small portion of this mutant was purified by measuring the protein absorbance at 280 nm, and by following the overexpressed band that was observed on SDS gels in the position corresponding to that of the native enzyme throughout the protein purification. The highly purified protein sample that was obtained was observed to be devoid of measurable enzymatic activity, suggesting a crucial role for this residue in the catalytic activity of aspartase.

4. *Recovery of Catalytic Activity*

An examination of the recently determined high-resolution structure of aspartase (discussed in Section VIII below) shows that Lys-327 is in the active site cavity, but Lys-55 is located on the surface and is not in the vicinity of the putative enzyme active site. It is possible that the conservative lysine to arginine mutation that was generated at this site causes a gross structural change in aspartase that results in the loss of activity. To test this hypothesis a variety of conditions were examined in an attempt to resolubilize and reconstitute this inactive protein, and then to determine if catalytic activity can be recovered.

Initial polyacrylamide gel electrophoresis of the K55R mutant in the presence of sodium dodecyl sulfate (SDS-PAGE) shows that the protein exists as a high molecular weight oligomer. Incubation of this mutant in the presence of guanidine.HCl showed the slow appearance of a low level of catalytic activity after several hours at high levels of this denaturant (Fig. 8). The increase in activity upon incubation with guanidine reached a limiting value of 20–25% of the wild-type activity after incubation under these conditions for 10 hours or more. Surprisingly, continued incubation at 4–5 M guanidine resulted in further increases in activity (Jayasekera and Viola, 1999). In the most favorable cases this mutant recovered nearly 75% of the wild-type activity after 90 hours in the presence of these denaturant levels (Fig. 8). Treatment with either lower or higher concentrations of guanidine.HCl does not lead to this further enhancement in catalytic activity. Examination of the reactivated K55R mutant by nondenaturing PAGE shows that after 5 hours the higher molecular weight oligomers had dissociated, and the primary band that is now observed coincides with the tetramer form of the enzyme. When incubated with guanidine. HCl for extended periods of time, some lower molecular weight bands are also observed that may correspond to either dimers or trimers. The presence of catalytic activity in these enzyme forms has not been established, although dimers of aspartase have been reported to be active (Murase et al., 1993b). In all cases, removal of the denaturant from these solubilized enzymes resulted in reaggregation and loss of activity.

Encouraged by the appearance of catalytic activity in the presence of denaturant in this previously inactive mutant, conditions were explored to allow

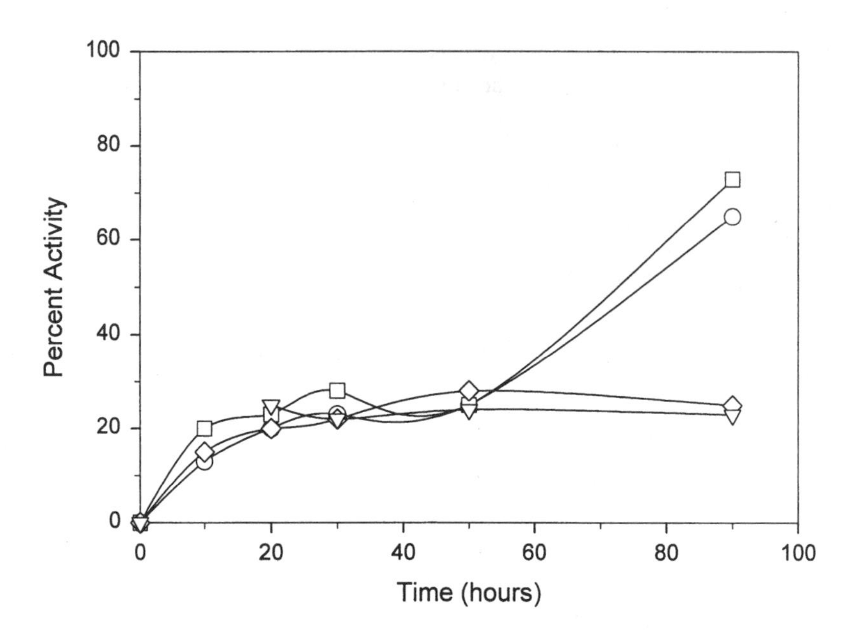

Figure 8. Reactivation of the K55R mutant of aspartase in the presence of high concentrations of guanidine.HCl. Aliquots were removed at the indicted times and assayed for catalytic activity. Incubations were conducted in the presence of 3 M (▽), 4 M (○), 5 M (□), and 6 M (◇) guanidine.HCl (Jayasekera and Viola, 1999).

the formation of a stable, active enzyme under nondenaturing conditions. Several approaches were attempted that were not successful, including changes in cell growth conditions, the addition and subsequent removal of a variety of denaturants, and the screening of a variety of protein refolding conditions. An artificial chaperone system, with exposure to detergents and their subsequent removal by binding to β-cyclodextrin, has been used to successfully refold thermally denatured carbonic anhydrase and guanidineHCl-denatured citrate synthase (Rozema and Gellman, 1995). This approach, using either Triton X-100 or cetyltrimethylammonium bromide (CTAB) as the detergents, was applied in an attempt to refold the inactive and aggregated K55R mutant. Exposure to Triton did not result in efficient resolubilization, however the cationic detergent CTAB did lead to solubilized mutant enzyme. Treatment of the native and the mutant enzymes with β-cyclodextrin or CTAB separately had little effect on either the loss or the recovery of catalytic activity (Table 4). However, subsequent treatment of the K55R mutant-CTAB complex with β-cyclodextrin resulted in recovery of enzyme activity

TABLE 4
Reactivation of an Inactive Mutant of Aspartase

Additive[a]	Percent Activity	
	Native	K55R
None	100	0
CTAB	73	0
β-Cyclodextrin	100	0
CTAB + β-cyclodextrin	81	40
Triton + β-cyclodextrin	0	0

[a]Addition of 1 mM CTAB, 5 mM β-cyclodextrin, or 1% Triton X-100 where indicated.

(Jayasekera and Viola, 1999). For the K55C mutant, extended incubation with CTAB before its removal by β-cyclodextrin leads to activity levels that are greater than 50% of the wild-type aspartase.

VIII. Structural Characterization

A. ENZYME STRUCTURE

The structure of aspartase has been solved in collaboration with Dr. Greg Farber at Penn State University. The *E. coli* enzyme was crystallized by microdialysis against Tris buffer, pH 8.5, with sodium acetate and polyethylene glycol as the precipitant (Shi et al., 1993). Aspartase crystals usually appeared after three days and attained maximum size in 7 to 10 days. The crystals have $P2_12_12$ symmetry and contain a tetramer in the asymmetric unit. X-ray diffraction data was collected on a flash frozen crystal at –178°C using an area detector. The structure of aspartase was solved by molecular replacement using the structure of fumarase (Weaver et al., 1995) as the search model, and the final density is continuous and well-defined for most of the main chain atoms (Shi et al., 1997). Each monomer of aspartase is composed of three domains oriented in an elongated S-shape (Fig. 9), and has approximate dimensions of 40 Å × 40 Å × 110 Å. The N-terminal domain is composed of residues 1–141. It consists of a short, two-stranded antiparallel β-sheet followed by five helices. Helix 1 and helix 4 are antiparallel to each other. Similarly, helices 2 and 3 are antiparallel to each other and together are oriented almost perpendicularly to the helix 1/4 pair. Helix 5 is a short stretch of five residues at the end of this domain.

Domain 2 of aspartase, consisting of residues 142–396, has almost all α-helical structure. This domain contains more than half of the total residues

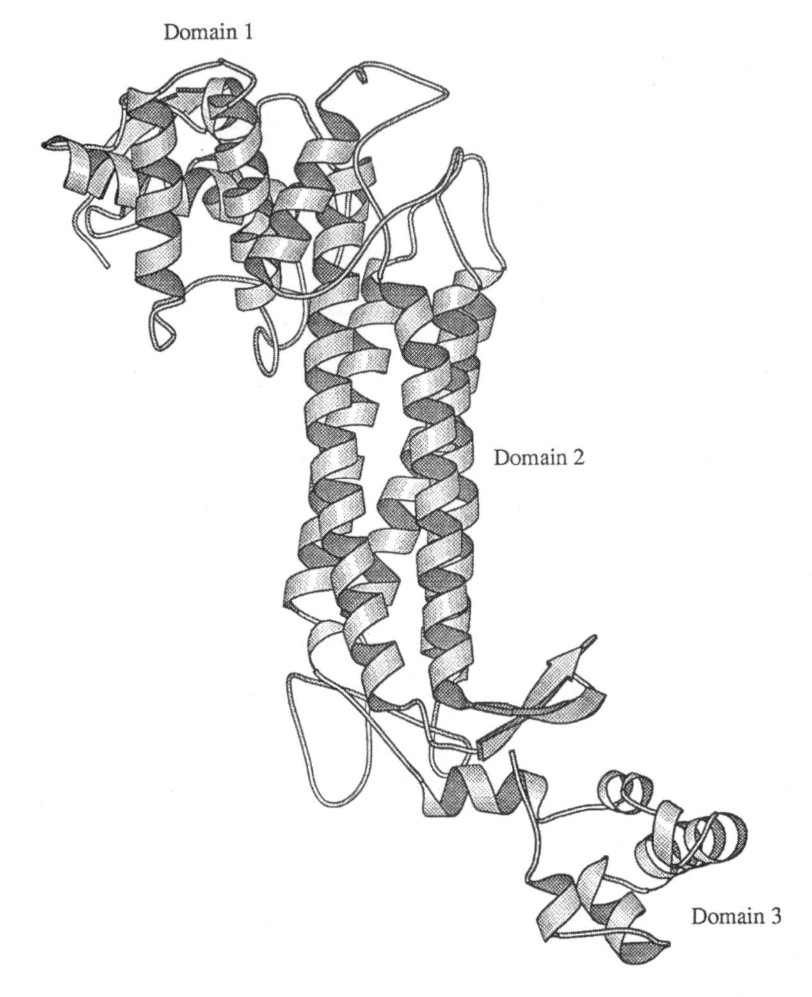

Figure 9. Molscript (Kraulis, 1991) diagram of an aspartase monomer with the three domains of the monomer labeled (Shi et al., 1997).

and is the most conserved domain in the aspartase-fumarase structural family. The central core of the domain is made up of five helices, and these helices are all slightly bent and are nearly parallel to each other to form a five helix bundle. The individual helices range from 25 to 38 residues in length and make domain 2 very long (about 50 Å). A relatively short helix connects two of the helices in this bundle by two long loops, making this part of the structure very flexible. This central domain contains the major portion of the subunit interface in the catalytically active tetramer of aspartase. When the tetramer is assembled, the central core of the enzyme consists of an unusual 20 helix bundle (Fig. 10), which has been observed only in the structures of enzymes in the fumarase-aspartase family (see Section VIII.B below).

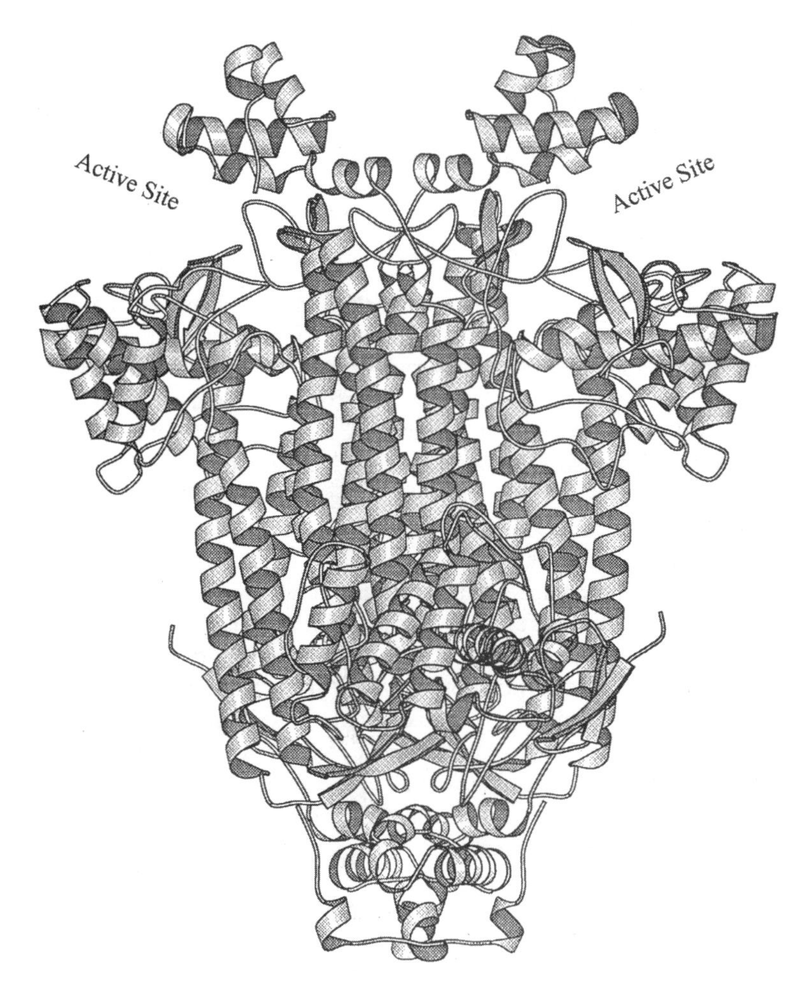

Figure 10. Molscript (Kraulis, 1991) diagram of the tetramer of *Escherichia coli* aspartase. Two of the four active sites in the aspartase tetramer are labeled (Shi et al., 1997).

The carboxyl terminus domain includes residues 397–478 and is the smallest domain in the subunit. It has a similar arrangement to the amino terminal domain, consisting mainly of two helix-turn-helix motifs oriented approximately 90° relative to each other. There is considerable disorder in this carboxyl terminus domain, such that there is very little well-defined density in the crystal from residue 461 to the carboxyl terminus (residue 478). This highly flexible region is of considerable interest because of work (reported in Section IX) that has shown a correlation between this region of the enzyme and the biological activity of aspartase. The four subunits of aspartase are arranged with the point symmetry 222 (Fig. 10). There are three unique dimer interfaces in the tetrameric structure. These are analogous to the interfaces present in fumarase (Weaver et al., 1995), although the subunit pairing is different.

B. STRUCTURAL HOMOLOGY

The amino acid sequence homology within each group of enzymes in the fumarase-aspartase family is generally quite high. However the sequence identity across the entire family is very low, even between different enzymes from the same organism. For example, aspartase and fumarase from *E. coli* have only 40% sequence identity, and human fumarase and argininosuccinate lyase are less than 15% identical. High-resolution structures have now been solved for four different enzymes in this family, with sequence identities that are as low as 15% when compared to aspartase. Despite this low overall sequence homology, all of the members of this family whose structures have been determined show high structural homology. Examples of high-resolution structures are now available for δ-crystallin (Simpson et al., 1994), fumarase (Weaver et al., 1995), aspartase (Shi et al., 1997), and argininosuccinate lyase (Turner et al., 1997). In each case the enzyme subunit consists of a central five helix bundle connecting smaller *N*- and *C*-terminal domains. This five helix bundle structure is arranged in an alternating up-down topology common to all members of the aspartase-fumarase structural family, and packs together to form a central 20 helix bundle as the core of the tetrameric structures.

C. ACTIVE SITE FUNCTIONAL GROUPS

Based on the high-resolution structure of aspartase we have been able to identify the active site cleft and, from an examination of that site, suggest several amino acid functional groups that may play a role in substrate binding or in catalysis. Lys-327 has been previously identified both by chemical modification studies (Karsten and Viola, 1991) and by site-directed mutagenesis studies (Saribas et al., 1994) as playing a significant role in the catalytic activity of aspartase. This loci was then the focal point in a search for potential amino acid targets that are located within a 15 Å radius. At least 30 functional amino acids on three of the four subunits were found that are proximal to Lys-327. This list was narrowed to a preliminary set of targets based on their relative proximity, sequence homology within the fumarase-aspartase family, and possession of a reasonable structure to provide a chemical basis for the proposed role. Other prospective candidates were eliminated from this initial consideration because they did not optimally conform to these selection criteria. This list of potential targets consists of amino acids that come primarily from the subunit adjacent to that which contributes Lys-327 to this intersubunit cleft.

His-26 was examined as a possible acid/base catalyst, but the identity of the amino acid at this position was found to be irrelevant for catalysis. Arg-15 and Arg-29 were tested as possible substrate binding groups. The kinetic characterization of these mutants supports such a role for Arg-29, with k_{cat} unaffected by replacement with an alanine but with the K_m increased by a factor of over 40 (Table 5). A similar substitution at Arg-15 has only a minimal effect on catalysis. Replacement of Ser-143 leads to a significant reduction in catalysis and, while substitution of an asparagine for Asp-10 only causes a fivefold reduction, replacement with an alanine leads to an enzyme form with barely detectable activity (Jayasekera et al., 1997).

TABLE 5
Kinetic Parameters of Aspartase Mutants

Mutant	Homology %	k_{cat} (min^{-1})[a]	Percent k_{cat}	K_m (mM)	k_{cat}/K_m (M^{-1} s^{-1})
Wild-type	—	40.6	100	1.8	3.8×10^2
H26Q	20	40.9	101	5.7	1.2×10^2
R29A	70	55.3	136	80	1.2×10^1
R15A	20	52.5	129	4.9	1.8×10^2
S143G	100	4.4	11	7.3	1.0×10^1
S143T	100	0.6	1.4	5.3	1.9×10^0
D10N	100	7.5	19	6.3	2.0×10^1
D10A	100	0.003	0.007	—	—

[a]Standard errors on k_{cat} are typically less than 10%, except for the D10A mutant, which has a standard error of about 40%.

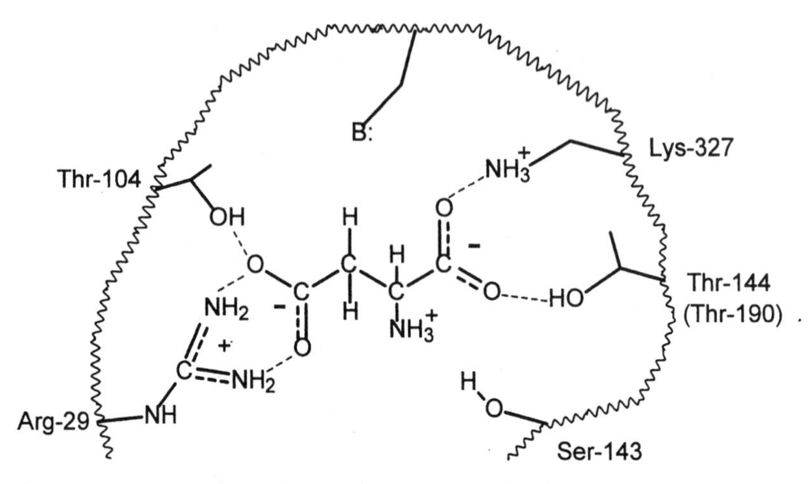

Figure 11. A cartoon of the active site of aspartase. Arg-29 and Lys-327 have been assigned to binding roles, with arginine interacting with the β-carboxylate of the substrate and lysine proposed to bind to an oxygen of the α-carboxylate. Structural studies have implicated threonines-104, -144, and -190 in substrate binding. The identity of the catalytic base is uncertain and Ser-143 is suggested as the catalytic acid.

Based on these results putative roles have been assigned to several of these active site amino acids. A cartoon of the active site structure (Fig. 11) shows the proposed substrate binding roles for Arg-29 and Lys-327, and a potential catalytic role for Ser-143. Asp-10 is most likely involved in a hydrogen-bonding network that is required to stabilize the active site structure. Preliminary structural studies on the binding of the competitive inhibitor 3-nitropropionate to aspartase (Dunbar, 1998) have suggested roles for several additional amino acids. Thr-104 is in a position to hydrogen-bond to the β-carboxyl group of the substrate L-aspartic acid, and Thr-144 (and possibly Thr-190) also are oriented to participate as substrate binding groups. The identity of the catalytic base that must abstract the *C-3* proton to generate the carbanion intermediate has not yet been conclusively determined.

While there are still some uncertainties that remain to be resolved, the functional and structural studies on aspartase have come to a reasonable consensus on the identity of many of the active site groups that participate in the catalytic activity of the enzyme. However, this consensus is not easily extended to the other members of the fumarase-aspartase family that have been studied. Banaszak and coworkers also had identified Lys-324 (the equivalent position to Lys-327) as being essential for catalysis in *E. coli* fumarase (Weaver et al., 1995). In addition, they assigned His-188 as the catalytic base that assists a bound water molecule in removing the C-3 proton from L-malate (Weaver et al., 1997). They propose that this histidine is helped by forming an ion pair with Glu-331. An analogous argument has been made for the catalytic mechanism of the argininosuccinate lyase activity of δ-crystallin II. In this case His-162 is proposed to be the catalytic base, but now acting directly rather than through an intervening water molecule (Vallée et al., 1999). Glu-296 is suggested to play the same charge relay role as the corresponding glutamic acid in fumarase. Unfortunately, these proposed active site assignments do not translate well to the current picture that we have of the aspartase structure. While Glu-334 is found in aspartase at the equivalent position to the Glu-331 in fumarase, the putative active site histidine base is a glutamine at the equivalent position in aspartase. Also, an inborn error leads to a fumarase deficiency in humans in which an essential and conserved glutamate (Glu-315) has been mutated to glutamine (Bourgeron et al., 1994). However, in aspartase this corresponding position is already a highly conserved glutamine (Gln-318), and this enzyme is fully functional. It would be unusual for a homologous enzyme family to catalyze chemically related reactions with a different set of catalytic residues. It is obvious that the detailed and definitive chemical mechanism of this family of enzymes has not yet fully emerged.

IX. New Tricks

Up to this point the studies that have been reported for aspartase describe a fairly "typical" enzyme. Typical in the sense that we have not indicated the presence of any unusual structural or functional properties. However, experience in enzymology has shown that there are really very few "typical" enzymes. Each enzyme that has been examined seems to have its own unusual properties, and aspartase is no exception. What follows is a presentation of some of these unusual properties, including the observed modulation of the catalytic activity of aspartase by both noncovalent and covalent means, and also a recently discovered nonenzymatic property of this enzyme.

A. SUBSTRATE ACTIVATION

When examined in the deamination reaction, *E. coli* aspartase displays non-Michaelis-Menten kinetics at higher pH (Fig. 12) (Karsten et al., 1986),

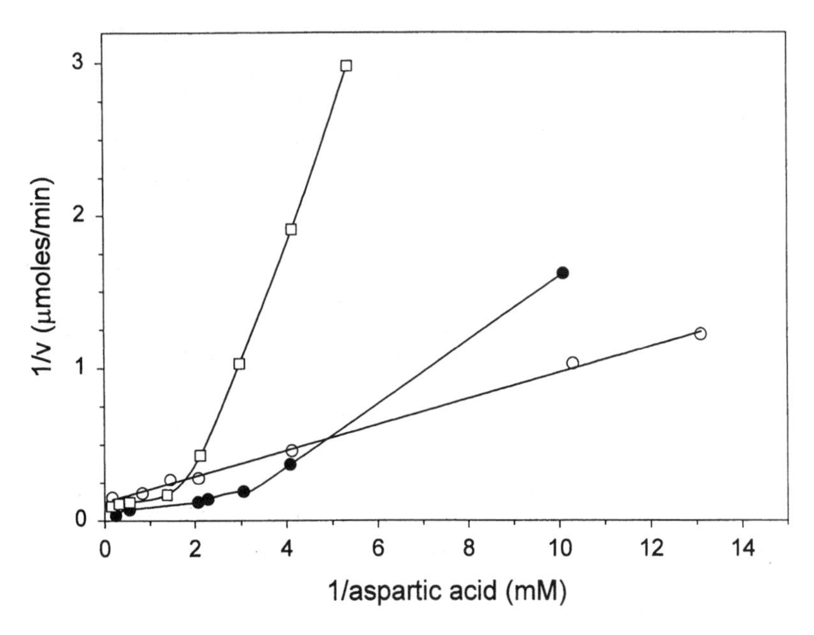

Figure 12. Reciprocal plot of aspartase activity at different pH values. The reaction mixtures contained 50 mM Hepes, pH 7 (○), Hepes, pH 8 (●), or Tris, pH 8.4 (□), 5 mM Mg^{2+}, and the indicated concentrations of L-aspartic acid (Karsten et al., 1986).

but linear kinetics are seen near neutral pH and below (Williams and Lartigue, 1967; Rudolph and Fromm, 1971). The time course of the reaction at higher pH in the amination direction shows a lag before a linear steady state rate is achieved (Ida and Tokushige, 1985). The lag in the amination direction and the non-Michaelis-Menten kinetics in the deamination direction have been shown to be related phenomena that are a consequence of the binding of the substrate L-aspartic acid to an activator site that is distinct from the enzyme active site (Karsten et al., 1986). In fact, the high pH form of the enzyme appears to be inactive unless L-aspartic acid and also a divalent metal ion are bound to the activator site.

Activators such as D-aspartate or *O*-phospho-D-serine have been shown, by NMR relaxation studies, to be bound close to the divalent metal ion binding site, with the carbon backbone of these molecules located about 3–4 Å from the aspartase-bound Mn^{2+} (Falzone et al., 1988). These distances are suggestive, but are not definitive proof, of direct coordination between the activator and the divalent metal ion. However, similar distances have been measured for the binary interaction between Mn^{2+} and L-aspartic acid in the absence of the enzyme (Khazaeli and Viola, 1984). In contrast, the relaxation of the product fumaric acid, which does not bind at this activator site, is not affected by the unpaired electrons of the divalent metal ion. Thus, the active site, where fumaric acid binds exclusively, must be greater than 10 Å away from this activator site (Falzone et al., 1988). By examining the results of these studies we can conclude that L-aspartic acid plays a dual role, both as a substrate and an activator for aspartase, with the metal complex of aspartic acid serving as the activator and the free amino acid functioning as the substrate.

Currently, however, there are few details concerning the location or the structure of the activator site in aspartase. Cysteine-430, which is solvent accessible since it can be modified with the fluorescent reagent DACM, has been proposed to be located near the activator site (Ida and Tokushige, 1985). Replacing this cysteine with a tryptophan leads to a decrease in the binding affinity of the activator site, while the overall enzyme activity is maintained. Also, the fluorescence spectrum of the Trp-430 mutant (the wild-type enzyme is devoid of tryptophans) is modified upon binding an activator, whereas the fluorescence from the 16 tyrosines that are distributed throughout the structure is unchanged even when the activator site is fully saturated (Murase et al., 1993a). A comparison with fumarase can provide some additional insights into the location of the activator site in aspartase. In the crystal structure of fumarase a second site in the enzyme, labeled site B, is observed to bind the substrate analog β-(trimethylsilyl)malate. By analogy with aspartase this second site has been suggested to be an activator site. The amino acids that form the putative activator site in fumarase include Arg-126 and Asn-135, located in a short helix segment (Weaver et al., 1997). The corresponding residues in aspartase are Gln-129 and Asn-138, also found in a helix (Shi et al., 1997), and these similarities suggest that this region in aspartase may contain the activator site.

Since the activator site in aspartase has not been identified from the crystal structure, and because the protein is too large for solution NMR characterization, we are currently investigating the activator site location in aspartase by using solid-state NMR. The aspartase samples have been lyophilized in the presence of cryo- and lyoprotectants, and were found to retain greater than 90% of their enzyme activity (M. Espe, unpublished results). By using the recently developed solid-state NMR dipolar recoupling techniques, atoms that are in close proximity (< 8 Å apart) can be identified

(McDowell and Schaefer, 1996). With both uniform and specific isotope labeling the atoms of the protein (such as ^{13}C and ^{15}N) that are near a bound activator (for example, ^{31}P in *O*-phospho-D-serine) can be determined. This information, along with the ^{13}C and ^{15}N chemical shifts, will allow the determination of the types of amino acids that are interacting with the activator. From these identities the location of this site will be mapped onto the existing three-dimensional structure.

B. PROTEOLYTIC ACTIVATION

In addition to the noncovalent activation, aspartase is also activated by proteolytic cleavage (Mizuta and Tokushige, 1975b, 1976). Cleavage of one or more peptide bonds near the carboxyl terminus results in a several-fold activation of aspartase activity (Yumoto et al., 1980). This proteolytic activation has been reported to be nonspecific, with the action of a number of different proteases including trypsin, subtilisin, and pronase each leading to enhanced catalysis by aspartase (Yumoto et al., 1982). Several primary and secondary sites of proteolysis were identified by mapping the carboxy-terminal peptides that are released (Fig. 13). The proteolytic activation of aspartase is quite transient, with the rate of peptide bond cleavages that result in enzyme activation competing with those cleavages that lead to inactivation. Changes in incubation times, or alterations in the protease to aspartase ratios can result in a failure to activate the enzyme and, under many conditions, only to decreases in activity. Based on these studies it was not clear whether cleavage at a unique site, or at several sites, is responsible for the enhanced catalytic activity of aspartase. Also, it had not been established whether the activated forms of aspartase were nicked subunits in which the carboxy-terminal peptides remained associated or if the cleaved enzyme is the activated species. To address these questions, we undertook a systematic study of the carboxy-terminal region of aspartase with the goal of identifying the amino acids in this region that influence the activity of the enzyme.

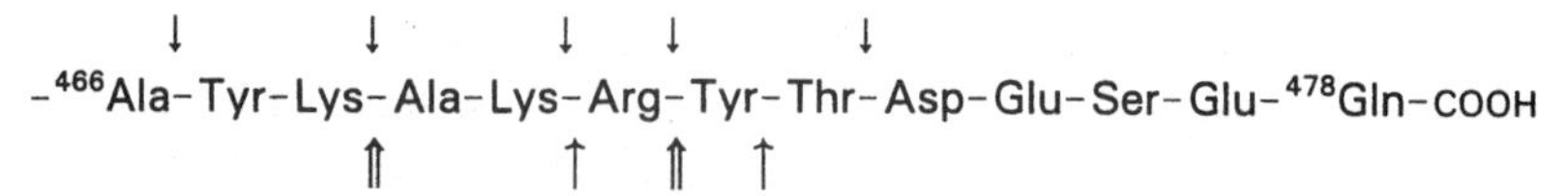

Figure 13. Carboxyl terminus sequence of aspartase with the truncation sites (↓) and the proposed sites of proteolytic cleavage by either trypsin (⇑) or by subtilisin (↑) indicated.

C. FORMATION AND CHARACTERIZATION OF TRUNCATION MUTANTS

Stop codons have been introduced at specific positions in the carboxyl terminus region that had previously been identified, from peptide mapping studies of partially proteolyzed aspartase, as having an effect on enzyme activity. The initial truncation site was located after Ala-466. Truncation at this position eliminates a 12-amino acid peptide that contains three positively charged and three negatively charged residues, and its removal was therefore expected to affect the structure of the enzyme in this region. However, this truncation mutant did not have enhanced catalytic activity, but instead showed a significant decrease in activity (Jayasekera et al., 1997a). By further examining the sequence in this region we observed that these charged amino acids are clustered, with the three negatively charged amino acids located nearest to the carboxyl terminus. Introduction of a stop codon after Thr-473 eliminates only the negative aspartate and glutamate residues without removing the positive lysines and arginine in this region. However this truncated form of aspartase has kinetic parameters that are essentially unchanged from those of the native enzyme.

Clearly, since truncation at position 474 has no effect on activity and truncation at position 467 causes substantial loss of activity, any cleavage sites that lead to enhanced catalytic activity must lie somewhere between these two positions. The location of subsequent truncation mutations was then designed to sequentially extend to the next charged or polar amino acid in this region when compared to the previous mutation site. Each of these truncation mutants was expressed, purified, and characterized in order to ascertain the impact of the amino acid that has been included. The kinetic parameters of this family of truncation mutants are summarized in Table 6. As

TABLE 6
Kinetic Parameters of Truncated Aspartases

Mutant	k_{cat} (min^{-1})	Percent k_{cat}	K_m (mM)	k_{cat}/K_m (M^{-1} s^{-1})
Wild-type	40.5	100	1.8	3.8×10^2
D474stop	42.7	105	2.8	2.5×10^2
Y472stop	97.2	240	3.3	4.9×10^2
R471stop	70.7	175	4.5	2.6×10^2
A469stop	39.6	98	10.2	6.5×10^1
Y467stop	2.5	6.1	31	1.3×10^0

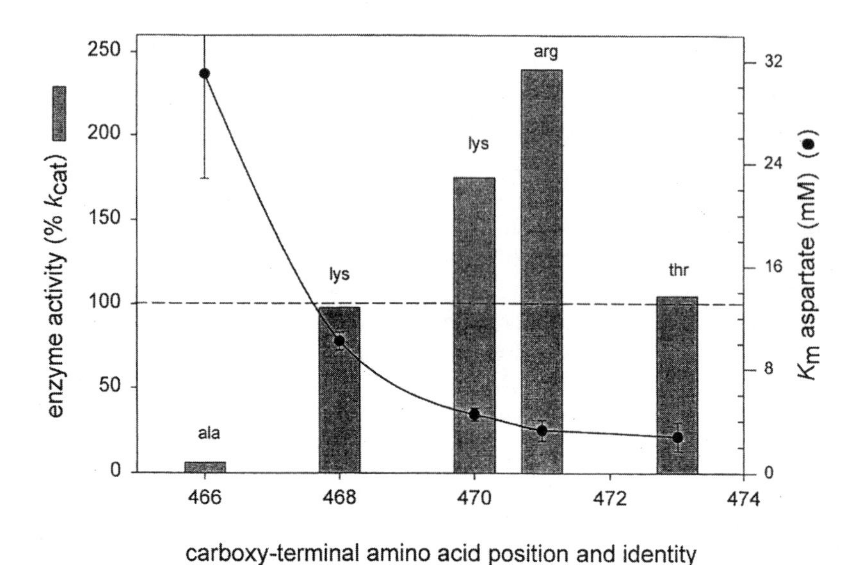

Figure 14. Relationship of the kinetic parameters of aspartase to the identity and position of the carboxyl terminus amino acid. The K_m values of the truncation mutants (●) are shown along with the associated standard errors, and the catalytic activity of the mutants ($\%k_{cat}$) are compared in a bar graph to the activity of the native enzyme (Jayasekera et al., 1997a).

described above, elimination of a 12–amino acid peptide from the carboxyl terminus (Y467stop) lead to an enzyme with significantly weaker affinity for its substrate and with decreased catalytic efficiency. Extending this truncated enzyme by two amino acids to a carboxyl terminus lysine (A469stop) leads to recovery of the catalytic activity to near native levels, but leaves an elevated Michaelis constant for aspartic acid. Addition of another dipeptide keeps the carboxyl terminus amino acid as lysine, but now results in an enzyme (R471stop) that has significantly enhanced catalytic activity over that of native aspartase (Fig. 14). Extending this enzyme by a single arginine (Y472stop) leads to a further enhancement in catalysis, and a tightening of the K_m for aspartic acid, to yield a modified aspartase with a k_{cat}/K_m that has been optimized beyond that which has been achieved by the native enzyme (Jayasekera et al., 1997a). Extending the carboxyl terminus by another dipeptide leads to an enzyme (D474stop) that terminates in a threonine, and results in a catalyst that has returned to native-like kinetic parameters.

Thus, extension of the truncated aspartase by the addition of one or two residues to yield an enzyme that is terminated by a positively charged amino acid results, in each case, in a significant enhancement in catalysis and a decrease in the Michaelis constant for aspartic acid (Fig. 14). What remains to be examined is whether the identity of the amino acids in this carboxyl terminus region are position sensitive (by placing lysine or arginine residues at other positions), or are context sensitive (by placing other amino acids at or adjacent to the sites where activation has been observed).

D. ACTIVATION OF PLASMINOGEN

The most recent surprise revealed by aspartase is its potential involvement in blood clotting. The activation of plasminogen can be influenced by the presence of a number of different bacterial plasminogen-binding proteins. During an examination of a strain of *Haemophilus influenzae* a binding protein was isolated by affinity chromatography to a bound kringle fragment of plasminogen. This purified plasminogen-binding protein was shown to stimulate the tissue plasminogen activator (tPA) catalyzed activation of plasminogen by about 300–fold (Sjöström et al., 1997). This protein specifically enhances tPA-activation and has no effect on the urokinase-activation of plasminogen. The gene that encodes for this protein was isolated and sequenced from a phage library and consists of an open reading frame corresponding to a 472 amino acid protein. A search of the *H. influenzae* genome revealed a 95% sequence identity with the aspartase gene of this bacterium, and 79% identity with the amino acid sequence of *E. coli* aspartase. Examination of the purified plasminogen-binding protein confirmed the presence of high levels of aspartase activity (Sjöström et al., 1997). These surprising results raised the question of whether other bacterial aspartases can also function as plasminogen-binding and tPA activating proteins. However, an examination of *E. coli* aspartase, conducted in collaboration with Drs. Sjöström and Wiman at the Karolinska Hospital in Sweden, did not detect any plasminogen binding or activation with this enzyme.

A comparison of the amino acid sequences between the aspartases from *H. influenzae* and *E. coli* showed that the *H. influenzae* enzyme is truncated by five amino acids and ends in a carboxyl terminus lysine. Numerous studies have shown that this carboxyl terminus lysine is a critical structural component for binding to the kringle domains of plasminogen (Plow et al., 1995). Since we had already shown (Jayasekera et al., 1997a) that specific truncations at the carboxyl terminus of aspartase lead to enhanced catalytic activity, our attention turned to the question of whether this enhanced activity would correlate with enhanced plasminogen binding activity. The truncation mutants that had previously been examined kinetically (Table 6) were tested for their ability to bind and activate plasminogen. Preliminary results indicated that the *E. coli* truncation mutants that end in a terminal lysine (A469stop and R471stop) bind to plasminogen with affinities that are comparable to that of the *H. influenzae* enzyme (I. Sjöström and B. Wiman, personal communication). Of these two enzyme forms, the A469stop mutant of *E. coli* aspartase has now been shown to stimulate plasminogen activation. This activation is not as efficient as that observed with the *H. influenzae* enzyme. However, we have been able, by the removal of a small carboxyl terminus peptide, to introduce a completely new biological activity into *E. coli* aspartase (Viola et al., 1999). The observation that the aspartase in *H. influenzae* is present at about 3% of the total soluble protein (Sjöström et al., 1997), in contrast to less than 0.1% in *E. coli*, lends credence to the argument that this protein is present to carry out an additional, nonenzymatic role in this organism.

It now appears that aspartase can be added to the growing list of proteins that have been recruited, in certain organisms, to carry out a separate, nonenzymatic function. There have been numerous other examples of enzymes that have been discovered to possess additional biological functions (Jeffery, 1999). Argininosuccinate lyase, another member of this fumarase-aspartase enzyme family, has been shown to be highly homologous with δ-crystallin, a major structural protein of the eye lens (Matsubasa et al., 1989), and δ-crystallin has been shown to possess residual lyase activity (Lee et al., 1992). Further studies will undoubtedly reveal many additional examples of the dual use of enzymes as a form of biological conservation.

X. Conclusions

The bacterial enzyme aspartase catalyzes the reversible deamination of L-aspartic acid with virtually absolute specificity. This high specificity does not arise exclusively from binding discrimination, but also from the enzyme's ability to select the correct binding orientation for catalysis. The deamination of aspartic acid catalyzed by aspartase has been shown to occur via a carbanion mechanism, and the involvement of divalent metal ions in this reaction takes place not at the active site, but at a separate activator site on the enzyme.

Structural and mutagenic studies of aspartase have established the location of the active site, and have identified a set of amino acids that are involved in substrate binding and in catalysis. The requirement for residues from several different subunits to participate in the formation of each active site provides a firm structural explanation for the lack of catalytic activity in an aspartase monomer. However, while there have been some intriguing hints, the structural elements that enforce the absolute specificity of aspartase remain to be elucidated.

Aspartase is activated noncovalently through the binding of the metal complex of its substrate at a separate activator site. Covalent activation is also observed as a consequence of the specific removal of seven to eight amino acids from the carboxyl terminus. Truncations in this region also introduce a new biological activity into aspartase, the ability to specifically enhance the activation of plasminogen to plasmin by tissue plasminogen activator. There is no obvious connection between the catalytic activity of aspartase and its possible role in enhancing the dissolution of blood clots. There clearly are a number of mysteries concerning this well-studied enzyme that still remain to be solved.

Acknowledgments

I would like to express my appreciation to the former students in my research group, especially Drs. Bill Karsten, Chris Falzone, Sami Saribas, Francesco Giorgianni, and Maithri Jayasekera, whose efforts have contributed to our improved understanding of aspartase. Also, a special thanks to Dr. Greg Farber and his co-workers for their persistence and tenacity in solving the structure of this enzyme. Much of this work has been supported over the years by grants from the National Institutes of Health (GM30217, GM34542, DK47838) and this financial assistance is gratefully acknowledged.

References

Acuna G, Ebeling S, Hennecke H (1991): Cloning, sequencing, and mutational analysis of the *Bradyrhizobium japonicum fumC*-like gene: evidence for the existence of two different fumarases. J Gen Microbiol 137: 991–1000.

Aimi J, Badylak J, Williams J, Chen Z, Zalkin H, Dixon JE (1990): Cloning of a cDNA encoding adenylsuccinate lyase by functional complementation in *Escherichia coli*. J Biol Chem 265: 9011–9014.

Baer HH, Urbas L (1970): Activating and directing effects of the nitro group in aliphatic systems. In Feuer H (ed): "The Chemistry of the Nitro and Nitroso Groups, Part 2." New York: Interscience Publishers, pp 75–200.

Berberich MA (1972): A glutamate-dependent phenotype in *E. coli* K12: the result of two mutations. Biochem Biophys Res Commun 47: 1498–1503.

Botting NP, Gani D (1989): Probing the mechanism of methylaspartase. In Roberts SM (ed): "Molecular Recognition: Chemical and Biochemical Problems." Cambridge: Royal Society of Chemistry, pp 134–151.

Bourgeron T, Chretien D, Poggi-Bach J, Doonan S, Rabier D, Letouze P, Munnich A, Rotig A, Landrieu P, Rustin P (1994): Mutation of the fumarase gene in two siblings with progressive encephalopathy and fumarase deficiency. J Clin Invest 93: 2514–2518.

Bright HJ (1965): On the mechanism of divalent metal activation of β-methylaspartase. J Biol Chem 240: 1198–1210.

Bright HJ (1967): Divalent metal activation of β-methylaspartase. The importance of ionic radius. Biochemistry 6: 1191–1203.

Chen HH, Chen JT, Tsai H (1996): Site-directed mutagenesis of cysteinyl residues in aspartase of *Escherichia coli.* Ann NY Acad Sci 799: 70–73.

Chen HH, Chen JT, Tsai H (1997): Site-directed mutagenesis of histidinyl residues in aspartase of *Escherichia coli.* Protein Eng 10: 60.

Chibata I, Tosa T, Sato T (1986): Continuous production of L-aspartic acid. Improvement of productivity by both development of immobilization method and construction of new *Escherichia coli* strain. Appl Biochem Biotechnol 13: 231–240.

Cook RP, Woolf B (1928): The deamination and synthesis of L-aspartic acid in the presence of bacteria. Biochem J 22: 474–481.

Dougherty TB, Williams VR, Younathan ES (1972): Mechanism of action of aspartase. A kinetic study and isotope rate effects with ^{2}H. Biochemistry 11: 2493–2498.

Dunbar JL (1998): Structural Studies of Enzymes Involved in Amino Acid Metabolism: Aspartokinase III, Human Branched-chain Amino Acid Aminotransferase, and L-Aspartate Ammonia Lyase: Ph.D. Dissertation, Pennsylvania State University.

Ellfolk N (1953a): Studies on aspartase. I. Quantitative separation of aspartase from bacterial cells, and its partial purification. Acta Chem Scand 7: 824–830.

Ellfolk N (1953b): Studies on aspartase. II. On the chemical nature of aspartase. Acta Chem Scand 7: 1155–1163.

Ellfolk N (1954): Studies on aspartase. III. On the specificity of aspartase. Acta Chem Scand 8: 151–156.

Emery TF (1963): Aspartase-catalyzed synthesis of *N*-hydroxyaspartic acid. Biochemistry 2: 1041–1045.

Englard S (1958): Studies on the mechanism of the enzymatic amination and hydration of fumarate. J Biol Chem 233: 1003–1009.

Englard S, Colowick SP (1956): On the mechanism of an anaerobic exchange reaction catalyzed by succinic dehydrogenase preparations. J Biol Chem 221: 1019–1035.

Englard S, Colowick SP (1957): On the mechanism of the aconitase and isocitric dehydrogenase reactions. J Biol Chem 226: 1047–1058.

Falzone CJ, Karsten WE, Conley JD, Viola RE (1988): L-Aspartase from *Escherichia coli*: substrate specificity and the role of divalent metal ions. Biochemistry 27: 9089–9093.

Farrar TC, Gutowsky HS, Alberty RA, Miller WG (1957): The mechanism of the stereospecific enzymatic hydration of fumarate to L-malate. J Am Chem Soc 79: 3978–3980.

Fusee MC (1987): Industrial production of L-aspartic acid using polyurethane-immobilized cells containing aspartase. Methods Enzymol 136: 463–471.

Gawron O, Fondy TP (1959): Stereochemistry of the fumarase and aspartase catalyzed reactions and of the Krebs cycle from fumaric acid to *d*-isocitric acid. J Am Chem Soc 81: 6333–6334.

Giorgianni F, Beranová S, Wesdemiotis C, Viola RE (1995): Elimination of the sensitivity of L-aspartase to active-site-directed inactivation without alteration of catalytic activity. Biochemistry 34: 3529–3535.

Giorgianni F, Beranová S, Wesdemiotis C, Viola RE (1997): Mapping the mechanism-based modification sites in L-aspartase from *Escherichia coli*. Arch Biochem Biophys 341: 329–336.

Goda SK, Minton NP, Botting NP, Gani D (1992): Cloning, sequencing, and expression in *Escherichia coli* of the *Clostridium tetanomorphum* gene encoding β-methylaspartase and characterization of the recombinant protein. Biochemistry 31: 10747–10756.

Guest JR, Roberts RE, Wilde RJ (1984): Cloning of the aspartase gene (aspA) of *Escherichia coli*. J Gen Microbiol 130: 1271–1278.

Halpern YS, Umbarger HE (1960): Conversion of ammonia to amino groups in *Escherichia coli*. Biochem J 80: 285–288.

Hanson KR, Havir EA (1970): L-Phenylalanine ammonia-lyase. IV. Evidence that the prosthetic group contains a dehydroalanyl residue and mechanism of action. Arch Biochem Biophys 141: 1–17.

Hanson KR, Havir EA (1972): The enzymic elimination of ammonia. In Boyer PD (ed): "The Enzymes." Vol. 3. New York: Academic Press, pp 75–166.

Harden A (1901): The chemical action of *Bacillus coli communis* and similar organisms on carbohydrates and allied compounds. J Chem Soc 79: 610–628.

Hernandez D, Stroh JG, Phillips AT (1993): Identification of Ser^{143} as the site of modification in the active site of histidine ammonia-lyase. Arch Biochem Biophys 307: 126–132.

Higashi Y, Isa N, Tokushige M (1988): Fumaraldehydic acid-induced inactivation of aspartase. Biochem Int 17: 103–109.

Ichihara K, Kanagawa H, Uchida M (1955): Studies on aspartase. J Biochem (Tokyo) 42: 439–447.

Ida N, Tokushige M (1984): Chemical modification of essential histidine residues in aspartase with diethylpyrocarbonate. J Biochem (Tokyo) 96: 1315–1321.

Ida N, Tokushige M (1985a): Assignment of catalytically essential cysteine residues in aspartase by selective chemical modification with N-(7-dimethylamino-4-methylcoumarynyl)maleimide. J Biochem (Tokyo) 98: 793–797.

Ida N, Tokushige M (1985b): L-Aspartate-induced activation of aspartase. J Biochem (Tokyo) 98: 35–39.

Imaishi H, Yumoto N, Tokushige M (1989): Characterization of intermediate species during the molecular assembly of aspartase. Physiol Chem Phys Med NMR 21: 221–228.

Jacobsohn KP, Pereira FB (1936): L'action du magnesium sur le systeme de l'aspartase. Comp Rend Seanc Soc Biol Lisb 120: 551–554.

Jayasekera MM, Saribas AS, Viola RE (1997a): Enhancement of catalytic activity by gene truncation. Activation of L-aspartase from *Escherichia coli*. Biochem Biophys Res Commun 238: 411–414.

Jayasekera MM, Shi WX, Farber GK, Viola RE (1997b): Evaluation of functionally important amino acids in L-aspartate ammonia-lyase from *Escherichia coli*. Biochemistry 36: 9145–9150.

Jayasekera MM, Viola RE (1999): Recovery of catalytic activity from an inactive aggregated mutant of L-asparatase. Biochem Biophys Res Comm 264:596–600.

Jeffery CJ (1999): Moonlighting proteins. Trends Biochem Sci 24: 8–11.

Karsten WE, Viola RE (1991): Kinetic studies of L-aspartase from *Escherichia coli*: pH dependent activity changes. Arch Biochem Biophys 287: 60–67.

Karsten WE, Hunsley JR, Viola RE (1985): Purification of aspartase and aspartokinase-homoserine dehydrogenase I from *Escherichia coli* by Dye-Ligand Chromatography. Anal Biochem 147: 336–341.

Karsten WE, Gates RB, Viola RE (1986): Kinetic studies of L-aspartase from *Escherichia coli*: substrate activation. Biochemistry 25: 1299–1303.

Khazaeli S, Viola RE (1984): A multinuclear NMR relaxation study of the interaction of divalent metal ions with L-aspartic acid. J Inorg Biochem 22: 33–42.

Kimura K, Takayama K, Ado Y, Kawamoto T, Masunaga I (1983): Processes for producing thermophilic aspartase. U.S. Patent No. 4,391,910.

Kraulis PJ (1991): Molscript: a program to produce both detailed and schematic plots of protein structures. J Appl Crystallogr 24: 946–950.

Langer M, Reck G, Reed J, Rétey J (1994): Identification of serine-143 as the most likely precursor of dehydroalanine in the active site of histidine ammonia-lyase. A study of the overexpressed enzyme by site-directed mutagenesis. Biochemistry 33: 6462–6467.

Lartigue DJ (1965) Purification and Characterization of Aspartase: Ph.D. Dissertation, Louisiana State University.

Lee HJ, Chiou SH, Chang GG (1992): Biochemical characterization and kinetic analysis of duck δ-crystallin with endogenous argininosuccinate lyase activity. Biochem J 283: 597–603.

Matsubasa T, Takiguchi M, Amaya Y, Matsuda I, Mori M (1989): Structure of the rat argininosuccinate lyase gene: close similarity to chicken δ-crystallin genes. Proc Natl Acad Sci USA 86: 592–596.

McDowell LM, Schaefer J (1996): High-resolution NMR of biological solids. Curr Opin Struct Biol 6: 624–629.

Mizuta K, Tokushige M (1975a): Studies on Aspartase. II. Role of Sulfhydryl Groups in Aspartase from *Escherichia coli*. Biochim Biophys Acta 403: 221–231.

Mizuta K, Tokushige M (1975b): Trypsin-catalyzed activation of aspartase. Biochem Biophys Res Commun 67: 741–746.

Mizuta K, Tokushige M (1976): Studies on aspartase. III. Alteration of enzymatic properties upon trypsin-mediated activation. Biochim Biophys Acta 452: 253–261.

Murase S, Kawata Y, Yumoto N (1993a): Use of hybridization for distance measurement by fluorescence energy transfer in oligomeric proteins: distance between two functional sites in aspartase. Biochem Biophys Res Commun 195: 1159–1164.

Murase S, Kawata Y, Yumoto N (1993b): Identification of an active dimeric form of aspartase as a denaturation intermediate. J Biochem (Tokyo) 114: 393–397.

Nishida Y, Sato T, Tosa T, Chibata I (1979): Immobilization of *Escherichia coli* cells having aspartase activity with carrageenan and locust bean gum. Enzyme Microb Technol 1: 95–99.

Nuiry II, Hermes JD, Weiss PM, Chen CY, Cook PF (1984): Kinetic mechanism and location of rate-determining steps for aspartase from *Hafnia alvei*. Biochemistry 23: 5168–5175.

O'Brien RJ, Fox JA, Kopczynski MG, Babior BM (1985): The mechanism of action of ethanolamine ammonia-lyase, an adenosylcobalamin-dependent enzyme. J Biol Chem 260: 16131–16136.

Plow EF, Herren T, Redlitz A, Miles LA, Hoover-Plow JL (1995): The cell biology of the plasminogen system. FASEB J 9: 939–945.

Porter DJ, Bright HJ (1980): 3-Carbanionic substrate analogues bind very tightly to fumarase and aspartase. J Biol Chem 255: 4772–4780.

Quastel JH, Woolf B (1926): The equilibrium between L-aspartic acid, fumaric acid and ammonia in the presence of resting bacteria. Biochem J 20: 545–555.

Rozema DB, Gellman SH (1995): Artificial chaperones: protein refolding via sequential use of detergent and cyclodextrin. J Am Chem Soc 117: 2373–2374.

Rudolph FB, Fromm HJ (1971): The purification and properties of aspartase from *Escherichia coli*. Arch Biochem Biophys 147: 92–98.

Sakano K, Hayashi T, Mukouyama M (1996): Process for the production of L-aspartic acid. U.S. Patent No. 5,541,090.

Saribas AS, Schindler JF, Viola RE (1994): Mutagenic investigation of conserved functional amino acids in *Escherichia coli* L-aspartase. J Biol Chem 269: 6313–6319.

Schindler JF, Viola RE (1994): Mechanism-based inactivation of L-aspartase from *Escherichia coli*. Biochemistry 33: 9365–9370.

Shi W, Kidd R, Giorgianni F, Schindler JF, Viola RE, Farber GK (1993): Crystallization and preliminary X-ray studies of L-aspartase from *Escherichia coli*. J Molec Biol 234: 1248–1249.

Shi WX, Dunbar JL, Jayasekera MM, Viola RE, Farber GK (1997): The structure of L-aspartate ammonia-lyase from *Escherichia coli*. Biochemistry 36: 9136–9144.

Shim JB, Kim JS, Yoon MY (1997): Chemical modification of cysteine residues in *Hafnia alvei* aspartase by NEM and DTNB. J Biochem Mol Biol 30: 113–118.

Simpson A, Bateman O, Driessen H, Lindley P, Moss D, Mylvaganam S, Narebor E, Slingsby C (1994): The structure of avian eye lens δ-crystallin reveals a new fold for a superfamily of oligomeric enzymes. Nat Struct Biol 1: 724–733.

Sjöström I, Gröndahl H, Falk G, Kronvall G, Ullberg M (1997): Purification and characterisation of a plasminogen-binding protein from *Haemophilus influenzae*. Sequence determination reveals identity with aspartase. Biochim Biophys Acta Bio-Membr 1324: 182–190.

Spencer ME, Lebeter VM, Guest JR (1976): Location of the aspartase gene (*AspA*) on the linkage map of *Escherichia coli* K12. J Gen Microbiol 97: 73–82.

Sun D, Setlow P (1991): Cloning, nucleotide sequence, and expression of the *Bacillus subtilis ans* Operon, which codes for L-asparaginase and L-aspartase. J Bacteriol 173: 3831–3845.

Suzuki S, Yamaguchi J, Tokushige M (1973): Studies on aspartase. I. Purification and molecular properties of aspartase from *Escherichia coli*. Biochim Biophys Acta 321: 369–381.

Takagi JS, Fukunaga R, Tokushige M, Katsuki H (1984): Purification, crystallization, and molecular properties of aspartase from *Pseudomonas fluorescens*. J Biochem (Tokyo) 96: 545–552.

Takagi JS, Ida N, Tokushige M, Sakamoto H, Shimura Y (1985): Cloning and nucleotide sequence of the aspartase gene of *Escherichia coli* W. Nucleic Acids Res 13: 2063–2074.

Takagi JS, Tokushige M, Shimura Y (1986a): Cloning and nucleotide sequence of the aspartase gene of *Pseudomonas fluorescens*. J Biochem (Tokyo) 100: 697–705.

Takagi JS, Tokushige M, Shimura Y, Kanehisa M (1986b): L-Aspartate ammonia-lyase and fumarate hydratase share extensive sequence homology. Biochem Biophys Res Commun 138: 568–572.

Takagi T, Kisumi M (1985): Isolation of a versatile *Serratia marcescens* mutant as a host and molecular cloning of the aspartase gene. J Bacteriol 161: 1–6.

Taylor RG, Lambert MA, Sexsmith E, Sadler SJ, Ray PN, Mahuran DJ, McInnes RR (1990): Cloning and expression of rat histidase. Homology to two bacterial histidases and four phenylalanine ammonia-lyases. J Biol Chem 265: 18192–18199.

Tokushige M (1985): Aspartate ammonia-lyase. Methods Enzymol 113: 618–627.

Tokushige M, Eguchi G, Hirata F (1977): Studies on aspartase. IV. Reversible denaturation of *E. coli* aspartase. Biochim Biophys Acta 480: 479–488.

Tosa T, Sato T, Mori T, Chibata I (1974): Basic Studies for Continuous Production of L-Aspartic Acid by Immobilized *Escherichia coli* Cells. Appl Microbiol 27: 886–889.

Tosa T, Sato T, Nishida Y, Chibata I (1977): Reason for higher stability of aspartase activity of immobilized *Escherichia coli* cells. Biochim Biophys Acta 483: 193–202.

Turner MA, Simpson A, McInnes RR, Howell PL (1997): Human argininosuccinate lyase: a structural basis for intragenic complementation. Proc Natl Acad Sci U S A 94: 9063–9068.

Vallée F, Turner MA, Lindley PL, Howell PL (1999): Crystal structure of an inactive duck δ II crystallin mutant with bound argininosuccinate. Biochemistry 38: 2425–2434.

Vender J, Rickenberg HV (1964): Ammonia metabolism in a mutant of *E.coli* lacking glutamate DH. Biochim Biophys Acta 90: 218–220.

Vender J, Jayaraman K, Rickenberg HV (1965): Metabolism of glutamic acid in a mutant of *Escherichia Coli*. J Bacteriol 90: 1304–1307.

Viola RE, Sarbas AS, Jayasekera MM (1999): Truncated aspartase enzyme derivatives and uses thereof. U.S. Patent No. 5,993,807.

Virtanen AI, Ellfolk N (1955): Aspartase. Methods Enzymol 2: 386–390.

Virtanen AI, Tarnanen J (1932): Die enzymatische Spaltung und Synthese der Asparaginsaure. Biochem Z 250: 193–211.

Weaver TM, Levitt DG, Donnelly MI, Wilkens-Stevens PP, Banaszak LJ (1995): The multisubunit active site of fumarase C from *Escherichia coli*. Nat Struct Biol 2: 654–662.

Weaver TM, Lees M, Banaszak LJ (1997): Mutations of fumarase that distinguish between the active site and a nearby dicarboxylic acid binding site. Protein Sci 6: 834–842.

Wilkinson JS, Williams VR (1961): Partial purification of bacterial aspartases by starch electrophoresis. Arch Biochem Biophys 93: 80–84.

Williams SE, Woolridge EM, Ransom SC, Landro JA, Babbitt PC, Kozarich JW (1992): 3-Carboxy-*cis,cis*-muconate lactonizing enzyme from *Pseudomonas putida* is homologous to the class II fumarase family: a new reaction in the evolution of a mechanistic motif. Biochemistry 31: 9768–9776.

Williams VR, Lartigue DJ (1967): Quaternary structure and certain allosteric properties of aspartase. J Biol Chem 242: 2973–2978.

Williams VR, Lartigue DJ (1969): Aspartase. Methods Enzymol 13: 354–361.

Wong OS, Sternson LA, Schowen RL (1985): Reaction of *o*-phthalaldehyde with alanine and thiols: kinetics and mechanism. J Am Chem Soc 107: 6421–6422.

Woods SA, Miles JS, Roberts RE, Guest JR (1986): Structural and functional relationships between fumarase and aspartase. Biochem J 237: 547–557.

Woods SA, Miles JS, Guest JR (1988): Sequence homologies between argininosuccinase, aspartase, and fumarase: a family of structurally-related enzymes. FEMS Microbiol Lett 51: 181–186.

Woolf B (1929): Some enzymes in *B. coli communis* which act on fumaric acid. Biochem J 23: 472–482.

Yoon MY, Thayer-Cook KA, Berdis AJ, Karsten WE, Schnackerz KD, Cook PF (1995): Acid-base chemical mechanism of aspartase from *Hafnia alvei*. Arch Biochem Biophys 320: 115–122.

Yumoto N, Mizuta K, Tokushige M, Hayashi R (1982): Studies on aspartase. VIII. Protease-mediated activation: comparative survey of protease specificity for activation and peptide cleavage. Physiol Chem Phys 14: 391–397.

Yumoto N, Murase S, Imaishi H, Tokushige M (1992): Determination of the subunit contact region of aspartase. Biochem Int 28: 413–422.

Yumoto N, Tokushige M, Hayashi R (1980): Studies on aspartase. VI. Trypsin-mediated activation releasing carboxy-terminal peptides. Biochim Biophys Acta 616: 319–328.

AUTHOR INDEX

Note: Page numbers followed by *t* and *f* indicate tables and figures, respectively.

SUBJECT INDEX

Note: Page numbers followed by *t* and *f* indicate tables and figures, respectively.